Cisco Self-Study:
Building Cisco Metro Optical Networks (METRO)

Dave Warren
Dennis Hartmann

Cisco Press

Cisco Press
800 East 96th Street, 3rd Floor
Indianapolis, IN 46240 USA

ii

Cisco Self-Study:
Building Cisco Metro Optical Networks (METRO)

Dave Warren

Dennis Hartmann

Copyright© 2004 Cisco Systems, Inc.

Published by:
Cisco Press
800 East 96th Street, 3rd Floor
Indianapolis, IN 46240 USA

Printed in the United States of America 1 2 3 4 5 6 7 8 9 0

First Printing August 2003

Library of Congress Cataloging-in-Publication Number: 2002104851

ISBN: 1-58705-070-6

Warning and Disclaimer

This book is designed to provide information about metro optical networks. Every effort has been made to make this book as complete and as accurate as possible, but no warranty or fitness is implied.

The information is provided on an "as is" basis. The author, Cisco Press, and Cisco Systems, Inc., shall have neither liability nor responsibility to any person or entity with respect to any loss or damages arising from the information contained in this book or from the use of the discs or programs that may accompany it.

The opinions expressed in this book belong to the author and are not necessarily those of Cisco Systems, Inc.

The Cisco Press self-study book series is as described, intended for self-study. It has not been designed for use in a classroom environment. Only Cisco Learning Partners displaying the following logos are authorized providers of Cisco curriculum. If you are using this book within the classroom of a training company that does not carry one of these logos, then you are not preparing with a Cisco trained and authorized provider. For information on Cisco Learning Partners please visit:www.cisco.com/go/authorizedtraining. To provide Cisco with any information about what you may believe is unauthorized use of Cisco trademarks or copyrighted training material, please visit: http://www.cisco.com/logo/infringement.html.

Trademark Acknowledgments

All terms mentioned in this book that are known to be trademarks or service marks have been appropriately capitalized. Cisco Press or Cisco Systems, Inc., cannot attest to the accuracy of this information. Use of a term in this book should not be regarded as affecting the validity of any trademark or service mark.

Feedback Information

At Cisco Press, our goal is to create in-depth technical books of the highest quality and value. Each book is crafted with care and precision, undergoing rigorous development that involves the unique expertise of members from the professional technical community.

Readers' feedback is a natural continuation of this process. If you have any comments regarding how we could improve the quality of this book, or otherwise alter it to better suit your needs, you can contact us through e-mail at feedback@ciscopress.com. Please make sure to include the book title and ISBN in your message.

We greatly appreciate your assistance.

Publisher	John Wait
Editor-in-Chief	John Kane
Executive Editor	Brett Bartow
Cisco Representative	Anthony Wolfenden
Cisco Press Program Manager	Sonia Torres Chavez
Manager, Marketing Communications, Cisco Systems	Scott Miller
Cisco Marketing Program Manager	Edie Quiroz
Production Manager	Patrick Kanouse
Acquisitions Editor	Michelle Grandin
Development Editors	Allison Beaumont Johnson
	Dayna Isley
Project Editor	San Dee Phillips
Copy Editor	Keith Cline
Contributing Editor	Robert Elling
Technical Editors	Joe Abley, Matthew J. "Cat" Castelli,
	Sanjay Kapoor, Roberto Narvaez, Drew Rosen
Team Coordinator	Tammi Ross
Book Designer	Gina Rexrode
Cover Designer	Louisa Adair
Composition	Mark Shirar
Indexer	Brad Herriman

CISCO SYSTEMS

Corporate Headquarters
Cisco Systems, Inc.
170 West Tasman Drive
San Jose, CA 95134-1706
USA
www.cisco.com
Tel: 408 526-4000
800 553-NETS (6387)
Fax: 408 526-4100

European Headquarters
Cisco Systems International BV
Haarlerbergpark
Haarlerbergweg 13-19
1101 CH Amsterdam
The Netherlands
www-europe.cisco.com
Tel: 31 0 20 357 1000
Fax: 31 0 20 357 1100

Americas Headquarters
Cisco Systems, Inc.
170 West Tasman Drive
San Jose, CA 95134-1706
USA
www.cisco.com
Tel: 408 526-7660
Fax: 408 527-0883

Asia Pacific Headquarters
Cisco Systems, Inc.
Capital Tower
168 Robinson Road
#22-01 to #29-01
Singapore 068912
www.cisco.com
Tel: +65 6317 7777
Fax: +65 6317 7799

Cisco Systems has more than 200 offices in the following countries and regions. Addresses, phone numbers, and fax numbers are listed on the
Cisco.com Web site at www.cisco.com/go/offices.

Argentina • Australia • Austria • Belgium • Brazil • Bulgaria • Canada • Chile • China PRC • Colombia • Costa Rica • Croatia • Czech Republic
Denmark • Dubai, UAE • Finland • France • Germany • Greece • Hong Kong SAR • Hungary • India • Indonesia • Ireland • Israel • Italy
Japan • Korea • Luxembourg • Malaysia • Mexico • The Netherlands • New Zealand • Norway • Peru • Philippines • Poland • Portugal
Puerto Rico • Romania • Russia • Saudi Arabia • Scotland • Singapore • Slovakia • Slovenia • South Africa • Spain • Sweden
Switzerland • Taiwan • Thailand • Turkey • Ukraine • United Kingdom • United States • Venezuela • Vietnam • Zimbabwe

About the Authors

Dave Warren, CCIP Optical, CCNP, CCDP, certified Cisco Systems instructor, is president and owner of Dynamic WorldWide Training (www.dwwtc.com) specializing in Network and Security training. With almost 30 years of experience in the computer and networking industry, he has designed, built, and consulted on the implementation of many different technologies, including Frame Relay, ATM, and optical. When Cisco initially began to deploy optical networks using the ONS 15454, Dave was chosen as one of the few instructors to deliver training to internal Cisco system engineers and account managers.

Dennis Hartmann is a service provider consultant/instructor who provides training services for Skyline Computer. Dennis has worked for various Fortune 500 companies, including AT&T, Sprint, Merrill-Lynch, KPMG Consulting, and Cabletron Systems. Dennis has knowledge and experience with a variety of Cisco optical platforms (including ONS 15327, ONS 15454, ONS 15216 series, ONS 15500 series, ONS 15200 series, DPT, and PoS). Dennis holds various certifications (including Cisco Optical Qualified Specialist, certified Cisco Systems instructor, CCIP in MPLS and Optical, Cisco IP Voice Support Specialist, Cisco IP Voice Design Specialist, CCNP, CCDP, CCNA, CCDA, and MCSE). Dennis co-wrote Chapters 4 through 12. Dennis can be reached at dh8@pobox.com.

About the Contributor

Robert Elling has worked at Cisco Systems since 1998. During that time, he has supported Fortune 1000 customers and service provider customers. He is currently working as the optical specialist in the Internal Learning Solutions Group (ILSG). Prior to his work at Cisco, Robert worked as a networking engineer for Arthur Andersen; he also worked as a senior networking engineer for Verizon (formerly called Bell Atlantic) in the Harrisburg, Pennsylvania, Network Management Center-Network Operations Center. Robert holds various certifications (including Cisco Qualified Specialist in Optical, CCNA, certified Cisco Systems instructor, Enterprise Certified Novell Engineer, and MCSE). Robert contributed significantly to the final versions of Chapters 1 through 3 and Chapter 13.

About the Technical Reviewers

Joe Abley has worked as a lead engineer in numerous projects, including backbone architecture, design, and operation of the AboveNet/MFN global IP network. He was also the technical lead in the early design work that led to the rollout of the Telstra Saturn (now TelstraClear) broadband packet VPN infrastructure in New Zealand. Joe participates actively in NANOG, NZNOG, and IETF.

Matthew J. "Cat" Castelli has more than 14 years of experience in the telecommunications networking industry. He started as a cryptologic technician (communications) in the U.S. Navy. Cat has since been working as a principal consultant for a Cisco Professional Services partner as a senior technical consultant/enterprise network design engineer, a global telecommunications integrator, and most recently as the information assurance IA liaison for a 360,000+ user network. Cat has broad exposure to LAN/WAN, Internet, and alternative technologies (VoX, for instance) for service provider and enterprise networks of all sizes. His exposure and experience includes implementation, application, configuration, integration, network management, and security solutions.

Sanjay Kapoor is a manager in software development working in the Cisco Systems Optical Networking Group.

Roberto Narvaez, CCIE No. 4439, is a leading network architect at RedNetworks, Inc., a consulting company. He has extensive experience working with optical, network security, MPLS, VoIP, and wireless projects with Fortune 500 companies. He holds a bachelor of science degree from Maryland University and currently is pursuing a master's degree in computer science at Johns Hopkins University.

Drew Rosen, CCIE No. 4365, is a product-marketing manager in the Cisco Internet Learning Solutions Group (ILSG). In his present role, Drew manages a team of technical consultants focusing on educational products for service providers. Previously, Drew spent four years as a systems engineer working on large named accounts in the enterprise space. He has been involved in the production and launch of numerous ILSG products, including Building Scalable Cisco Internetworks (BSCI), Configuring BGP on Cisco Routers (CBCR), Configuring Cisco Routers for IS-IS (CCRI), Advanced MPLS VPN Solutions (AMVS), Building Cisco Optical Networks (METRO), and Implementing QoS (QoS). Drew lives in Florida with his wife Meredith and daughter Chelsea.

Dedications

Dave Warren: This book is dedicated to my wife Lorrie and her incredible patience while I have spent every extra hour working on this book. To my family who continues to grow up and make a father proud. And finally to my friend Mark; every person needs a friend, and he has been the best.

Dennis Hartmann: This book is dedicated to my wife Missy. Thank you very much for your support and understanding. Thank you to the rest of my family for their love and support: Little Dennis, Chile, and China.

Special thanks to Mom, who steered me in the right direction.

Acknowledgments

Dave Warren: This book could not have made without the input and support of many people.

I would like to give special recognition to my technical editors, who have provided a tremendous amount of help and insight to make the book even better and more informative.

Also, I would like to recognize those people who gave me my start in Networking and have helped me along the way—Carolyn Cutler and Doug Hall—who continue to support my thirst for knowledge in new technologies and continue to help me stretch.

A huge thank you to Michelle Grandin and Dayna Isley, whose patience and help was essential as I discovered exactly what it takes to write a book.

Dennis Hartmann: Thanks to all my family and friends. The list is too long to include here, but you know who you are. Thanks to all the instructors who have taught me a lot. Thanks to Mike Z. and Wendell Odom for their recommendation. Thanks to Ali Nooriala, Yong Kim, Seth Higgins, and Robert Elling for their assistance. Thanks to Ken Peterson, Marty Nusbaum, and Chuck Terrien for the opportunity to work on the project that provided me my first optical training opportunity.

Thanks to Allison, Michelle, Dayna, and everyone at Cisco Press for helping me through this book.

Contents at a Glance

Contents

Introduction

The use of optical fiber as a primary means of transporting voice and high-speed data continues to increase. Supporting bandwidth-intensive data applications and voice is a challenge for service providers. The need for qualified personnel who not only understand the technologies involved but also have an understanding of how it should be implemented has helped drive the need for Cisco to develop courseware and a certification exam on the use of optical equipment in the metro area.

Objectives

Cisco Self-Study: Building Cisco Metro Optical Networks (METRO) is designed to help you obtain knowledge about what it takes to build a metro area optical network using Cisco metro equipment, which includes the ONS 15454, 15327, and 15216 platforms as well as the Cisco 12000 series switch and routers. This self-study book follows the form and content of the Cisco course of the same name.

Who Should Read This Book?

This book is geared toward the networking professional desiring to obtain a greater level of expertise in the technology and Cisco equipment used in the metro area. This book assumes the reader is already familiar with the following technologies:

- Internetworking concepts and fundamentals

- Routing concepts

- Router configuration knowledge and experience

- Optical network fundamentals

- Basic knowledge in the areas of SONET, TLS, DPT, DWDM, and PoS

- Experience in configuring the Cisco ONS 15454 and 15327

Readers building, planning to build, or designing optical networks, as well as people working with the ONS 15454, 15327, 15216, or Cisco 12000 series switch/routers in the metro area, can use this book as additional reference material.

How This Book Is Organized

This book is divided into 13 chapters. The first three chapters are designed to provide background information so that you can better understand the technologies used in the metro area. Chapters 4 and 5 deal specifically with the ONS 15454 and 15327 products. Chapters 6 and 7 delve into metro Ethernet services. Chapter 8 talks about DWDM products, and Chapters 9 through 12 cover the Cisco 12000 series switch and routers technologies. Chapter 13 finishes with some design exercises.

A brief description of each chapter follows:

- **Chapter 1, "Metro Area Optical Networks"**—This chapter introduces you to the characteristics of a metropolitan-area network (MAN) and discusses the application and services used. It introduces the Cisco products and relates them to the applications and services they support. You learn about the components of a service network architecture, the physical topologies that support metro networks, and the key service traffic patterns.

- **Chapter 2, "Voice History"**—This chapter provides a brief history on voice technology.

- **Chapter 3, "SONET Overview"**—This chapter takes you into the workings of SONET. Starting with a brief history of why SONET was developed, this chapter discusses SONET frame structures, SONET components, and SONET network design. This includes linear and ring designs and the various protection schemes available with SONET.

- **Chapter 4, "ONS 15454 and ONS 15327 Optical Platforms"**—This chapter reviews the Cisco ONS 15454 and Cisco ONS 15327 platforms and shows you how they would work in a SONET MAN network. Details on the hardware available and their capabilities are provided. An introduction to the CTC management interface to the ONS products completes the chapter.

- **Chapter 5, "Configuring ONS 15454 and ONS 15327"**—This chapter covers the steps required to initially configure and make an ONS platform part of an optical network. Sections cover controlling access to the node, as well as configuring UPSR and BLSR rings, with special attention paid to configuring clocking for the rings.

- **Chapter 6, "Metro Ethernet Services"**—This chapter examines metro Ethernet services and why this has become such an important topic for customers in the metro area. You are shown the architecture used to build a TLS system and explore the capabilities of the Cisco equipment that supports this environment.

- **Chapter 7, "Configuring Metro Ethernet"**—This chapter takes you through the steps required to build a TLS system using the ONS 15454 and 15327 products. Singlecard EtherSwitches, multicard Ethergroups, point-to-point and point-to-multipoint circuits, and shared packet rings are described, and step-by-step instructions are provided for their implementation.

- **Chapter 8, "Implementing DWDM in the Metropolitan-Area Networks"**—This chapter covers not only why you need DWDM but also what it is and how it works. The Cisco equipment that supports DWDM, specifically the 15216 products, is covered.

- **Chapter 9, "Packet over SONET"**—This chapter describes in broad terms the current, popular uses for PoS technologies and how PoS compares with other technologies, such as ATM and DPT. You learn how PoS is structured and how it can be used in a network. You also learn about how PoS protection is implemented using SONET protection and about the Cisco proprietary Protection Group Protocol (PGP).

- **Chapter 10, "Configuring Packet over SONET"**—This chapter describes the Cisco equipment that supports PoS and the steps required to implement PoS in a network. Clocking issues, troubleshooting tips, and screen samples are included.

- **Chapter 11, "Dynamic Packet Transport"**—This chapter examines the technology behind DPT and why DPT is a useful solution for today's networks. You learn the concepts behind DPT and see its inner workings through the packet structure of the control messages. Issues concerning bandwidth usage, quality of service, and protection are addressed.

- **Chapter 12, "Configuring Dynamic Packet Transport"**—This chapter discusses the procedures required to create a DPT ring. You identify the hardware that is required to build a Cisco DPT ring. You learn how to implement DPT over an existing SONET ring when used in conjunction with an ONS 15454. Finally, you review the commands needed to activate and send traffic across a DPT ring.

- **Chapter 13, "Metro Optical Case Studies"**—This chapter provides an opportunity for you to be creative and apply the concepts learned in this book. Two ficticious case studies ask you to choose equipment and produce a rough network design.

Icons Used in This Book

Throughout this book, you will see the following icons used for networking devices:

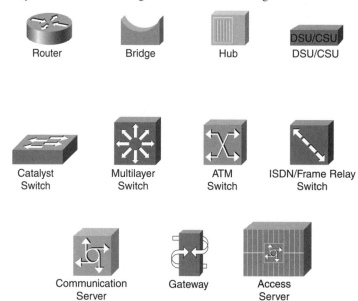

The following icons are used for peripherals and other devices:

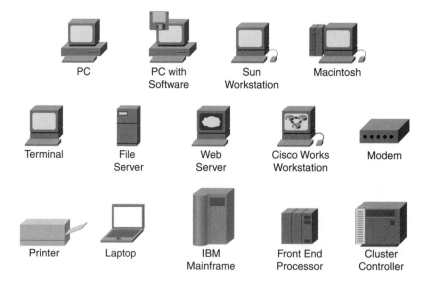

The following icons are used for networks and network connections:

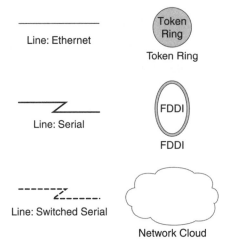

Line: Ethernet

Token Ring

Line: Serial

FDDI

Line: Switched Serial

Network Cloud

Conventions and Features

The conventions used to present command syntax in this book are the same conventions used in the IOS Software Command Reference. The Command Reference describes these conventions as follows:

- Vertical bars (|) separate alternative, mutually exclusive elements.

- Square brackets [ ] indicate optional elements.

- Braces { } indicate a required choice.

- Braces within brackets [{ }] indicate a required choice within an optional element.

- **Boldface** indicates commands and keywords that are entered literally as shown. In actual configuration examples and output (not general command syntax), boldface indicates commands that are manually input by the user (such as a **show** command).

- *Italics* indicate arguments for which you supply actual values.

This chapter covers the following topics:

- The MAN environment within the total network architecture
- Key services that are needed in the MAN
- Technologies that are used in the MAN
- Cisco products and the services they support

CHAPTER 1

Metro Area Optical Networks

Service providers (SPs) today face the greatest set of opportunities and challenges ever. Enterprise customers have started to outsource their management and connectivity services to other providers. At the same time, to meet the challenges presented by a fluctuating economy and changing customer needs, SPs must explore new solutions and services.

This chapter introduces you to the characteristics of a metropolitan-area optical network (MAON) and discusses the applications and services in that environment. Because this book focuses on the optical nature of metropolitan-area networks (MANs), this book uses the terms *MAON* and *optical* interchangeably; even when referring to MAN, an optical solution is the focus.

This chapter presents the Cisco optical products that support optical services in a MAON, including the following:

- Synchronous Optical Networking (SONET)
- Synchronous Digital Hierarchy (SDH)
- Dense wavelength division multiplexing (DWDM)
- Transparent LAN Services (TLS) or metro Ethernet services
- Packet over SONET (PoS)
- Dynamic Packet Transport (DPT) or Resilient Packet Ring (RPR)
- Time-division multiplexing (TDM)

To understand the MAN, you need to learn about the components of a service network architecture, MAN services, the physical topologies that support metro networks, and the four key service traffic patterns. With that background, you will see the service opportunities that MANs make possible.

Components of a MAN

Figure 1-1 shows how the metro area architecture is divided into the following three key components:

- Metro
- Service point of presence (POP)
- Core/backbone

Figure 1-1   *Three Components of a MAN Architecture*

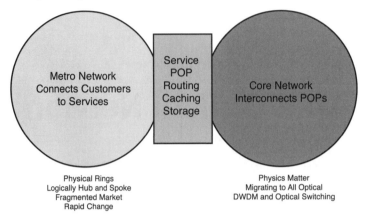

Starting from the left side of the figure, you can see the relationship between the metro network, service POP, and the core network. Each area has its own function.

Metro

The metro optical space is an evolving space with the next-generation SONET/SDH equipment penetrating this arena (Cisco ONS 15454, for instance). The metro optical space is where companies want to maximize their return on investment (ROI) for existing infrastructures, thereby reducing the cost of deployment and increasing productivity. The metro optical space is where the customer premises equipment (CPE) can connect to the access ring.

The metro optical network components have the following features:

- The metro optical network is where the customer's network connects to the optical network, usually at the access ring.

- Metro optical networks are physical rings, often deployed in the form of fiber daisy chained between buildings.

- Metro optical networks have a clear master node that is usually co-located in the IP service's POP. As a result, metro optical transport networks are designed with rings but look like logical hub-and-spoke networks.

An important part of the metro/POP is the aggregation and grooming of multiservice traffic to and from different user locations into a higher-speed core or long-haul infrastructure network.

Service POP

Service POPs act as interconnects, historically seen as central offices. These interconnects were not extremely intelligent—just fast and easy to manage at the management interface. Nowadays, the service POP acts as a common connection point between the metro optical network and the core/backbone/long-haul networks. Service POPs usually handle application services, caching, distributed load balancing, multimedia content, and aggregated TDM services.

The service POP component offers the following features:

- The service POP controls local content and provides switching to the backbone (core/long-haul) optical networks.
- The service POP provides IP services, such as access to content servers, connections to Internet service provider (ISP) networks, and connections to and from virtual private network (VPN) services.

Core

The core or long-haul backbone area is where the aggregation of high-volume traffic is transferred between remote cities or remote metro optical networks. This network usually runs multiple OC-192 (optical carrier) rings connecting with DWDM technology. You learn more about DWDM in Chapter 8, "Implementing DWDM Metropolitan-Area Networks".

Common aspects of core/long-haul backbones include the following:

- It provides focused high-speed throughput—usually at OC-192 speeds.
- The core area can be a set of rings connected or point-to-point connections in a physical full/partial mesh topology.

MAN Services

MAN network architecture supports a variety of services that can be divided into four distinct groups, as follows:

- Access services
- Metro optical transport
- Service POP services
- Core/long-haul transport services

Figure 1-2 identifies potential revenue opportunities for SPs that offer these services to their customers.

Figure 1-2   *Four Key Service Groupings*

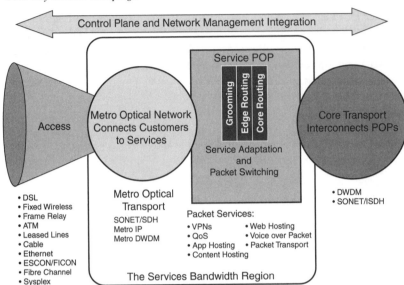

Access Services

The access network provides the connectivity and transport to the end customer. Access services include such technologies as digital subscriber line (DSL), fixed wireless, Frame Relay, and Asynchronous Transfer Mode (ATM).

Metro Optical Services

Metro optical services are important to both the SP and its customers. The customers are looking for the SP to offer solutions that cover their needs. These services can and should include anything from legacy-based Enterprise Systems Connection (ESCON), TDM voice lines, and private lease lines to transparent Ethernet-based services such as metro Ethernet or Transparent LAN Services (TLS). An SP that provides customers with DWDM-based metro services can offer many transparent services in a high-bandwidth environment.

SPs want to offer as many services over a common platform as their customers demand and technology allows. Metro optical services offer a viable platform, giving the SP operational efficiencies and enhanced operating margins.

The metro optical network aggregates customer traffic for transport to another location. Metro optical network transport might stay in the metro network, might cross to the service provider POP and terminate at a provisioned service, or might travel over the core/long-haul network to another metro ring to terminate at the CPE. The MAON uses technologies such as SONET, SDH, TLS/metro Ethernet, DPT/RPR, PoS, DWDM, and even coarse wavelength division multiplexing (CWDM). Wavelength division multiplexing (WDM) has two groups: DWDM and CWDM. Chapter 8 covers DWDM. The next revision of this book will cover CWDM.

NOTE   TLS is described in more detail in Chapter 6, "Metro Ethernet Services." Enterprise customers are requesting and SPs are providing Ethernet services that extend the WAN throughout the MAON.

Service POP Services

The service POP performs service adaptation and packet switching. The service POP performs three functions:

- Grooming traffic from the metro network
- Edge packet switching, where IP services are enabled
- Core packet switching, where POPs are interconnected over the IP backbone

The service POP is the hub of high-value Internet services. Modern service POPs can host a wide array of services, including the following:

- Web content
- Domain Name System (DNS) servers
- Connections to ISP networks
- VPN services
- Application services

To reduce response time, service POPs act as local repositories hosting tremendous amounts of local content for application service providers (ASPs) and a variety of e-businesses.

Core Services

The core network, where optical technologies dominate, is the domain of the long-haul carrier. This high-speed transport fabric interconnects service POPs and has traditionally been built as SONET ring architectures.

The core is undergoing tremendous change. SPs are moving to a more efficient architecture for optical connectivity. As IP networks grow, the role of the optical core will focus on management, switching, and protection. Cisco optical core products enable all network elements to communicate with each other. This common communication means simplified, cost-effective provisioned bandwidth, which will result in quicker revenue realization.

To provide the core service in the MAON, certain physical topologies must be used. The following section covers some of them.

MAON Physical Topologies

MAONs are configured as a variety of network topologies, including the following:

- Ring topology
- Point-to-point topology
- Mesh topology
- Linear topology

The most commonly used metro optical topology has been the ring configuration, brought on by the extensive use of SONET/SDH in SP optical networks. Rings provide fault tolerance with universal connectivity and self-healing. Unidirectional path switched rings (UPSRs) are generally used as an access/collector ring. You most often find bidirectional line-switch rings (BLSRs) used in the metro core and long-haul networks. Chapter 3, "SONET Overview," covers more about UPSR and BLSR ring configurations. See Figure 1-3 for an example of the most common MAN ring topologies.

Figure 1-3   *MAN Ring Topologies*

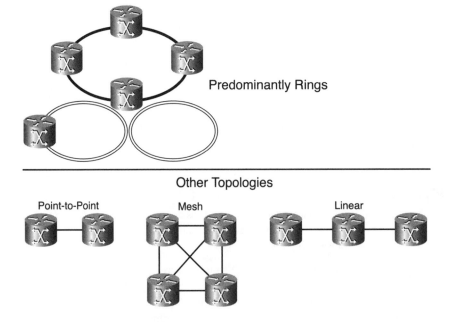

The ring topology offers advantages over the other MAON topologies. Figure 1-4 shows the transport service offerings in a MAON:

- **Point-to-point**—High-service density
- **Point-to-point**—Low- to medium-service density
- **Point-to-multipoint**—Medium- to high-service density
- **Multipoint-to-multipoint**—Medium- to high-service density

Figure 1-4   *Types of Communication Traffic in Different Ring Topologies*

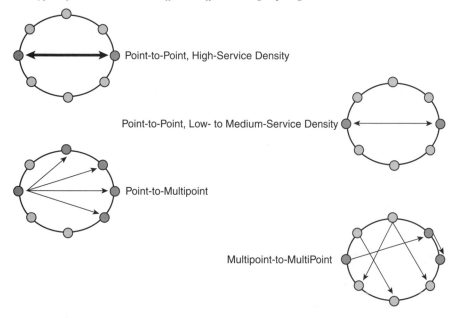

The various MAON topologies provide support for services that customers need and carry the traffic to its desired destination. Customer traffic patterns can be categorized into four key service patterns, as discussed in the following section.

Four Key Service Traffic Patterns

In a MAN, the four key service traffic patterns as are follows:

- **Point-to-point with high-service density and broadband requirements**—This type of traffic pattern is usually handled by Hot Standby Router Protocol (HSRT) remote servers, high-speed video server backup, and replication service.

- **Point-to-point with low- to medium-service density**—This type of traffic pattern is most common in small- to medium-size businesses, and it is sometimes called a linear pattern. Customers often request this type of service to allow nightly backups and replication to occur during off-peak hours.

- **Point-to-multipoint**—This type of traffic pattern typically occurs in a hub-type environment, as from corporate headquarters to remote sites, and it is sometimes referred to as a hub-and-spoke topology. This is usually designed around a multi-cast feed such as IPTV or some type of application that is sent out, such as videoconferencing.

- **Multipoint-to-multipoint**—This type of traffic pattern generally involves a VPN-type application, and it is sometimes called *a mesh* or *full-mesh topology*. It is used in enterprises where everyone needs to talk to and communicate with everyone else.

Customer service traffic patterns drive the need for an end-to-end solution. Corporate network traffic must be carried in a metro region as well as it is in a core/long-haul network. This need challenges SPs to build networks that can support many and various company requirements that might dramatically change as businesses expand or downsize. Metro optical networks must be designed to meet these requirements as they arise, both quickly and efficiently.

Now that you have seen the architecture of the metro optical network, its topologies, and key traffic patterns, this discussion shifts its focus to the MAON service offerings and opportunities.

Service Offerings and Opportunities

MAONs can provide corporate customers with, among many others, the following services:

- **Private line aggregation**—This refers to the consolidation of multiple lines that get multiplexed into a higher-speed circuit that usually traverses the access ring/metro core ring and sometimes the core backbone or long-haul rings. At the other end or the circuit, they usually get multiplexed back out into the original private line services. (Examples include OC3/OCx/10/100/1000 Ethernet or basic T1/E1/DS3 services.)

  - **Service opportunity**—Corporate customers can consolidate multiple OC-3 to OC-48 circuits, 10/100 Gigabit metro Ethernet services, and have storage-area networking connectivity all running over a single fiber pair.

- **Transparent lambda**—This refers to the raw broadband connectivity, such as OC-48 to OC-192 circuits, between different points in the MAON. This technology allows multiple circuits to be created with multiple customers running over the same fiber at different wavelengths.

- **Data networking**—This refers to broadband service, usually based on an IP layer connectivity, but it also includes Layer 2 metro Ethernet solutions. This would include TLS or metro Ethernet services and would also include multiple Multiprotocol Label Switching (MPLS) VPN solutions.

- **Service opportunity**—Customers might extend the existing Ethernet technology out of the existing building into and over the metro access ring, metro core rings, metro service POP rings, and over the core/long-haul rings. These opportunities start to address support of VPN services for intranet, extranet, and Internet business needs and might allow customers to extend local Ethernets to almost any location close to the access rings.

  Customers can purchase virtual "dark fiber" in the metro region. A dark-fiber purchase enables a customer to have a dedicated circuit running between different sites with a high amount of security.

This section covers in detail the types of services that transit an MAON. It provides the foundation for further learning in metro optic design.

Time-division multiplexing (TDM) services can be found in the MAON. TDM is a mechanism for dividing the bandwidth of a link into separate channels or timeslots. TDM services are mostly made up of voice, dialup, broadband, and private-line services. Generally, TDM revenue generation results from two sources: residential and business. TDM service is typically sold on a full-time (24×7) basis with no overbooking. Bandwidth allocated to a customer is removed from the SP's inventory regardless of the SP's customer usage.

Wavelength and transparent wavelength services are also found in the MAN. Wavelength services are equivalent to offering a customer a "dark fiber," where the entire bandwidth of the fiber is available to the customer. Transparent wavelength services allow access to some of the fiber's bandwidth for some traffic and other parts of the bandwidth for other traffic. Transparent wavelength services provide for bandwidth aggregation. By using technologies such as DWDM and bandwidth aggregation, traffic can be made to flow more effectively over existing fiber with no need to acquire additional fiber runs.

Service aggregation allows multiple services to share the existing bandwidth. Efficient multiplexing makes it possible to have LAN traffic, DS1, DS3, and OC-n traffic share the same bandwidth. This differs from DWDM, where individual services can be given different lambdas (wavelengths) within a single fiber to keep the services separate.

If the customer's 1 wavelength gets full, the SP or customer can expand into DWDM and make the 1 fiber pair look like 8, 16, 32, 64 or more individual fibers, even though physically it is just 1 fiber pair.

DWDM, CWDM, and service aggregation allow SPs to access a large amount of bandwidth, which enables them to carry additional customer services. This translates to more sales and profit.

Corporate Network Requirements

The various corporate network transmission requirements can be divided into four types of transmission services, as follows:

- **Transport services (Layer 0 and 1)**—These include SONET/SDH services, DS1/DS3/E1 services, and dark-fiber/lambda services.

- **Data services (Layer 2)**—These include VPNs, MPLS VPNs, Ethernet campus-area networks, and Ethernet WANs (which are beginning to emerge now that there has been a ratification of the IEEE 802.3ae [10 Gigabit Ethernet] standard).

- **IP services (Layer 3)**—These include e-commerce, web hosting, caching, and Internet access.

- **Content and applications (Layer 4 and up)**—These include content and software applications.

Figure 1-5 depicts corporate network requirements along with their expected level of complexity.

Figure 1-5   *Corporate Network Requirements*

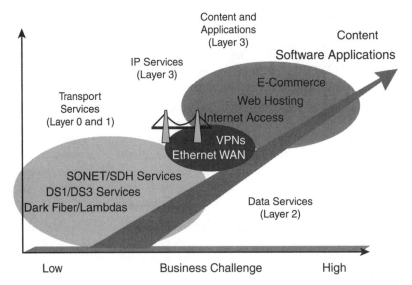

At Layer 0 and 1, you find transport services. Here the customer provides and maintains all his equipment. Therefore, the customer wants to acquire raw bandwidth from the SP and not necessary own the physical equipment between all the buildings scattered throughout multiple cities. In this case, the SP provides the bandwidth and charges according to the guaranteed level of service or Service Level Agreement (SLA) that the customer desires and pays for. The customer might acquire all or part of the available bandwidth of a given circuit. The biggest issue with providing services at this level is that the amount of income an SP can get from resources tends to be the amount it can charge that customer. The SP has less control of the bandwidth and cannot oversubscribe its network. The customer also has more work: The customer is responsible for its side of the equipment, which means that not only does the customer pay more for the service, but the customer also has to maintain qualified support personnel.

Services sold at Layers 0 and 1 include technologies such as DS1/E1/DS3, and SONET/ SDH circuits. Service providers can generate additional revenue through the use of service aggregation and DWDM.

As you move up the layers in Figure 1-5, to Layer 2 and Layer 3, VPNs, MPLS VPNs, Ethernet WAN, Internet access, web hosting, and e-commerce can be provisioned services. At this point, the SP can take the burden of some of the management issues from the customer, provide secure transmissions through the network, and have more control over the use of the bandwidth. The customer can save money by not reserving unnecessary bandwidth and offloading some of the management functions to the SP instead of maintaining an internal staff.

At Layer 4 and higher, you find content services, such as content mining and sharing, and a number of software application support services.

As you can see, SPs can offer and deploy many types of services, from the physical layer (Layer 1) service offerings to the application solutions in Layers 4 and higher (such as content mining).

Ethernet as a Service

Ethernet, ubiquitous in the LAN, is now extending its efficiencies into the MAON in an attempt to obviate the need for higher-cost WAN solutions. The effect, metro Ethernet, enables enterprise customers to have a single LAN, MAN, and WAN based on a familiar cost-effective interface. Metro Ethernet also allows for troubleshooting tools such as sniffers to be used across the WAN, reducing training and test-equipment costs.

Compared to WAN technologies such as ATM, Frame Relay, and SONET/SDH, Ethernet can be relatively inexpensive. Ethernet is scalable with speeds from 10 Mbps to 10 Gbps and is provisionable in a much more granular fashion, down to 1-Mbps increments. For example, some SPs provision in 1-Mbps increments up to 20 Mbps, 5-Mbps increments from 20 to 50 Mbps, and 10-Mbps increments above 50 Mbps. Compare that granularity to the jump between an OC-3 and an OC-12, for instance, or between a T1 and a T3.

An extension of Ethernet across the MAON gives the user geographical independence beyond the building without the user having to resort to a WAN technology. Ethernet can run over optical, IP, or MPLS.

Metro Ethernet allows SPs to offer services such as IP telephony, storage-area networks (SANs) over IP, and new managed services, such as firewalls and intrusion detection systems.

From a management perspective, SPs can reduce provisioning from days to minutes and can allow customers to manage their accounts locally through their web browser.

Ethernet technology has taken over the LAN footprint with the emergence of Gigabit over Token Ring. The ability to directly extend Ethernet connectivity through the MAN and WAN is extremely attractive to corporate clients for a number of reasons, including the following:

- Everyone is familiar with Ethernet; it is standard on the LAN and within the POP.

- Ethernet offers the most bandwidth at the cheapest cost.

- Ubiquity: Ethernet is everywhere.

- Ethernet is inexpensive.

Ethernet is used in almost all LANs. Its low cost and high performance makes it an ideal candidate for optical connectivity.

Metro Ethernet is the next-generation (NGen) WAN technology. Whereas ATM and Frame Relay are at the end of their evolutionary cycles, Ethernet technologies are evolving to provide lower-cost alternatives for the enterprise to expand its LAN. SPs recognizing this trend can provide a customer with a tremendous value when compared to traditional data transport services. A customer's router port cost is far lower with a Fast Ethernet port than it is with a traditional OC3 ATM port. In addition, the customer bandwidth increments are variable from 1 Mbps to 100 Mbps or even 1000 Mbps without changing the local access interface.

As a result of these trends, Ethernet-only carriers, such as Cogent and Yipes, are offering metro Ethernet services. You can buy 100-Mbps Ethernet circuits in the North American footprint, as long as you are close to these providers.

Corporate customers seeking to implement direct Ethernet transport across the MAN or WAN will expect SLAs from the SP. Corporate customers need to have a sense of security in the network services on which their businesses rely. SLAs can be tied to throughput guarantees and uptime of the services being purchased.

SPs can offer Ethernet service in many different service types, including IP VPNs, TLS, and lambda transport, as well as direct access to SANs.

Enterprises view metro LAN connection services as highly attractive alternatives to existing SONET/ATM, private-line, and Frame Relay services for a number of fairly obvious and a few not-so-obvious reasons. Metro LAN connection services offer enterprises the following advantages, among others:

- **Cost**—At list price, metro LAN connection services can save enterprises anywhere from 30 to 80 percent over "legacy" Frame Relay, private-line, or SONET-based Internet services in a 3-location implementation (4 to 30 Mbps). Additionally, enterprise hardware switch investments are maximized in an Ethernet environment. (Gigabit Ethernet LAN switch ports are one-quarter of the cost of ATM LAN switch ports.)

- **Simplicity**—Ethernet is the clear winner in the enterprise LAN environment. (In 2000, three times more ports of Gigabit Ethernet were shipped than ATM ports.) Utilization of a metro LAN connection service no longer requires network management staff to understand and implement the intricacies of ATM or even Frame Relay. The only operational expertise required of the enterprise is an understanding of the various VLAN types (port, MAC, protocol, or policy based), IEEE 802.1Q for interswitch VLAN communication, and how the SP and enterprise VLAN implementations will interoperate. In addition, metro LAN connection services enable network managers to use existing network management software, systems, and processes to manage their metropolitan networking infrastructures. This not only simplifies their network implementations, but also reduces capital and operational investment by integrating with systems already in place.

- **Flexibility and speed**—Most emerging metro LAN connection services allow customers to either rapidly request bandwidth upgrades in 1-Mbps increments to 1-Gbps maximums or "self-provision" bandwidth (as noted previously) through an easy-to-use web interface. No longer constrained by static telephony bandwidth hierarchies, metro LAN connection services offer enterprise customers the "bandwidth-on-demand" capabilities that enable them to investigate cost-saving and revenue-generating streaming and real-time video applications.

SPs can use new technologies to expand their service offerings. Additional services might result in additional revenue streams for the SP and might provide the SP opportunities to increase usage and revenues of the SP's MAN network. Three such technologies are storage-area networks (SANs), storage service providers (SSPs), and metro and remote mirroring, which are discussed in the following subsections.

Storage-Area Network

A SAN is defined as a Fibre Channel fabric connecting servers and storage devices. Within the industry, the term SAN generally refers to a Fibre Channel fabric with multiple devices attached, which provide the following features:

- Storage consolidation (lower total cost of ownership)

- Centralized management (lower administration costs)

- High availability and fault tolerance (HA or multiple paths)

- Modular scalability (independently add compute or storage devices as needed)

- Shared data access via Silicon Graphics's CXFS (Clustered neXt-generation File System) software

- High bandwidth (compared to standard SCSI devices)

Figure 1-6 shows how a SAN system works. The figure shows a server farm with redundant storage servers connecting into the server farm sharing the storage across the servers.

Figure 1-6   *A SAN System*

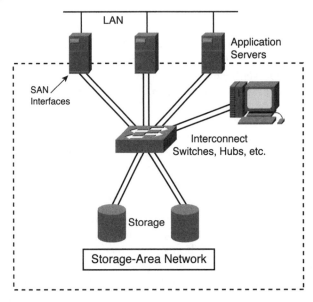

Storage Service Providers

A new market emerging for SPs is managed storage services. This market is the technology associated with storage-area networking. It allows SPs to sell remote/network disk storage and other types of content-driven applications such as video entertainment, e-learning, and so on.

Figure 1-7 shows a server farm of multiple servers—they could be mini, mainframe, imaging servers, or SuperServers—connecting as a storage offering over the DWDM metro network.

As you can see from the figure, different types of connectivity might be used (for instance, Gigabit Ethernet, Fibre Channel, or iSCSI). You can see that many types of solutions could be provisioned and sold, such as remote storage backup from mainframe or mini, LAN server replication, remote server, or hot standby. With the emerging SAN technology supplemented by optical speeds, you can start to drive the solution that customers need. For instance, this could be a hospital network with multiple hospitals all talking to each other. You could video surgeries and send them over to the teaching hospital with live feeds to enable other doctors to help or review the progress.

Figure 1-7   *Storage Service Providers*

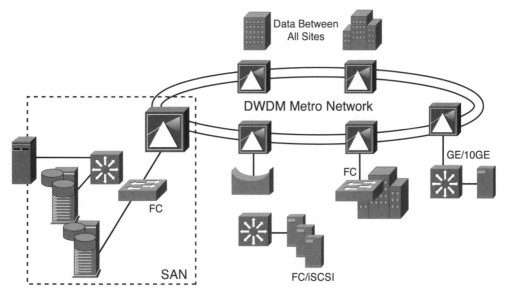

The challenge for the SP is to provide enough bandwidth between its corporate customer and the remote storage system. For access technologies, customers might be using technologies such as the following:

- **Gigabit Ethernet**—Ethernet transport running at 1 Gbps.

- **10 Gigabit Ethernet**—Ethernet transport running at 10 Gbps.

- **Fibre Channel**—A technology for transmitting data between computer devices at rates of up to 1 Gbps. Fibre Channel is used to connect computer servers' shared storage devices and for interconnecting storage controllers and drives.

- **Internet Small Computer System Interface (iSCSI)**—A new IP-based storage networking standard for linking data storage facilities. It is used to facilitate data transfers over intranets and to manage storage over long distances.

- **Enterprise Systems Connection (ESCON)**—A set of products that interconnect attached storage, workstations, and other devices. The signaling rate of an ESCON I/O interface is 20 Mbps, but the maximum channel data rate is 18.6 Mbps up to 5 miles (8 km) and 17.6 Mbps at 8.5 miles (9 km).

An MAON can meet this challenge. An MAON can provide gigabits of bandwidth wavedivision multiplexed over a single fiber to provide customers with broadband connectivity to their remote storage system.

Metro and Remote Mirroring

The term *mirroring* refers to the ability to maintain an exact copy of data in more than one location in real time. Although mirroring can be accomplished within a single site, this does not provide the protection that is needed in the event of a major catastrophe, such as a fire, that could destroy an entire building or complex. Metro mirroring provides for this process within a metro area, and remote mirroring traverses greater distances.

An SP can provide metro and remote mirroring as a service to fulfill disaster-recovery and peak-loading needs of its corporate customers. Extensive bandwidth is required to meet the needs of today's broadband content. An MAON can provide the amount of bandwidth needed at a price point that makes this an affordable and useful service.

In Figure 1-8, the primary data center in New York is mirrored in the metro area by a New Jersey location. The New Jersey center is also mirrored through a WAN to London.

Figure 1-8   *Metro and Remote Mirroring*

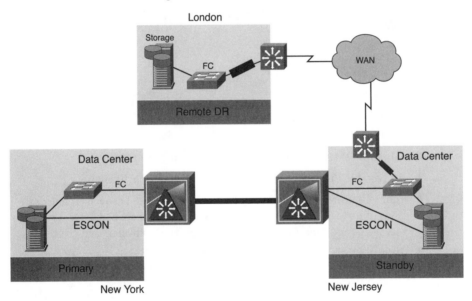

With the potential for a disaster—either from nature or man—to occur at any time, mirroring is one of the most important options a business can consider.

Cisco Support for Metro Services

Cisco provides equipment that supports all of these services and more.

Figure 1-9 depicts the capability of the Cisco ONS 15454 and 15327 products to work together as customer premises equipment (CPE) for customer-level aggregation and to work as a metro aggregation device.

Figure 1-9 *Metro Area with Cisco ONS and Optical Router Products*

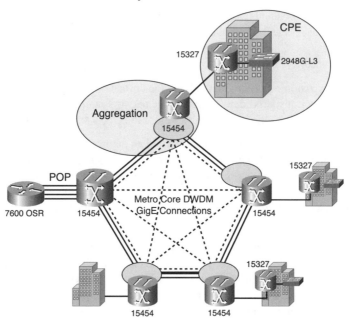

Now that you understand corporate network requirements, it is time to take a look at the various service technologies available in the MAN environment.

Service Technologies

To meet the needs of corporate clients, SPs must use various broadband optical technologies. These technologies provide the foundation for the broadband services that are offered over the metro/core optical network.

To provide the private-line aggregation service, the SP needs to provide multiservice aggregation using either SONET/SDH or DWDM. To provide transparent lambda service, which is the capability to transport any protocol over individual wavelengths within a single fiber, the SP must use DWDM. And finally, to provide data networking services, it is necessary to use Ethernet switches, routers, DPT, PoS, SONET/SDH, and DWDM. (DPT, PoS, SONET, and DWDM are discussed in more detail in the next sections.)

Implementing these technologies requires an understanding of what the industry has used in the past, what it is using now, and what it needs to use to meet the needs of the future.

Early Voice and Data Network Architectures

Early voice and data network architectures were based on TDM circuits established across a SONET access and backbone network. The network was primarily designed to support voice switches (Class 4/5 switches). Circuits between rings were switched using a digital cross-connect system.

Circuit provisioning was a relatively tedious process. This was fine in this environment because data transport did not dominate the network, and voice transport was automatically switched "through" the backbone by way of the Class 4/5 switches.

Advances in data technologies have created an environment where data connectivity and bandwidth have become an issue. As more customers require higher bandwidths to handle their data services and applications, the component-based SONET architecture becomes inefficient in handling these demands. As a result, more SONET rings, digital cross-connects, and SONET fiber need to be provisioned, and SPs need a new solution that enables them to service this demand.

The expansion of the global Internet and dominance of the IP-based applications are the impetus behind the efforts of SPs to offer exciting new services to their customers. Traditionally, packets over DS3/DS1 services have been the predominant access methods offered by SPs in dense metropolitan markets. Because applications are becoming increasingly complex and network intensive, however, the demand for higher-bandwidth services at a lower cost is increasing at a rapid pace among various customers. As a result, local exchange carriers (LECs) face the challenge of providing higher-bandwidth services in a cost-effective manner to maintain a competitive edge.

The changes in business customer networks have made this new challenge even more formidable for SPs. Enterprises want to increase outsourcing for cost savings and access to networking expertise. Service providers who seize this market opportunity will enjoy financial success, with new revenue streams and increased profits.

SONET

To deal with the challenge of providing high-bandwidth services in a cost-effective manner, LECs are carefully considering MAONs. In North America, MAONs are configured in physical ring topologies. The predominant technology deployed by both incumbent local exchange carrier (ILECs) and competitive local exchange carrier (CLECs) to these fiber rings is SONET.

SONET provides the high speed, high capacity, and rapid-recovery times that are required for an SP environment. As for high speed and high capacity, SONET basically starts at 51 Mbps and can transport traffic at up to 10 Gbps from readily available commercial equipment. The SONET integrated optical platform can also support multiservice data transport. Within a single fiber, multiple services of different protocols and speeds can be transported. A single fiber can support multiple wavelengths called *lambdas*. Each lambda can be used as an open pipe to support a single service or can be segmented to support multiple services, such as voice, data, and video, packaged in different transport systems, such as ATM or IP. Figure 1-10 shows the capability of support for multiple services over an optical circuit.

Figure 1-10 *Multiple Service Data Transport with SONET*

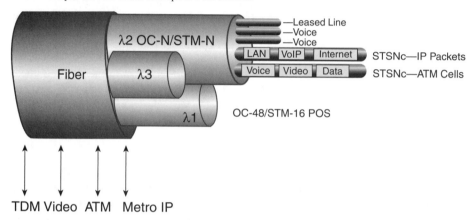

SONET reduces the overall backbone cost and equipment requirements, and at the same time simplifies the provisioning process for backbone services. The net result is a high-efficiency, low-cost, and scalable broadband service.

The Cisco ONS 15454 and Cisco 12xxx series routers are the Cisco optical-based platforms that can support SONET speeds up to OC-192, and the Cisco ONS 15327 can support SONET speeds up to OC-48. These products provide both the TDM support for legacy data and the high SONET speeds of today. Supporting multiple service types, the Cisco ONS and Cisco 12xxx series products can mix multiple service types onto the same wavelength to maximize bandwidth revenues. The ONS 15454 and Cisco 1200 series products can also offer support of bandwidth over subscription and SLA enforcement.

SONET is discussed in detail in Chapter 3. The Cisco ONS products are covered in Chapter 4, "ONS 15454 and ONS 15327 Optical Platforms," and Chapter 5, "Configuring ONS 15454 and ONS 15327."

Transparent LAN Services or Metro Ethernet Services

TLS brings simplicity to the customer and extra revenue potential to the SP by providing an Ethernet interface for customer access into the metro network. Customers today have moved beyond the need for basic point-to-point Internet circuits and legacy voice tie-lines. With the steadily growing acceptance of Voice over IP (VoIP) systems, increased demand for VPNs, and expanding distributed mesh networking, SPs must put into place new systems to support these infrastructures.

Transparent LAN services is where customers with distributed facilities want to operate their LANs and interconnect without having to purchase, configure, and manage extra equipment. It is up to the SP to figure out how to make this happen (and at a cost that is acceptable to the customer). The customer just hands off an Ethernet connection, and the rest is managed for him. The ONS products can accept native Ethernet and provisioning circuits through the SONET metro network quickly and easily. With the use of shared packet rings on the ONS, Ethernet traffic can be multiplexed for even more efficient transport.

Cisco's TLS enables the SP to support both native LAN services at LAN speeds over metro and wide-area links to the customer and to support legacy customers with framed traffic. You can find more details about TLS and the ONS products in Chapter 6 and Chapter 7, "Configuring Metro Ethernet."

Packet over SONET

Packet over SONET (PoS) is a router-based technology found in Cisco 12xxx series routers and other routers. Operating at up to OC-192 speeds, PoS provides high-speed transport to data-centric networks. PoS has the distinct advantage of a larger maximum transmission unit (MTU) than the ATM 53-byte cell, providing it with more efficiency (less overhead) at higher transmission speeds. PoS is also a SONET/SDH-compliant interface that supports SONET/SDH-level alarm processing, performance monitoring, synchronization, and protection switching. This support enables PoS systems to seamlessly interoperate with existing SONET infrastructures and migrate slowly to packet-based optical networks, thus eliminating the need for expensive "forklift" upgrades to existing equipment as demand for services escalates.

PoS is best used in an environment where the majority of the traffic is data that must be transported across a WAN. With PoS, data can be efficiently encapsulated into IP traffic inside a SONET synchronous payload envelope (SPE) and pass across just about any vendor's SONET circuits until it reaches the far side. Voice and video can still be carried within the IP packets and given priority with Layer 3 controlling the quality of service.

PoS can be used in a number of applications and with a number of technologies, including ATM and DPT. PoS is covered in detail in Chapter 9, "Packet over SONET," and Chapter 10, "Configuring Packet over SONET."

Dynamic Packet Transport

DPT is a newer ring technology that currently works with Cisco higher-capacity routers. DPT can work across a SONET ring but can utilize more of the ring's bandwidth for working traffic than the standard SONET ring. Although it is Layer 1 independent, DPT can maximize the SONET ring for packet transport. DPT uses both of the rings' bandwidth for working traffic. When an IP packet comes into a router, it is switched into the inner or outer ring immediately. Control packets are sent one way and data packets are sent another. In addition to having both working fibers, the Cisco routers also have built-in protection-switching technology called *intelligent protection switching*. This switching supports restoration in less than 50 milliseconds. Figure 1-11 shows a DPT ring.

Figure 1-11   *A DPT Ring*

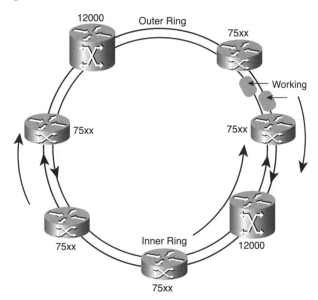

Most of the applications for DPT are POP connectivity and metro. Supported OC-192 and OC-48 speeds will drive more backbone ring architectures.

DPT operation in an MAON can be supported over a DWDM transport system. DPT is used on the metro access ring portion of the MAON. DPT provides the following:

- Fault-tolerant switching and a less than 50-millisecond recovery from ring failure
- Efficiency, because there is no need to put half the SONET ring in "protect" (unused as a spare)
- Fair allocation of bandwidth through fairness algorithm

DPT is an ideal access protocol in that it is highly robust, fault tolerant, and provides broadband connectivity. It reduces the amount of fiber feeds for fault-tolerant connectivity, while providing maximum efficiency because the entire bandwidth of the ring can be used. DPT does not require "protection" (unused) bandwidth.

DPT enables an SP to offer services such as broadband IP. SPs can sell broadband IP as Internet access, VPN, storage-area networking, VoIP, and so on. The DPT metro network transports IP over optical in a fault-tolerant, high-quality environment.

DPT is covered in Chapter 11, "Dynamic Packet Transport," and Chapter 12, "Configuring Dynamic Packet Transport."

DWDM

Dense wavelength division multiplexing (DWDM) allows a single fiber to carry multiple wavelengths simultaneously. This one fiber acts as many. DWDM enables an SP to install once and sell many times, essentially selling wavelengths or lambdas each time.

The benefits of metro DWDM include the following:

- Rapid service creation and deployment.

- Customers are interested in how soon it can be up, not how much it costs to build.

- Transparent services: With protocol transparency, customers can transport any type of service (ESCON, Ethernet, IP, TDM, or proprietary applications/protocols) without protocol conversion complexities and cost.

- Optimal reliability and protection.

- Sub-50-millisecond restoration that monitors photons.

The metro DWDM strategy is to build an optical network out as fast and with as much bandwidth as possible. DWDM makes adding bandwidth as easy as activating a new wavelength. Now it becomes cost effective to sell customers OC-48s. In fact, instead of the long-term leases normally associated with SONET, DWDM can provide bandwidth on short notice with "rental" terms.

DWDM can be used to transport a number of broadband services over a single fiber. This allows an SP to "build it once; sell it many times." Gigabit Ethernet for IP transport, DS3 for private-line and voice transport, and OC-n transport for "virtual dark fiber" are all service possibilities for the SP.

DWDM is detailed in Chapter 8.

Summary

This chapter covered what constitutes the metro area and which type of services and traffic patterns are found in the respective locations. In addition, you saw the types of applications that can generate revenue and the technologies and equipment that make it possible.

Review Questions

1 Which of the following is not an optical metro region service provider service?

    A Private-line aggregation

    B Voice line switching

    C Transparent lambda

    D Data networking

2 Which portion of the network focuses on routing and caching?

   A   Metro network

   B   Service POP

   C   Core network

   D   None of the above

3 Which portion of the network is focused on connecting to the customer?

   A   Metro network

   B   Service POP

   C   Core network

   D   None of the above

4 What are the four main elements of a metropolitan environment?

   A   Metro network, core packet switching, web content, core network

   B   Access network, core packet switching, edge packet switching, service POP

   C   Access network, metro network, service POP, core network

   D   Metro network, DNS servers, VPN services, core network

5 Which four physical topologies are used in a metro network?

   A   Ring, mesh, fault, point-to-point

   B   Point-to-multipoint, bus, mesh, ring

   C   Point-to-multipoint, point-to-point, mesh, bus

   D   Bus, mesh, point-to-point, ring

6 Which one of the following functions is generally not associated with a service POP?

   A   Grooming of traffic

   B   Edge packet switching

   C   Core packet switching

   D   Accessing technologies

7 Which type of physical topology is predominantly used in metro networks?

   A   Point-to-point

   B   Mesh

   C   Bus

   D   Rings

8 How is time-division multiplex service typically sold?

   A Demand basis

   B Full-time basis

   C Usage-based basis

   D None of the above

9 Which of the following is *not* a service traffic pattern?

   A Point-to-point, high-service density

   B Point-to-point, low- to medium-service density

   C Multipoint-to-point

   D Multipoint-to-multipoint

   E Point-to-multipoint

10 Wavelength service offering is equivalent to which of the following?

   A Wireless

   B Dark fiber

   C Multiple fibers

   D Gigabit Ethernet

11 SAN is defined as which of the following?

   A Fibre Channel fabric connecting servers and servers

   B Fibre Channel fabric connecting servers and storage

   C Super high-speed access network

   D None of the above

12 Metro and remote mirroring can be offered to fulfill _____ needs. Choose all that apply.

   A Disaster-recovery

   B Peak-loading

   C Video-on-demand

   D Bandwidth-protection

13 Transparent lambda wavelengths will be enabled by which of the following?

   A SONET

   B DWDM

   C DPT

   D PoS

14 Which is the predominant broadband technology deployed by service providers in the local metro market?

   A Wireless

   B DSL

   C SONET/SDH

   D ATM

15 DPT is a _____-based technology.

   A Point-to-point

   B Hub-and-spoke

   C Ring

   D Point-to-multipoint

16 What are the capabilities of next-generation SONET multiservice?

   A It reduces the overall backbone cost.

   B It reduces the requirement for equipment.

   C It simplifies the provisioning process for backbone services.

   D All of the above.

This chapter covers the following topics:

- A short history of voice and data transmission
- The digital hierarchy
- Developing a digital hierarchy outside the United States
- Expanding the digital hierarchy

CHAPTER 2

Voice History

This chapter examines the history of the telco network with voice. You examine the original digital hierarchy scheme, which provides an overview of the speeds normally used to transport traffic over copper. You discover the multiplexing functionality in the telco world, and you examine the differences between analog and digital signaling.

A Short History of Voice and Data Transmission

When the telephone was first introduced in 1876, Alexander Graham Bell used sound waves to create an electrical current that was transported across a wire and then converted back into sound waves at the other side. The term *analog* describes this method of voice transmission because of the continuous representation of the acoustical wave through an electrical current. Through the 1960s, the global telephone network was almost exclusively analog. Figure 2-1 shows an analog signal.

Figure 2-1 *Analog Signal*

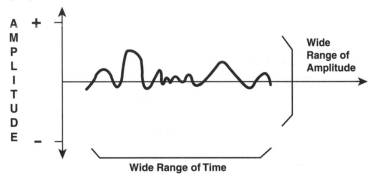

Voice or Modem/Fax Carrier Signal

Purely analog voice systems do raise some issues. A current being passed through a wire attenuates (degrades) over distance. If the distance is too great, the receiving system cannot re-create the signal. Before the signal becomes unrecognizable, the signal needs to be amplified and returned to its original strength. The distance that a signal can be transmitted before it needs amplification varies depending on multiple factors, such as what device is producing the signal and at what strength as well as the quality of the line. Figure 2-2 provides an example of signal amplification and attenuation.

Figure 2-2   *Analog Signal Amplification and Attenuation*

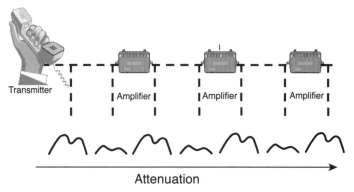

Line quality is another issue limiting analog signals. Nearby electrical fields can distort analog signals by changing the shape and characteristics of the signal. Electrical devices or wires, other phone lines, or even electromagnetic energy in nature, such as lightning, can generate these electrical fields. The phone system cannot distinguish distortion to the analog signal from the original signal itself. If the traffic passes through an amplifier, it is amplified and passed on to the receiver. Distortion comes across a voice line as static and crackles, which is why this distortion is often called "noise." Figure 2-3 provides an example of analog signal distortion.

Figure 2-3   *Analog Signal Distortion*

Attenuation:
Reduces Power Level with Distance

Dispersion and Nonlinearities:
Erodes Clarity with Distance and Speed

Signal Detection and Recovery is an <u>Analog</u> Problem

Fiber offers the capability to traverse long distances without repeaters. Repeaters "clean up" and resend the signal on a fiber span. The distance between repeaters on a digital link is purely a function of the signal loss over the distance. Greater signal loss means that a greater number of repeaters is needed. Figure 2-4 provides an example of analog signal regenerators.

Figure 2-4   *Analog Signal Regenerators*

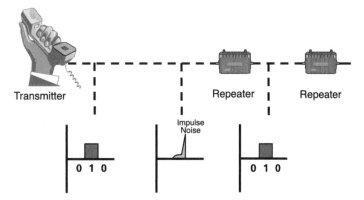

The result is a "clean signal."

The ratio of transmitted-to-received signal strength is known as *attenuation*. Engineers measure attenuation in decibels (dB), which is easier to work with. The formula is dB = 10 log SNR, where SNR is the signal-to-noise ratio. By this scheme, a send/receive ratio of 1000 to 1 causes a 30-dB signal loss.

One of the major reasons that fiber-optic cable is popular and effective is that the loss of signal strength over a given distance is much less than with other transmission media. That is not to say that fiber has no signal loss, or attenuation. Much of the loss of signal strength on a fiber link happens right at the interface between the light source and the fiber. This is called the *injection loss* and is a function of what is known as the *numeric aperture* of the particular fiber.

Copper media, from the twisted copper pair to coaxial cable that is used in many telephone networks and cable networks, becomes basically long antennas. Noise is just an unwanted signal that must be distinguished from the send signal at the receiver.

Part A of Figure 2-5 shows an example of an analog signal. Part B shows the same signal after distortion. They look similar, but distortion can result in an unrecognizable voice, static sounds, or background noise on the line.

Figure 2-5   *Analog Signal Before and After Distortion*

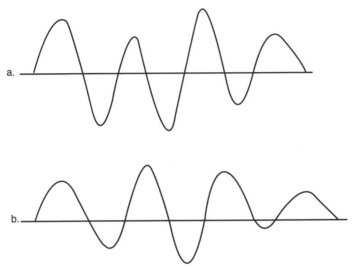

In 1928, Dr. Nyquist, working for AT&T, published a paper that postulated that by taking a digital sample of a frequency being transmitted at twice the highest signal frequency rate, you could capture the signal perfectly and reproduce it. Dr. Nyquist at this time was not specifically working on a process to digitize voice. This theorem was a result of trying to increase the speed of linear circuits. Figure 2-6 provides a representation of digital voice.

Figure 2-6   *Digital Voice*

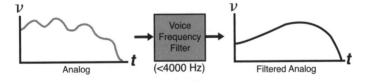

The analog wave must be sampled often enough and converted to create a stream of 1s and 0s that is precise enough to be re-created at the distant end, producing a signal that sounds like the original conversation. According to Nyquist rule, a digital signal should be created by sampling the analog circuit, delivered by the local telephone company. The frequency range on a twisted pair of wires is represented in hertz (Hz). Figure 2-7 shows a digital voice sample.

Figure 2-7   *Digital Voice Sample*

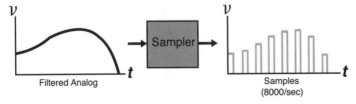

According to Nyquist's rule, the sampling of an analog wave at twice the maximum frequency of the line means that a minimum of 6600 samples per second should be sufficient to re-create the wave for the digital-to-analog and analog-to-digital conversion. To produce the quality of the higher range of frequencies of the human voice, the number of samples is rounded up to 8000 per second. Figure 2-8 shows an example of digital voice transmission.

Figure 2-8   *Digital Voice Transmission*

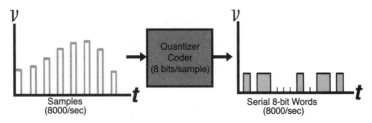

8000 Samples/sec X 8 bits/sample = 64 Kb/s

The sample now must be converted into a bit stream of 1s and 0s, as already stated. To represent the true tone and inflection of the human voice, enough bits need to be used to create a digital word. Using an 8-bit word creates enough different points on a wave to do just that.

The conversion of 1s and 0s is done by using pulse code modulation (PCM) to create the sample. Using PCM and the rules established, the transmission of the digital equivalent of the analog wave results in 8000 samples per second × 8 bits per sample or 256 amplitudes = 64,000 bits per second, which is the basis for the voice transmission in PCM mode.

In 1938, Alec Reeves patented PCM, which specifically addressed the voice-conversion process. This invention could take the varying voice signals and create a series of digitally coded pulses. Basically, this created a pattern of pulses, which could be construed as 1s and 0s.

Figure 2-9 shows an analog signal being sampled. The amplitude of the wave at a particular instant of time is measured and saved as a sample. The sample is quantized into discrete values, which are then coded into a binary form. That binary representation is then transmitted across the wire.

PCM was a huge step forward for the digitization of voice. The military used much of Reeves' work during World War II. However, more work needed to be done to improve its application in a commercial environment. Reeves' work was not to be more fully realized until after World War II.

Figure 2-9   *Pulse Code Modulation*

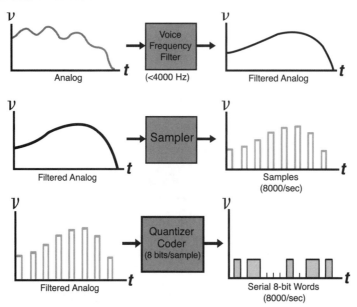

With the boom in the economy after World War II, the phone companies were pressed to keep up with the demand for new phones and the bandwidth needed to support them. The phone companies had developed ways to pass multiple analog signals across a single line using a method called *frequency-division multiplexing* (FDM), which expanded capacity by allowing up to 12 channels to be transmitted simultaneously across a single line. However, demand continued to outpace the available bandwidth. Either more cable was going to have to be laid or another solution found.

In 1948, Claude Shannon expanded PCM by more fully incorporating Dr. Nyquist's theorem and putting PCM into a more usable form. This was to aid in the development of a digital transmission system truly optimized for voice, and was later to include computer information systems.

With this expanded PCM, each sample was represented by an octet, which is 8 bits. This allowed for 256 discrete values to be used in the companding and quantizing of the voice signal. *Companding* is the *comp*ressing and ex*panding* algorithms applied to voice signals as they arrive and leave the voice network. *Quantizing* is the process of assigning a value to a companded signal, which could then be encoded into an 8-bit word. The 8 bits multiplied by the suggested 8000 samples per second resulted in the basic 64-kbps digital voice channels. So 8000 samples per second is explained by taking the 8-bit sample every 1/125 microseconds, which equates to 8000 samples every second, thus providing a 64-kbps voice channel.

Figure 2-10 shows a single digitized voice example.

Figure 2-10 *Digital Voice Transmission*

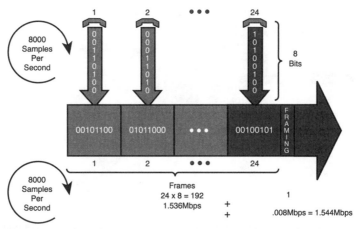

8000 Samples/sec x 8 bits/sample = 64 Kbps
64 Kbps * 24 per T1 = 1536 Kbps + framing

The conversion of voice to a series of 1s and 0s meant that voice signals could be more reliably re-created without the adverse effects of noise. The continuous wave pattern was replaced with discrete representations, which only had to be evaluated to a 0 or a 1. Noise on the line, which might affect the form of the digitized signal, would usually not be enough to change how the equipment would evaluate it. Signals could still be amplified, but regenerators could be used to return the signal to its original digitized state. This allowed voice traffic to be transmitted farther with greater clarity than ever before.

Digitalization brought its own set of issues, such as how to tell the difference between a long string of 0s and a loss of signal. Different systems were developed to resolve these problems. One system, called *Alternate Mark Inversion*, has a 0 represented as no signal and has a 1 represented as either an alternating positive or negative voltage. AMI easily handled density of 1s, but not of 0s. B8ZS and H3BD were designed to handle this. (Refer to *Cisco WAN Quick Start* [Cisco Press, June 2000] for more information about this and other coding techniques.)

The next big development came with the introduction of time-division multiplexing (TDM) for digital signals. TDM allowed 24 64-kbps channels to be multiplexed onto a single cable. This was twice as much bandwidth as was available through FDM. Figure 2-11 shows a TDM example.

Figure 2-11   *Time-Division Multiplexing (TDM)*

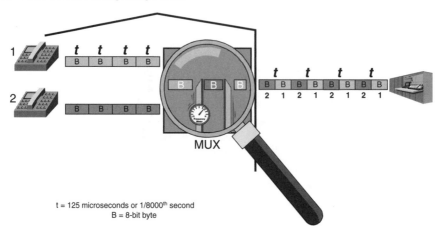

t = 125 microseconds or 1/8000th second
B = 8-bit byte

When compared to the cost of laying new cable, the more expensive equipment was less costly overall. The digital age of telephones really got off the ground in the 1960s. As the world converted to a digital telephone system, another technology arrived on the horizon: computers. As computers evolved and developed a need to communicate over long distances, there were two choices. One choice was that the computer companies could buy right of ways and lay the fiber cable themselves to interconnect the systems. The other choice was that companies could use a system that was already in place: telephone networks.

The problem was that the telco network had gone digital, but the connection from the CPE phone to the telco network was still analog. For computers to use the phone network, a device was needed to translate the digital signals of the computer into analog signals or sounds to be sent to the central phone office to be converted back into digital signals. This device was a modem (modulator/demodulator). A modem would basically take a 0 and give it one tone and then take a 1 and give it another tone, thus making it compatible with the telephone system. The framework for the transmission of voice and data was in place, and the digital hierarchy was born.

Digital Hierarchy

Like small streams of water merging into a larger river and being carried downstream, so are small phone circuits fed into a central office and multiplexed into larger traffic streams and sent across trunk lines to their destination.

The phone circuit is multiplexed into T1 circuits at the local central office and then multiplexed into T3/OC-3 trunk circuits that run between central offices. Then, the closest central office to the circuit is multiplexed back out to the phone line that goes to your house. Figure 2-12 shows an example of a digital hierarchy.

Figure 2-12 *Digital Hierarchy*

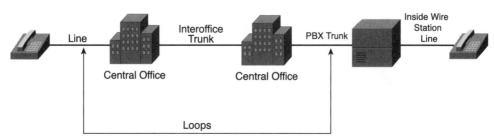

The phone circuits, which are digitized into 64-kbps channels, have become the base unit of telecommunication networks. They have become known as *digital signal level 0*, or DS0 for short. The 0 represents the phone circuit's position in a digital hierarchy, which is composed of multiple levels. Twenty-four DS0s and a framing bit are then multiplexed into a DS1, which is the first level above DS0. A DS1 specifies the structure and organization of the 1s and 0s that are transmitted across a T-carrier level 1 (T1) line. A T1 represents the transmitters, receivers, and wiring requirements needed to transport the DS1. Although the terms *DS1* and *T1* are often used interchangeably (in North America), they are not technically the same thing. E1 circuits are used in European countries, and J1 circuits are used in Japan. These circuits are not interchangeable; they all have different characteristics. Figure 2-13 shows how a T1 is created from 24 DS0s.

Figure 2-13 *Time-Division Multiplexing*

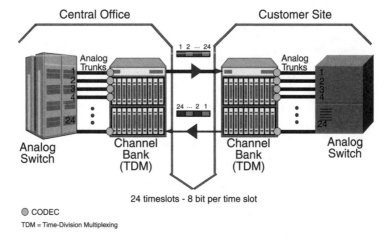

To create a T1, an octet (8 bits) is taken from each of the 24 DS0s in a round-robin sequence and placed in a matching timeslot. A single bit is then added to the front to complete the T1 frame. Framing is normally used to delimit the traffic, but framing bits can also carry

separate but related information to a traffic stream so that management, alarms, and maintenance information can be sent at the same time as the user traffic. This additional information is referred to as *overhead*. The process used to combine multiple signals one octet
at a time is called *time-division multiplexing* (TDM).

Because each signal is given a distinct timeslot, it is very easy to distinguish one DS0 from
another. You might wonder how a single bit could result in valid management information.
A single bit by itself doesn't say a lot; if you string bits from multiple frames together,
however, information can be passed. With 8000 samples of 1 byte each taken every second,
there are 8000 framing bits, which produce an 8-kbps communications channel that can be
used to pass management information. Because an 8-kbps communication channel was not
sufficient for voice traffic, additional methods were developed to pass control and
protection information.

To enhance the control information for voice traffic, extra bits were needed in a DS1. The
creation of the A and B signaling bits is done by robbing (RBS) the eighth bit from each of
the 24 timeslots on the sixth and twelfth frame. The overall structure defines the D4 framer.
Figure 2-14 shows an example of frame formats.

Figure 2-14   *Frame Formats*

- D1 (1965)
  - 8 bits/channel
  - Least significant bit robbed for signaling every 6th frame
    (A bit)
- Super Frame (1969)
  - 12 Frames = 1 Super Frame
  - Least significant bit robbed for signaling in 6th and 12th frames
    (A and B bits)
- Extended Super Frame (1985)
  - 24 Frames = 1 Extended Super Frame
  - Framing Bits
    - 6 used for frame synchronization (2 Kb/s)
    - 6 used for embedded operations channel (4 Kb/s)
    - 12 used for embedded operations channel (4 kb/s)
  - Least significant bit robbed for signaling in 6th, 12th, 18th and 24th frames (A, B, C, D bits)

Extended super frame (ESF) expands this by doubling the size of the frame unit to 24
frames, resulting in A, B, C, and D signaling bits in the eighth bit of each timeslot in the
sixth, twelfth, eighteenth, and twenty-fourth frame.

D4 and ESF were created to further define how to use the extra framing bit on each frame.
With a D4 frame, for instance, you have 12 bits for frame check sequence (FCS), sync, and
so on. ESF expands this to 24 bits. Currently, 4 functions are associated with the ESF
framing bits, giving 6 bits to each.

When 12 frames are grouped together in this fashion, it is called D4 framing or super frame.
Figure 2-15 shows an example of D4 frame format.

Figure 2-15   *D4 Frame Format*

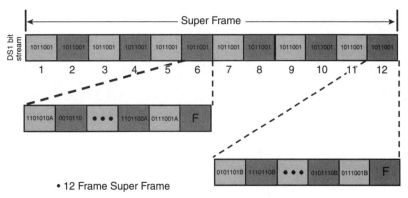

• 12 Frame Super Frame

• LSB of each channel of frame six used for A bits

• LSB of each channel of frame twelve used for B bits

When 24 frames are grouped together, this is called *extended super frame* (ESF). Under ESF, signal bits are conveyed in frames 6, 12, 18, and 24, with the bits labeled A, B, C, and D, respectively. Because T1 lines can be provisioned as either a 12-frame sequence known as D4 framing or a 24-frame sequence known as ESF, you must consider the compatibility between the signaling method used and the framing format when acquiring equipment to construct a voice network. Figure 2-16 shows an example of ESF frame format.

Figure 2-16   *ESF Frame Format*

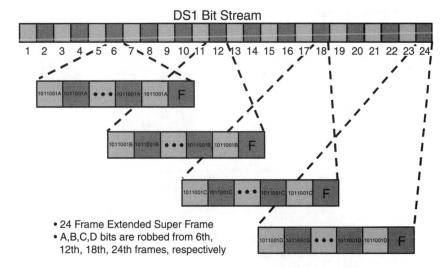

• 24 Frame Extended Super Frame
• A,B,C,D bits are robbed from 6th,
  12th, 18th, 24th frames, respectively

Figure 2-17 is another graph, showing an ESF.

Figure 2-17 *ESF Frame Format*

Frame* Counter																									
	←————————————— One Super frame —————————————→																								
Frame* Counter	1	2	3	4	5	6	7	8	9	10	11	12	13	14	15	16	17	18	19	20	21	22	23	24	Speed
ESF BIT: 001011				0				0				1				0				1				1	2 Kb/s
Data Link	D		D		D		D		D		D		D		D		D		D		D		D		4 Kb/s (12/24 X 8000)
ERROR DETECTION: (CRC-6)		E1				E2				E3				E4				E5				E6			2 Kb/s
OVERALL FRAMING:	D	E1	D	0	D	E2	D	0	D	E3	D	1	D	E4	D	0	D	E5	D	1	D	E6	D	1	

* Not to Be Confused with DSO Channel

Developing a Digital Hierarchy Outside the United States

While the United States developed the digital hierarchy, the international community worked on their own version of a digital hierarchy. Figure 2-18 shows an example of a North American digital hierarchy.

Figure 2-18 *North America Digital Hierarchy*

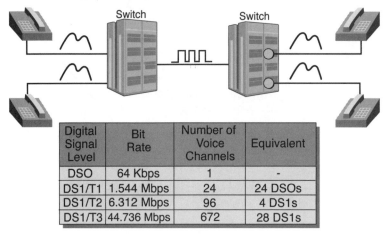

Digital Signal Level	Bit Rate	Number of Voice Channels	Equivalent
DSO	64 Kbps	1	-
DS1/T1	1.544 Mbps	24	24 DSOs
DS1/T2	6.312 Mbps	96	4 DS1s
DS1/T3	44.736 Mbps	672	28 DS1s

◯ CODEC

The international digital hierarchy developed by the ITU-T (formerly CCITT) was not happy with the management channel being limited to just 8000 bps-nor did they like that the U.S. version needed signaling information to be passed inside the circuit by robbing bits from the actual traffic. The ITU-T community wanted an alternative solution to the existing U.S. version, so they multiplexed 30 voice channels together rather than 24 and then added 2 additional 64-kbps channels—one for management and one for signaling. The first extra channel, channel 0, is used for timing synchronization. The second extra channel, channel 15, is used for signaling, alarms, and maintenance.

This combination of DS0s is transported over an E1 line, which gives this standard a rate of 2.048 Mbps rather than the 1.544 Mbps in the United States. Figure 2-19 shows an example of a European digital hierarchy. Figure 2-20 shows an example of E1 voice transmission.

Figure 2-19   *European Digital Hierarchy*

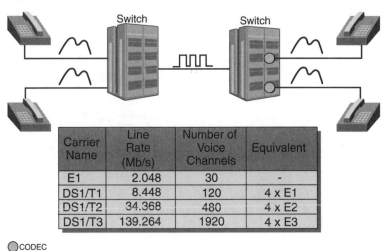

Carrier Name	Line Rate (Mb/s)	Number of Voice Channels	Equivalent
E1	2.048	30	-
DS1/T1	8.448	120	4 x E1
DS1/T2	34.368	480	4 x E2
DS1/T3	139.264	1920	4 x E3

⬤CODEC

If you work in North America, you might or might not have ever had exposure to an E1 circuit. E1 is the level 1 digital signal that is found in the ITU-T community. Most countries in the world use the E1 circuits, with the exception of North America and Japan. Parts of Korea, Taiwan, and Hong Kong also use T-carrier interfaces, but they are gradually using more and more E-carrier interfaces.

Admittedly, the E-carrier technology differs from its North American counterpart, but it still uses many of the same PCM principles. "Analog-to-digital" conversion is an area that E-carrier circuits use (A-Law) and T-carrier circuits use (μ-Law). Figure 2-21 shows an example.

Figure 2-20 *E1 Voice Transmission*

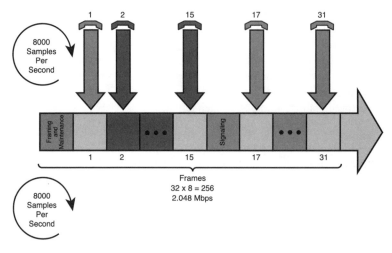

Figure 2-21 *1A-Law and μ-Law*

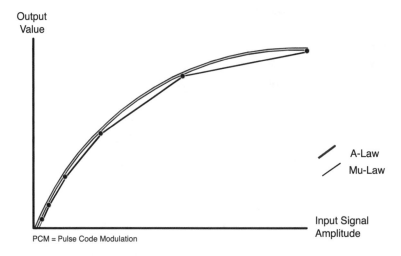

A-Law and μ-Law are similar in many ways. Figure 2-22 lists the similarities.

Figure 2-22 *A-Law and μ-Law Similarities*

A-Law and Mu-Law PCM
They Are Similar in Many Ways

— Both are linear approximations of a logarithmic input/output
  relationship.
— Both are implemented using 8-bit code words (256 levels)
  resulting in a 64 Kbps channel for voice.
— Both break the dynamic range into a total of 16 segments.
  • 8 positive and 8 negative segments.
  • Each segment is twice the length of the preceding one.
  • Uniform quantization is used within each segment.
— Both use a similar approach to coding the 8-bit word.
  • First (MSB) identifies polarity.
  • Bits 2,3,4 identify segment.
  • Final 4 bits quantize the segment.

PCM = Pulse Code Modulation
MSB = Most Significant Bit

A-Law and μ-Law differ significantly in many ways. Figure 2-23 lists the differences, and
Figure 2-24 illustrates them.

Figure 2-23 *A-Law and μ-Law Differences*

A-Law and Mu-Law PCM
They Are Different

— Different linear approximations lead to different segment lengths and
  slopes.
— The numerical assignment of bit positions in the 8-bit code word to
  segments and to quantization levels within segments are different.
— A-law provides a greater dynamic range than Mu-law.
— Mu-law provides better signal/distortion performance for low level
  signals than A-law.
— A-law requires 13 bits for a uniform PCM equivalent; Mu-law
  requires 14 bits for a uniform PCM equivalent.
— An international connection should use A-law: Mu to A conversion is
  the responsibility of the Mu-law country.

PCM = Pulse Code Modulation

Figure 2-24 *A-Law and μ-Law Differences*

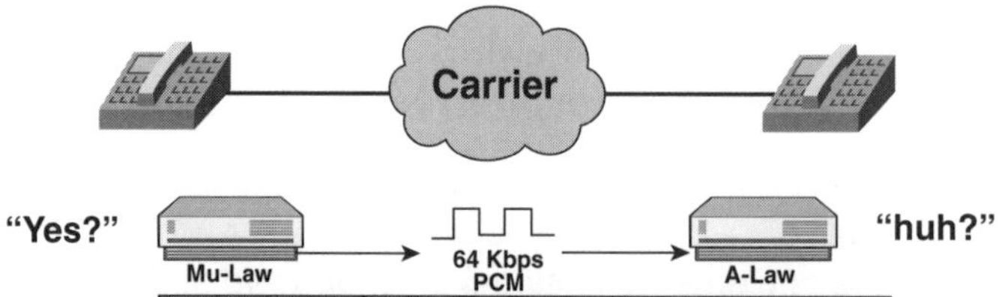

Coded	Mu-Law	A-Law
+127	10000000	11111111
+196	10011111	11100000
+64	10111111	11000000
+32	11011111	10100000
0	11111111	10000000
0	01111111	00000000
-32	01011111	00100000
-64	00111111	01000000
-96	00011111	01100000
-126	00000001	01111110
-127	00000000	01111111

Expanding the Digital Hierarchy

AT&T developed T1s in the early 1960s as interoffice trunks but tariffed their use to consumers in 1984.

The largest circuit the customers could get until 1984 was a 64-kbps circuit. As computers and their associated applications grew, customers started requiring higher-speed circuits.

Initially, the higher levels of multiplexing were proprietary, which meant users who wanted to adopt that method had to get all equipment from a single source. Deregulation made customers wary of single sources and made them want instead something based on standards, so the digital hierarchy was expanded.

At first two DS1s were just pasted together to form a DS1C, which operated at 3.152 Mbps. This was fairly simple, but the need for combining four and more DS1s resulted in new problems. Combining DS1s that were being clocked from different sources and arriving at different time rates—because of the different clock sources—made the process of multiplexing them together difficult. TDM did not provide the flexibility needed to account for these timing differences.

Bit stuffing occurs as the traffic enters the network to be combined with higher-rate traffic streams. Just 1 bit (rather than 1 byte used as a DS1) at a time is added to the frame from each channel in a round-robin fashion. If a channel does not have a bit ready to send because its rate is off from the others, a bit is stuffed in the frame to compensate for the difference. Other information is then placed in the frame as well to let the demultiplexing equipment know where the stuffed bits are. This allows for the higher-speed frame to be created but makes extracting an individual DS1 impossible without having to go through the entire demultiplexing process.

The timing difference of individual signals is why the multiple DS levels are called the *plesiochronous digital hierarchy* (PDH). Plesiochronous means "close time," reflecting the fact that the electronic equipment of the network does not acquire timing from the same source, but the timing of individual devices is nearly the same and within strict limits.

This new bit-stuffing technique was used to define a new level of multiplexing called the *DS2*. A DS2 was created by multiplexing four DS1s, providing a rate of 6.213 Mbps (see Figure 2-25). The equipment was called an M12 multiplexing.

Figure 2-25   *M12 Multiplexing: Four DS1s into a DS2*

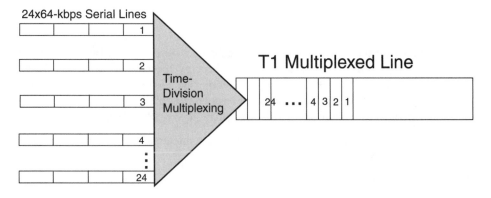

The next level in the hierarchy provided for seven DS2s to be multiplexed into a DS3, providing rates up to 44.736 Mbps. The equipment that multiplexed the signals was an M23 multiplexing. For more detail about M12 and M23 multiplexing, refer to the *Telecommunications Technologies Reference* (Cisco Press). Figure 2-26 shows an M23 multiplexing example.

Figure 2-26 *M23 Multiplexing: Seven DS2s into a DS3*

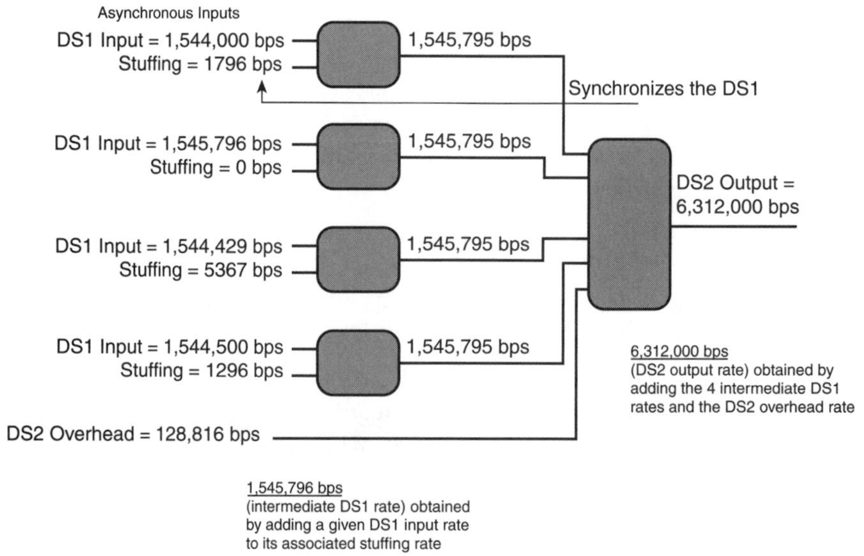

Asynchronous Inputs
DS1 Input = 1,544,000 bps — 1,545,795 bps
Stuffing = 1796 bps —

Synchronizes the DS1

DS1 Input = 1,545,796 bps — 1,545,795 bps
Stuffing = 0 bps —

DS2 Output =
6,312,000 bps

DS1 Input = 1,544,429 bps — 1,545,795 bps
Stuffing = 5367 bps —

DS1 Input = 1,544,500 bps — 1,545,795 bps
Stuffing = 1296 bps —

6,312,000 bps
(DS2 output rate) obtained by
adding the 4 intermediate DS1
rates and the DS2 overhead rate

DS2 Overhead = 128,816 bps —

1,545,796 bps
(intermediate DS1 rate) obtained
by adding a given DS1 input rate
to its associated stuffing rate

As with many things, these multiplexing techniques were both good and bad. Higher speeds were being made available, but the time and energy needed to recover an individual DS1 became a major issue, especially when multiplexing was occurring in stages (which meant multiplexing and demultiplexing had to be done multiple times). This led to the development of an M13 multiplexer, which could multiplex 28 DS1s into a DS3 directly.

A DS3 is transported over a T3 line. Table 2-1 summarizes the basic levels and bit rates of the plesiochronous digital hierarchy.

Table 2-1 *Plesiochronous Digital Hierarchy*

Signal Level	Digital Bit Rate	64-kbps Circuits	Carrier System
DS0	64 kbps	1	(None)
DS1	1.544 Mbps	24	T1
DS1C	3.152 Mbps	48	T1C
DS2	6.312 Mbps	96	T2
DS3	44.736 Mbps	672	T3

At this point, the DS3 (44.736 Mbps) was the largest standard available. The maximum user rate is 43.008 Mbps, due to overhead and bit stuffing on a channelized DS3. On an unchannelized DS3, you have 1.544 Mbps × 28 = 43.232 Mbps. (The 44.736 speed is with no overhead and no bit stuffing.)

Higher rates were available, but they were proprietary. The companies selling the bandwidth could sell customers one or more T1s, which could be transported over T3 trunk lines. The fact was that more customers than ever were requesting greater amounts of bandwidth. Not only was there a need to pass more DS0s and DS1s, but also there was a demand from customers to have their own DS3s.

If a customer wanted to obtain a T3 to carry a DS3's worth of traffic, all the bandwidth of the connecting devices through the communications network had to be dedicated to that one customer. This was not necessarily the most desirable option for the provider, so prices for a T3 were kept very high. High prices, however, did not stop the demand. Deregulation was in full force, and companies such as MCI were pushing for higher speed standards. The reason that standards were so important was that vendors seeking to compete would have a way to design equipment that could interconnect with other vendors' equipment. This would provide the customer with a choice, and these new vendors wanted the choice to be for their equipment. This point of interconnection was called the *mid-span meet*, and interoperability was very much a driving force for change.

Although higher copper speeds could be obtained, a new technology based on optical fibers held greater promise. Even though data computing needs for bandwidth were increasing, voice was still the prevalent use of North American networks. Because communications still dominated the majority of U.S. WANs, compatibility with existing technologies, especially the 64-kbps voice circuit, was a priority. This demand for compatibility resulted in the proposal to the American National Standards Institute (ANSI) of an optical standard referred to as *SONET*.

Summary

This chapter covered voice history and the design evolution of the telco network. You reviewed the differences between analog and digital signals. You reviewed the different ways to multiplex traffic, from PCM, FDM, and TDM. You reviewed the makeup of the digital hierarchy—both the North America hierarchy and the international hierarchy. In this chapter, you also learned the similarities and differences between A-Law and μ-Law.

Review Questions

1 Engineers measure attenuation in an optical network with which unit of measure?

   A   dB

   B   dBμ

   C   SNR

   D   Reflection

2 Who invented the voice-sampling method of digital voice?

   A Sir Isaac Newton

   B Bill Gates

   C Einstein

   D Nyquist

3 Digital voice transmission compared to analog transmission uses a sampling method of what?

   A 8000 samples per second × 64 bits per sample = 256 bits per second

   B 8000 samples per second × 8 bits per sample = 64,000 bits per second

   C 6400 samples per second × 8 bits per sample = 256 bits per second

   D 64 samples per second × 8000 bits per sample = 64 bits per second

4 What invention took the voice signal and created a pattern of pulses, which could be construed as 1s and 0s?

   A FDM

   B PCM

   C QCM

   D CPM

   E TDM

5 What invention allowed up to 12 channels to be transmitted simultaneously across a single line?

   A FDM

   B PCM

   C QCM

   D CPM

   E TDM

6 What device was created to take the digital signals of the computer and put them into analog signals to be sent to the central office phone system to be then converted back to digital signals?

   A Phone

   B Router

   C Switch

D   Modem

E   Multiplexer

7   How fast is a DS0 transmission rate?

   A   64 bps

   B   256 bps

   C   64 kbps

   D   256 kbps

   E   1.544 kbps

8   A DS1 has a transmission rate of what speed?

   A   1.544 Mbps

   B   43.008 Mbps

   C   43.232 Mbps

   D   44.736 Mbps

9   A DS3 with no overhead and no bit stuffing has a transmission of what speed?

   A   1.544 Mbps

   B   43.008 Mbps

   C   43.232 Mbps

   D   44.736 Mbps

10   A channelized DS3 with overhead and bit stuffing has a transmission rate of what speed?

   A   1.544 Mbps

   B   43.008 Mbps

   C   43.232 Mbps

   D   44.736 Mbps

11   An unchannelized DS3 with overhead and bit stuffing has a transmission rate of what speed?

   A   1.544 Mbps

   B   43.008 Mbps

   C   43.232 Mbps

   D   44.736 Mbps

This chapter covers the following topics:

- SONET/SDH overview
- SONET/SDH development and fundamentals
- SONET layers
- SONET framing
- SONET network elements
- SONET topologies
- SONET alarms
- SONET management
- Cisco SONET enhancements

CHAPTER **3**

SONET Overview

In this chapter, you learn about the history that led to the development of SONET. The digital hierarchy is then expanded with SONET, and the higher speeds are made possible through the use of optical fiber. To help you understand SONET and its inner workings, this chapter covers the architecture, layers, framing, and network elements that make up a SONET network. You learn the different ways a SONET network can be constructed and protected. Finally, you learn how SONET provides for alarm reporting and the ability for management information to pass through the network for operations, administration, maintenance, and provisioning (OAM&P).

SONET and SDH Overview

If you build an optical network in a metropolitan area, chances are good that you will use Synchronous Optical Network (SONET) as a transport mechanism somewhere in that network. SONET and Synchronous Digital Hierarchy (SDH, the SONET equivalent in Europe and Asia) have been the most widely used framing technique for transmitting optical signals. Any implementation that you deploy over SONET requires an understanding of its architecture, components, and topologies. That is the function of this chapter.

SONET was not adopted as the de facto standard for the entire world. The international community, which used E-carrier lines, was also interested in the benefits of optical transmission. However, they wanted the standard to be in increments, which was more suited for them. They wanted the base rate to start at 155 Mbps to be more in line with their E4 rates of 139.264 Mbps. The standard they proposed came to be known as the *Synchronous Digital Hierarchy* (SDH).

When dealing with SONET and SDH, remember a couple of points. SONET was developed by the Bellcore Labs in North America. The standards derived are used mainly in North America. SDH is the international community's SONET equivalent. SDH development began in Europe after the development of SONET in the United States. The SDH standard is used throughout most of the world outside of the United States. When discussing SONET, the terms *U.S.* and *North American standards* can be used interchangeably. When discussing SDH, the terms *international community* and *European standards* can be used interchangeably.

SONET is an optical-based digital transport system that can carry voice, video, and data traffic at very high speeds. Cisco currently ships products such as the ONS 15454 that can carry circuits of voice, video, and data traffic over an OC-n optical circuit.

You may ask yourself, what is SONET/SDH?

- SONET is a set of standard interfaces in an optical synchronous network of elements (NE) that conform to these interfaces.

- SONET is a synchronous network.

- SONET interfaces define the layers, from the physical to the application layer, referring to the OSI in data networking.

- SDH is a set of standard interfaces in an optical synchronous network of elements (NE) that conform to these interfaces.

- SDH is a synchronous network.

- SDH interfaces define the layers, from the physical to the application layer.

So you say they look the same, but they are not. You will see that many details make them different.

Some areas of similarities include the following:

- Frame format and bit rates

- Frame synchronization schemes

- Multiplexing rules

- Demultiplexing rules

- Error control

Some areas of differences include the following:

- The overhead bytes definition is similar but have some variations to accommodate differences between U.S. and European/Asia communication nodes and networks.

- In SDH, the photonic interface requires more parameters than SONET.

- Synchronous transport signal (STS) versus synchronous transport module (STM).

- Synchronous payload envelope (SPE) versus virtual container (VT).

The advantages of SONET/SDH are summarized here:

- Reduced cost both in operating and equipment

- Integrated network elements

- Remote operations capabilities, such as remotely provision, inventoried, customized, tested, and reconfigured

- Compatible with legacy and future networks

- Allows or offers network-survivability features

As with most technologies, SONET was developed to solve a number of problems that existed in networks at the time. These problems can be summarized as follows:

- SONET addressed the limitation of the existing plesiosynchronous digital hierarchy (PDH) in that as link speed increased, so did the overhead, thus making it not well suited for high-speed networks.

- As traffic was being multiplexed into higher bit-rate signals, it was not possible to directly access the underlying traffic without fully demultiplexing it first. This was expensive both in equipment and in the time it added to signal processing.

- Copper wire had inherent distance limitations.

- Because SONET/SDH standards were at an infant stage, most vendors designed their equipment to those standards, with added features and capabilities. Because of these added features in each vendor's products, most equipment had to be the same at each end of the fiber to operate correctly.

SONET was designed to address these problems and at the same time prepare for the future.

Developed during the 1980s by Bellcore, SONET/SDH was adopted as a standard in 1988 by the International Telecommunication Union-Telecommunication Standardization Sector (ITU-T) and then by the American Nation Standards Institute (ANSI). SONET/SDH has continued to evolve and has become a major factor in today's network design. (See Figure 3-1.)

When referring to SDH throughout the rest of this book, SONET is covered and SDH is referred to, unless otherwise stated.

Figure 3-1   *SONET/SDH Standardization*

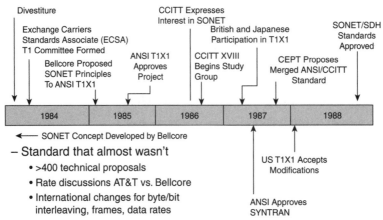

SONET Development and Fundamentals

Although some last-minute negotiations allowed SONET to be compatible with the international standard of SDH, SONET provides for a base rate that took up where the PDH network left off. The base rate of a SONET frame is 51.84 Mbps and reaches up into the gigabits per second range. Table 3-1 summarizes the SONET digital hierarchy.

Table 3-1   *The SONET Digital Hierarchy*

Signal Level Electrical Level	Digital Bit Rate	Voice Circuits	Carrier System
STS-1	51.84 Mbps	672	OC1
STS-3	155.52 Mbps	2016	OC3
STS-12	622.08 Mbps	8064	OC12
STS-24	1244.16 Mbps	16,128	OC24
STS-48	2488.32 Mbps	32,256	OC48
STS-192	9953.28 Mbps	129,024	OC192

Although SONET is considered an optical-based system, most of its work is done while it is in its electrical state. As non-SONET traffic enters the network, it is normally in an electrical state. The equipment processes the traffic from its original form, probably a DS0,

DS1, or DS3, and either multiplexes or places the traffic into a SONET frame, as appropriate, while the traffic is still in its electrical form. The equipment then passes the signal to an optical transmitter that converts the traffic into optical pulses to pass across the fiber. In the electrical state, the signals are referred to as *STS-n* where *n* is a multiple of the base rate (51.84 Mbps). When the signal is converted into optical signals and passed over fiber, it is referred to as *OC-n*. Many times, engineers make references to the STS-*n* signal and the OC-*n* signal as equal, but they are not, much like the confusion of DS1 and T1.

The bandwidth of an STS-1 signal is 51.840 Mbps. This includes a payload of 50.112 Mbps with 1.728 Mbps for overhead. As you can see, the STS frame provides more room for management information than a DS1 frame does, which means more control and management information can be used to pass more detailed information. This has given SONET both a good name and a bad reputation. On the good side, SONET has this abundance of management information and control. On the bad side, the critics say that SONET dedicates too many bytes for overhead. You explore this issue later in this chapter in the section "SONET Framing."

Each increment of the STS/OC layers is a multiple of the STS-1. So an OC-3 is 155.520 Mbps (51.840×3) and an OC-12 is 622.08 Mbps (51.84×12). When people ask whether there is an OC-9 or an OC-18, the answer is yes. Some vendors do provide these intermediate levels, but most vendors tend to offer those levels that provide service increases in factors of 4 to the base OC3, not just small increments, to justify the development costs. This is why you see the rates of OC-3, OC-12, OC-48, and OC-192 most often.

An STS-*n* is constructed from multiple STS-1s. Multiplexing can be in one stage or in multiple stages. For example, three STS-1s can be multiplexed into an STS-3. Those can then be combined with other STS-3s and even some STS-1s to form an STS-12. Each STS-1 payload is given a position in the SONET frame and can be extracted without the frame having to be fully demultiplexed. Each STS payload is still 50.112 Mbps and is separate from any other payloads.

Sometimes, this payload can be limiting, such as when the original traffic stream is IP, video, or ATM. In this case, multiple STSs can be combined, but the traffic is placed into a single larger payload envelope. An STS in this format is referred to as *STS-n*c. The small *c* indicates that the payload is concatenated and cannot be broken down into smaller units. This form of STS has less overhead and can carry larger amounts of information without the need to divide the information into different frames. The price is that the entire STS-Nc frame must be extracted and processed as a single unit. Figure 3-2 shows a concatenated and nonconcatenated frame.

SONET and SDH links most of the time are channelized—that is, a provision exists that allows for unchannelized transmission of information. Concatenated or unchannelized SONET and SDH links form a single large synchronous payload envelope (SPE) and do not specify any lower-level digital signals. Even though the circuit is concatenated, the same section, line, and path overhead are used. Section, line, and path overhead are covered in detail later in this chapter.

Concatenated frames are great for data circuits, whereas voice and non-IP video/voice nonconcatenated frames are better.

Figure 3-2 *Concatenated and Nonconcatenated SONET Traffic*

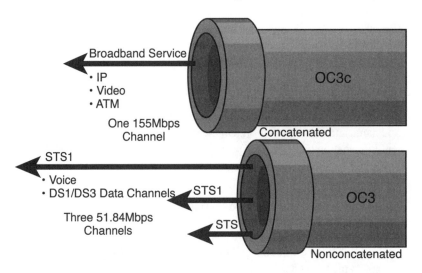

Table 3-2 shows the various SDH levels and the SONET equivalents.

Table 3-2 *SDH Levels and SONET Equivalents*

Signal Level	Digital Bit Rate	SONET Equivalent	Carrier System
STM-1	155.52 Mbps	STS-3/OC-3	STM1
STM-4	622.08 Mbps	STS-12/OC-12	STM4
STM-16	2488.32 Mbps	STS-48/OC-48	STM16
STM-64	9953.28 Mbps	STS-192/OC-192	STM64

SONET Advantages

The big advantage of SONET is that it was designed to provide the following functions needed in networking at that time:

- Single-step multiplexing

- Access to low-level signals directly

- Carry existing DS1, DS3, ATM, and packet traffic

- Synchronous timing to eliminate bit stuffing

- Overhead room for acceptable network management information

- Allow transmission of data at higher speeds (50 Mbps+)

The following sections describe these functions in more detail.

Single-Step Multiplexing

In single-step multiplexing, lower-rate traffic is combined into higher-rate traffic without having to go through stages. So if you have an STS1 and want to put it in an STS3 for transmission, you can do so in one step. If you want something to be transported inside an STS-48, again one step is needed. The process is provided through byte interleaving like that used in creating a DS1.

Access to Low-level Signal Directly

Direct access to low-level signals allows the receiving system to know exactly where an STS-1 is located within a STS-*n* and extract it for use at the local site. A new STS-1 can also be inserted into the traffic for delivery to another node. This can be done without having to fully demultiplex the signal, buffer the contents, extract the desired content, and then multiplex a new signal for transport across the wire.

Carry Existing DS1, DS3, ATM, and Packet Traffic

A single STS-1 is able to carry a payload of 50.112 Mbps. Placing a single DS1, which is 1.544 Mbps, into a single STS-1 frame would be a waste of space. Because backward-compatibility was always important in SONET, a system was devised as part of the standard to subdivide an STS-1 when necessary.

An STS-1 frame can be subdivided into seven virtual tributary groups (VTG) of 108 bytes each. Each VTG can carry one of four different types of traffic organized into virtual tributaries (VTs). VTs can differ in size based on the traffic they carry. Figure 3-3 shows a representation of virtual tributaries. VTs can be one of the following types:

- **VT 1.5(1.728 Mbps)**—A VT1.5 is 27 bytes and carries a full DS1. A DS1 contains 24 bytes plus 1 bit for framing. A VTG is 108 bytes, so one VTG can hold 4 VT1.5s, which carries 4 DS1s. The extra unused bytes are used for VT overhead.

- **VT 2(2.304 Mbps)**—A VT2 is used to transport the European E1 frame. A VT 2.0 can carry three E1s with 12 bytes left for overhead.

- **VT 3(3.456 Mbps)**—A VT3 is designed to carry a DS1C signal, which is 2 DS1s pasted together. Once again, the fit is such that 12 bytes are left over for VT overhead.

- **VT 6(6.912 Mbps)**—A VT6 can carry a DS2, which is 4 DS1s. Again, the leftover bytes are 12 bytes.

Figure 3-3   *Virtual Tributaries*

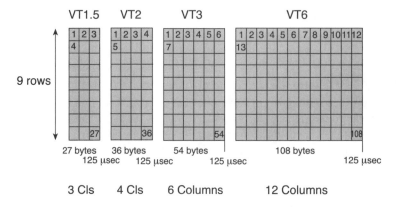

A SONET frame can carry any combination of VT groups, but a VT group can carry only one type of traffic. Because a VTG can carry 4 VT1.5s and there are 7 VTGs in an STS1, an STS1 can carry 28 DS1s, like a DS3. If a SONET frame carries VT groups, it cannot carry non-VT group traffic. If the traffic is not in the form of a DS1, E1, DS1C or DS2, the traffic can be mapped directly into a SONET frame. This includes DS3, ATM, High-Level Data Link Control (HDLC, which includes Dynamic Packet Transport [DPT] and packet over SONET [PoS]), and Ethernet. Figure 3-4 compares payloads.

Figure 3-4   *Comparison of Payloads*

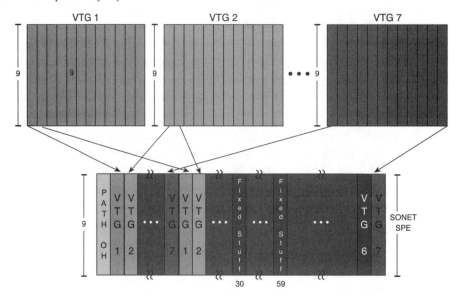

Synchronous Timing to Eliminate Bit Stuffing

SONET is synchronous. This requires a constant timing source for both the receiving and transmitting of traffic. The advantage of this common timing source is that it is possible to have byte interleaving rather than bit interleaving at these higher speeds. This results in being able to access the underlying traffic directly instead of having to demultiplex the entire signal.

In a plesiochronous system such as that used with DS1 and DS3 traffic, the clock of the sender is run independently of the receiver's clock. In a system running at 1.544 Mbps, an individual bit is sent every 648 nanoseconds. Electrically the voltage of a 1 bit only exists for half that time, so accuracy has to be down to 324 nanoseconds. Variations in the clocking rate of the sender and the receiver does not greatly impact the reception of the traffic at slower speeds because the receiver can adjust its clock by utilizing the sender's clock extracted from the signal. Issues develop when multiple senders are delivering traffic, such as DS1s, to be aggregated for passage across the network inside a larger multiplexed signal, such as a DS3. When the aggregating switch that is using its own clock source accumulates bits from each of the independently timed sources, it needs them there at precise intervals to successfully create the larger frame. The timing variances of each source can cause problems. If a bit is not ready to be processed because one of the originating systems was delayed, the switch still needs to send something to preserve its timing. The solution is for the switch to stuff a filler bit in the frame. Bit stuffing maintains the timing but prevents the demultiplexing device from accessing the underlying traffic directly because the position of any given bit cannot be guaranteed. In this case, the entire DS3 must be demultiplexed to get access to any of the DS1s. (See Figure 3-5.)

Figure 3-5   *Timing Source*

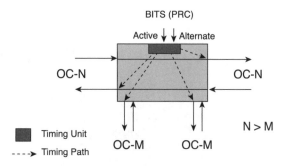

SONET uses a pointer system to account for these timing variances, which allows it to keep its synchronous nature. This avoids the use of bit stuffing, by allowing the frame to be created using the more efficient byte interleaving process. The pointer is called the SPE pointer located in the line overhead (LOH), which points to the location of the first byte of the STS-1 SPE. (See Figure 3-6.)

Figure 3-6 *Synchronous Payload Envelope (SPE)*

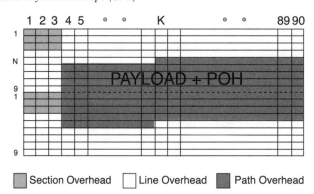

Overhead Room for Network Management Information

A SONET frame carries multiple types of overhead. A single STS-1 has 9 bytes of section overhead, 18 bytes of line overhead, 9 bytes of path overhead, and if the payload is packaged into VT groups, another 12 bytes of VT overhead for each of the 7 VT groups. All this overhead carries a wealth of management and other information between devices. Although the amount of space dedicated to overhead in SONET is considered excessive by some, as a percentage of the traffic, it is fixed at 3.45 percent. This is true at all speeds of SONET. SONET overhead is discussed in detail in the section "SONET Framing" later in this chapter. (See Figure 3-7.)

Figure 3-7 *STS-1 Frame format*

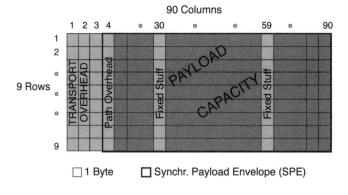

Allow Transmission of Data at Higher Speeds

At the time of the development of SONET, standards in the United States were maxed out at T3, which was approximately 45 Mbps. Other higher rates were available, but they were proprietary, which meant vendor interoperability was not an option. SONET basically picked up from this rate and moved upward. Refer to Table 3-1 for a recap of the speeds. Although not included in the table, work is being done on even higher speeds of OC-768 or approximately 40 Gbps across optical fiber.

Now that you have reviewed the benefits of SONET, the next section covers SONET's architecture.

SONET Layers

When discussing SONET, it is important to realize that SONET also conforms to standards put forth by the International Organization of Standards (ISO). The ISO defines a networking framework to be used for implementing protocols in seven layers. The model the ISO uses to show how to implement this framework is called the *Open System Interconnection (OSI) reference model*, which is shown in Figure 3-8.

Figure 3-8   *OSI Reference Model*

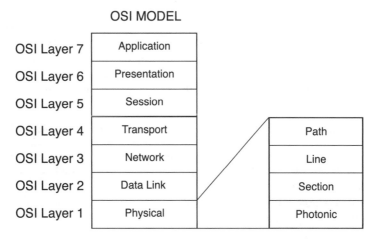

The OSI model was developed to provide for the interoperability and growth of systems that needed to interconnect. It represents all aspects of internodal communication from the application layer at the top to the physical layer at the bottom. The OSI model numbers the layers from 1 to 7, with 1 being the physical layer and 7 being the application layer. This model defines how each layer is to communicate only with the layer directly above or directly below. It was constructed in layers so that a single layer could be replaced as new technologies came into place without the need to redesign all the other layers. (See Figure 3-9.)

Figure 3-9 *SONET/SDH Protocol Stack*

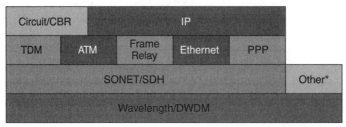

One Example, Many Layer Variations Are Possible

*eg., Gigabit Ethernet, Digital Wrapper, FICON

SONET is referred to as a Layer 1 technology because it works at Layer 1 of the OSI model. Within that layer, SONET is further subdivided into four separate layers, as illustrated in Figure 3-8.

The sections that follow describe the four layers of the SONET model in more detail.

Photonic Layer

At the lowest level, you find the photonic layer. This is the layer where the electrical STS signal is converted to an optical signal (OC-n) and photons are passed through the fiber to the next network element. The lasers are sending the traffic across the fiber as directed by the section layer. All devices in a SONET network interact at the photonic layer. The photonic layer is mainly concerned with such things as pulse shape, power level, and wavelength.

You have old-generation SONET equipment and new-generation SONET equipment.

ITU has defined an optical layer, which takes into account wavelength division multiplexing (WDM)-based networks (new physical transport networks) that could carry SONET/SDH/IP/etc. in a transparent or semitransparent manner. This new optical layer is itself divided into three sublayers: the optical channel (Och), optical multiplex section (OMS), and optical transport section (OTS).

Section Layer

A *section* is defined as the interconnection of any two adjacent SONET network elements. (See Figure 3-10.) Network elements that work at the section layer are called *section terminating equipment* (STE). STEs can be either regenerators or one of several forms of multiplexers. (These pieces of equipment are discussed in the section "SONET Network Elements" later in this chapter.) Because the STE controls the photonic layer, you will always see them depicted together.

Figure 3-10 *Section Layer*

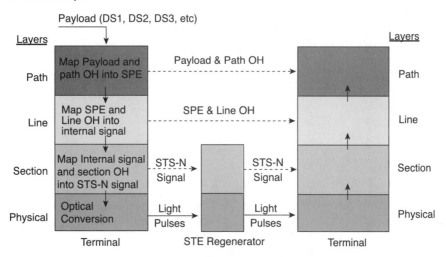

The section layer directly converses with the section layer of the adjacent network element and does not read or create any other layers itself. (See Figure 3-11.) It is here where STS-*n* blocks are encapsulated into frames and transported to the next device. The section layer is responsible for the final framing and scrambling of the traffic. It adds its own overhead to the traffic it has received before passing the traffic to the photonic layer, which transmits the traffic between systems as light pulses. The section layer communicates at the same level at both ends of the optical segment.

Figure 3-11 *Section Layer Communications*

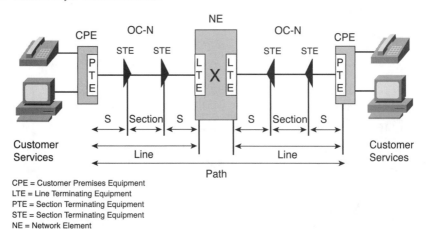

CPE = Customer Premises Equipment
LTE = Line Terminating Equipment
PTE = Section Terminating Equipment
STE = Section Terminating Equipment
NE = Network Element

There are $9 \times n$ bytes of overhead associated with section framing. These bytes are used for frame detection, section error monitoring, order wire communications, and section maintenance. You learn more about these functions in the section "Section Overhead" later in this chapter. (See Figure 3-12.)

Figure 3-12 *Section Overhead*

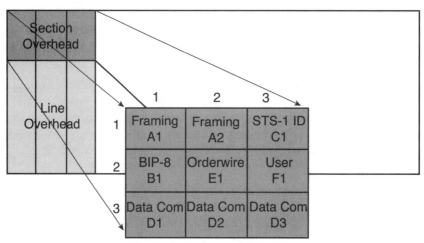

Line Layer

A line layer is used to support traffic flow between two adjacent SONET multiplexers. If regenerators exist between the multiplexers, the regenerator does not participate at the line layer. (See Figure 3-13.)

Figure 3-13 *Line Layer*

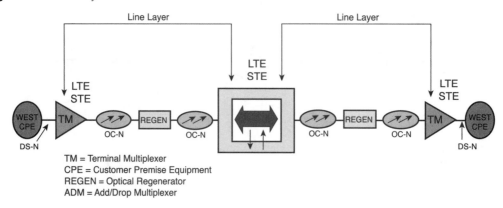

The multiplexer that terminates a line is called *line terminating equipment* (LTE). The LTE itself is also an STE and must support the functionality of both the STE and LTE. A line can be made up of multiple sections. Like the section layer, the line layer only talks to the line layer on other LTEs and is unaffected by the intermediate equipment. (See Figure 3-14.) Traffic at this level is in STS blocks. The SPE and line overhead are mapped into STS-*n*s.

Figure 3-14 *Line Layer Communications*

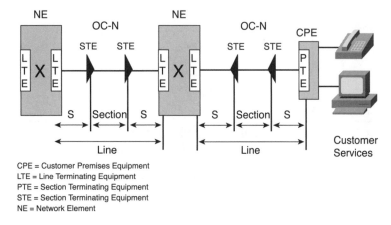

CPE = Customer Premises Equipment
LTE = Line Terminating Equipment
PTE = Section Terminating Equipment
STE = Section Terminating Equipment
NE = Network Element

The line layer adds its own overhead and is responsible for the reliable transport of the path layer payload across the physical medium. It is at this layer where multiplexing/demultiplexing of STSs into the required STS-*n* occurs. Line level protection switching is implemented here as well.

There are $n \times 18$ overhead bytes reserved in the STS-*n* frame for the line layer. Overhead bytes provide synchronization for the traffic between the two LTE elements using payload pointers as well as error checking, performance monitoring, order wire, and line maintenance and protection switching. (See Figure 3-15.)

The section "Line Overhead" later in this chapter discusses these bytes in greater detail.

Path Layer

The path layer serves as a logical connection between the point where a signal entering the SONET network is mapped into a SONET payload envelope and the point where it is delivered to its destination. (See Figure 3-16.) The path layer is responsible for the transport of payloads between SONET terminal multiplexing equipment.

Figure 3-15 *Line Overhead*

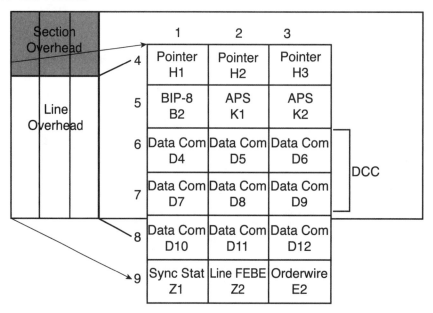

Figure 3-16 *Path Layer*

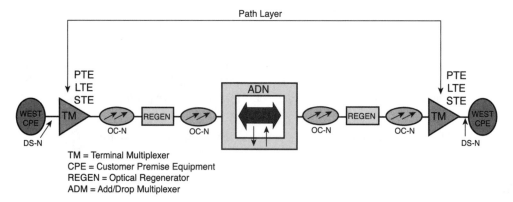

TM = Terminal Multiplexer
CPE = Customer Premise Equipment
REGEN = Optical Regenerator
ADM = Add/Drop Multiplexer

The devices (terminal multiplexers) that terminate a path are called *path terminating equipment* (PTE). The PTE is also an LTE and an STE and must support the functions of each. A path can be built up of multiple lines and sections. As with the line layer, the path layer talks only with other path layers. (See Figure 3-17.) It is here where the payload envelope is created and path overhead is added.

Figure 3-17 *Path Layer Communications*

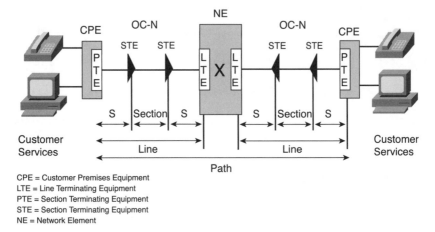

CPE = Customer Premises Equipment
LTE = Line Terminating Equipment
PTE = Section Terminating Equipment
STE = Section Terminating Equipment
NE = Network Element

The main function of the path layer is to map the incoming signals (DS1, DS3, E1, ATM cells, HDLC-framed IP) into the format required by the line layer (that is, STS-1, STS-3c, STS-12c, STS-48c, and STS-192c). This mapping can be accomplished by placing the traffic directly into payload envelopes, such as with DS3, ATM, or IP traffic, or by placing the traffic in a VT, such as a DS1, DS1C, DS2, or E1 signal, which finally ends up in an STS.

There are $n \times 9$ bytes of overhead reserved in the payload section of an STS frame for nonconcatenated traffic. For concatenated, there will be only 9 bytes of path overhead for the entire payload. The path overhead stores information used for error checking, connection verification, equipment type used, and payload type being transmitted. The path layer can provide information for protection switching in case of a failure or signal degradation. (See Figure 3-18.)

At this point, you should have a basic understanding of the layers used by SONET and that each layer has certain responsibilities. Overhead bytes are found at three of the four layers, and a greater understanding of these bytes will provide greater insight into the workings of SONET.

Figure 3-18 *Path Overhead*

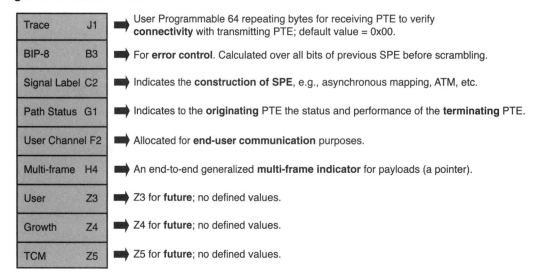

Trace	J1	➡	User Programmable 64 repeating bytes for receiving PTE to verify **connectivity** with transmitting PTE; default value = 0x00.
BIP-8	B3	➡	For **error control**. Calculated over all bits of previous SPE before scrambling.
Signal Label	C2	➡	Indicates the **construction of SPE**, e.g., asynchronous mapping, ATM, etc.
Path Status	G1	➡	Indicates to the **originating** PTE the status and performance of the **terminating** PTE.
User Channel	F2	➡	Allocated for **end-user communication** purposes.
Multi-frame	H4	➡	An end-to-end generalized **multi-frame indicator** for payloads (a pointer).
User	Z3	➡	Z3 for **future**; no defined values.
Growth	Z4	➡	Z4 for **future**; no defined values.
TCM	Z5	➡	Z5 for **future**; no defined values.

SONET Framing

Framing is used to provide organization to the 1s and 0s that comprise the information being transmitted. Information used in framing can indicate the beginning and ending of a stream of traffic and can include management, control, and signaling information.

A SONET frame regardless of its size is transmitted at 8000 frames per second or 1 frame every 125 microseconds. This makes SONET backward-compatible with voice even though it supports all traffic types.

It is easiest to understand a SONET frame by looking at it in its smallest unit: the STS-1. The STS-1 frame is 810 bytes, which is configured in 9 rows by 90 columns. This works out to a bit rate of 51.84 Mbps (810 bytes in a frame × 8 bits in a byte × 8000 frames per second). Each byte of a SONET frame represents a single 64-kbps channel. The SONET frame is sent out row by row, from left to right, which means the first row is sent out before the second row and so on. The SONET frame consists of two parts, which are the transport overhead (TOH) and the SPE, as illustrated in Figure 3-19.

In an STS-1, the TOH is 27 bytes and is further divided into 2 more parts:

- **Section overhead**—9 bytes, which comprise the first 3 bytes of the first 3 rows
- **Line overhead**—18 bytes, which comprise the first 3 bytes of the last 6 rows

Figure 3-19 *Basic SONET Frame*

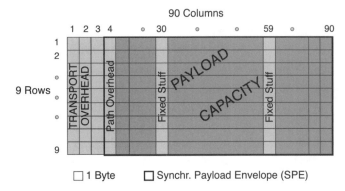

The rest of the frame is used for the payload envelope (SPE), which is used to carry the actual traffic. The SPE is 783 bytes $(810 - 27)$ providing a payload bit rate of 50.112 Mbps; depending on the payload, however, the SPE has its own overhead. As a minimum, the SPE contains 9 bytes of POH. When the SPE carries non-DS0-based traffic, such as IP or ATM, most of the remaining SPE bytes can be dedicated to the transport of traffic. If the SPE is transporting a DS3, only one DS3 can fit into the SPE and the leftover space is filled by SONET. If the SPE carries sub-DS3 traffic, the SPE is further subdivided into VTGs, each having its own overhead. The structure of the overhead depends on the type of VTG that is being used. Figure 3- 20 shows how the structure of a SONET frame is organized.

Figure 3-20 *SONET Frame Structure*

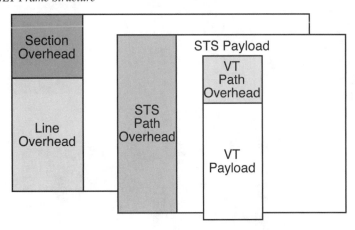

One STS-1 can contain seven VTGs. Each VTG can carry one of the following:

- 4 DS1s
- 3 E1s
- 2 DS1Cs
- 1 DS2

Figure 3-21 shows a representation of VTGs group.

Figure 3-21 *VTGs Group*

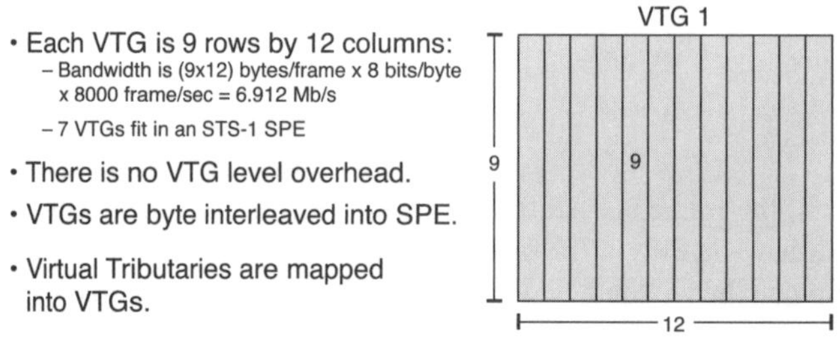

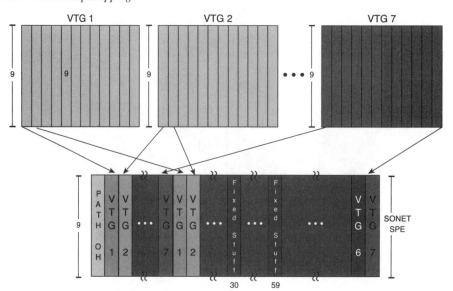

Figure 3-22 shows a representation of VTGs group mappings.

Figure 3-22 *VTGs Group Mappings*

When used, VTGs add overhead bytes for each VTG.

SONET frames are commonly depicted with the SPE and VTG floating outside the boundaries of the SONET frame. For SONET to be synchronous, it must account for the timing differences of the incoming traffic being multiplexed. Instead of using bit stuffing, it allows the SPE and VTG to float within and between SONET frames. Pointers are used in the TOH to point to the location in the first SPE byte—thus allowing the float. Pointers provide simple, dynamic phase alignment of both STS and VT payloads. This results in easy dropping, insertion, and cross-connection within the network. Transmission signal wander and jitter are also minimized with pointers. (See Figure 3-23.)

Figure 3-23 *VTGs Group Mappings*

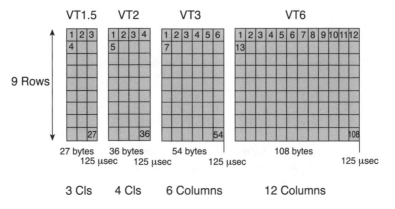

SONET frames are created using time-division multiplexing (TDM) and byte interleaving processes like a DS1 frame. This gives SONET its capability to access individual traffic streams without having to demultiplex the entire signal. The issue of the incoming asynchronous traffic being transmitted across the synchronous network is overcome with pointers in the line and path overhead.

SONET frames can be multiplexed to create larger frames. The multiplexing can be single stage or multistage. In single-stage multiplexing, multiple STS-1s are directly combined into a higher-level STS such as an STS-3 or STS-12. In multistage multiplexing, several STS-1s can be multiplexed into an STS-3, and then STS-3s and STS-1s can be multiplexed together into an STS-12.

When SONET frames consist of multiple STS-1s, the frame is referenced as STS-n where n equals the number of STSs that have been combined. The total number of TOH overhead bytes becomes $27 \times n$. The size of the payload becomes $783 \times n$ even though each SPE maintains its individuality. The overhead bytes of each STS are also maintained and placed into the TOH; however, some bytes become redundant and are not used. (See Figure 3-24.) Remember that an STS-1 is 90 columns, and hence STS-3 is $3 \times$ STS-1 = $90 \times 3 = 270$ columns.

Figure 3-24 *Multiplexed SONET Frames*

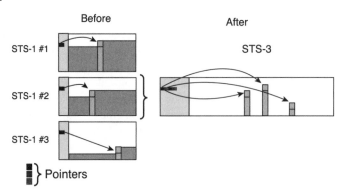

When multiple frames are multiplexed, both the overhead and the payloads are interleaved to create the larger frame. So the first byte is from the first STS-1, the second byte is from the second STS-1, and so on.

When non-DS1, or DS3, traffic needs to be transported such as ATM or IP and the rate is less than 50 Mbps, the traffic can be placed into a single STS-1. If it is higher than 50 Mbps, it is necessary to have a larger SPE. You can create a larger SPE by combining frames. Frames can be combined only in certain combinations, such as 3, 12, and 48, thus creating STS-3, STS-12, and STS-48. When frames are combined, they are said to be *concatenated*. These STS frames still have overhead bytes equal to the number of STSs, but there is only one much larger payload envelope for all the traffic. A concatenated STS is indicated by a small *c* at the end (that is, STS-3c, STS-12c). Like the nonconcatenated STS-*n* frame, the overhead bytes that are redundant are not used. Refer back to Figure 3-2, which shows the various ways traffic can be combined and multiplexed into a SONET frame and placed on fiber. Figure 3-25 shows the SONET frame hierarchy.

Now that you understand that SONET has overhead bytes to perform its job at three of its layers, it is now time to look at these overhead bytes in detail.

Section Overhead

Section overhead (SOH) comprises the first three columns of the first three rows of a basic SONET frame. Each overhead is given an alphanumeric designation such as A1, B2, and so on to identify each byte. Figure 3-26 illustrates the nine bytes used for overhead.

Figure 3-25 *SONET Frame Hierarchy*

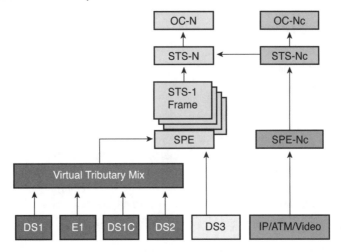

Figure 3-26 *Section Overhead Bytes*

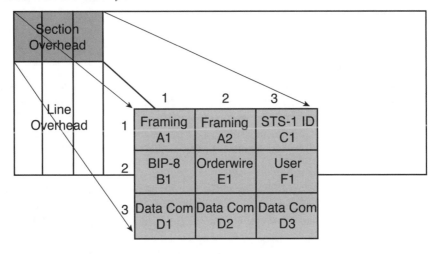

The following list describes these bytes in greater detail:

- **A1 and A2**—Framing alignment bytes used to ensure that the beginning of a SONET frame can be detected by the equipment. A1 is always set to 1111-0110, and A2 is always 0010-1000. These values are never scrambled during the frame-scrambling process. When four consecutive nonconforming frame patterns have been received, an OOF (out-of-frame) condition alarm is reported. When two consecutive conforming frame patterns have been received, an in-frame condition returns and the alarm is no longer reported.

- **J0/Z0/C1**—Formally defined as the C1 byte and called the STS-ID, this byte has been redefined as either the J0 or Z0 byte depending on the bytes position in an STS-n frame. If the byte is part of the first STS of the STS-n, this byte is defined as the J0 byte and is used for section trace purposes. If the byte is in the second through nth STS-1, it is defined as the Z0 byte and reserved for future section growth.

- **B1**—Bit interleaved parity (BIP-8) checks for error monitoring on the previous frame. The value of the B1 byte is calculated against the previous frame that has been scrambled and is inserted into the current section overhead before it is scrambled. On the receiving side, a new frame comes in. The section equipment calculates the B1 byte of the STS-n frame and compares it with the B1 byte received from the first STS-1 of the next STS-n frame. If they match, no error is reported. If they do not match and a defined error threshold is exceeded, an alarm indicator is set. The B1 byte is located in the first STS-1 of an STS-n. The corresponding byte locations in the second through nth STS-1s are currently undefined.

- **E1 orderwire**—One byte in a SONET frame is equal to a 64-kbps circuit, the same as a voice channel. This 64-kbps circuit can be used by maintenance personnel for voice access to other personnel plugged into the same orderwire circuit on the span. The E1 byte is located in the first STS-1 of an STS-n. The corresponding byte locations in the second through nth STS-1s are currently undefined.

- **F1 user defined**—This allows the vendor to add vendor-specific information within the section overhead without compromising the standard. This can give the vendor flexibility; when two vendors use the same byte for different purposes, however, interoperability may be an issue. The F1 byte is located in the first STS-1 of an STS-n. The corresponding byte locations in the second through nth STS-1s are currently undefined.

- **D1–D3**—Three bytes that comprise a 192-kbps channel have been designed for the passing of operations, administration, maintenance and provisioning (OAM&P) information. These bytes, which are located in the first STS-1 of an STS-n, are known as the *section DCC channel*. The corresponding byte locations in the second through nth STS-1s are currently undefined. Although full standards have yet to be defined, the D1 through D3 bytes are meant to be used by management systems for passing of alarms, maintenance, control, monitoring, administration, and communication information for managing the network. Many companies have used the OSI model to develop applications that use this channel. Cisco has opted to base the channel for its 15454 and 15327 product lines on the TCP/IP protocol stack, which differs from the generally accepted practice of vendors. Cisco products enable you to manage the network using browser technologies. Chapter 5, "Configuring ONS15454 and ONS15327," discusses this in more detail.

Line Overhead

Line overhead (LOH) resides just below the section overhead in the SONET frame. It is comprised of three columns and the last six rows of the frame. This provides for 18 bytes of overhead. The SOH and LOH are commonly referred to as a group called *transport overhead* (TOH). Figure 3-27 depicts the LOH bytes.

Figure 3-27 *Line Overhead*

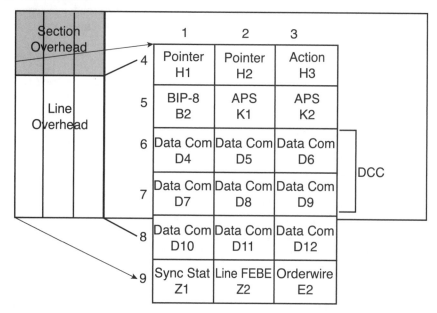

The following list describes the LOH bytes in greater detail:

- **H1 and H2 pointers**—Used together, but the 16 total bits are broken into distinct parts. The first 4 bits are referred to as the *new data flag* (NDF). When major changes in the starting position of the SPE occur, this flag inverts. The next 2 bits are unassigned. The final 10 bits are used as a pointer value, with any value between 0 and 782 being valid. This means the SPE can start at any location within the SONET frame. A value of 0 means the SPE starts at the 4 byte of the fourth row. This is the normal starting position of the SONET SPE. H1 and H2 are also used to indicate a concatenated payload by transmitting the pointer word 1001XX1111111111 in the second through *n*th STS-1 in and STS-*N*c. Another purpose is to detect STS path alarm indication signals (AIS-P).

- **H3 pointer action byte**—This byte is used to accommodate timing variations in the SPEs. Remember that an SPE is sent out in SONET frames every 125 microseconds. In a perfect world, only 783 bytes would arrive at the sending equipment for transmission every 125 microseconds, but occasionally the sending equipment has an extra byte ready to send or the SPE may be short a byte. When there is an extra byte, the byte can be placed in the H3 byte and indicated to the receiving equipment by modifications to the H1 and H2 bytes. This process is called a *negative timing adjustment*. Conversely, if an SPE does not have the full 783 bytes to send, a different adjustment needs to be made and indicated to the H1 and H2 bytes. This process is called a *positive timing adjustment*.

- **B2**—Bit interleaved parity (BIP-8) checks for error monitoring. This is similar to the B1 byte in the section overhead but with a couple of differences. The B2 byte is calculated against just the payload and line overhead sections of the previous frame. It is also calculated before scrambling rather that after. It is inserted into the frame containing the trailing part of the SPE.

- **K1 and K2 bytes**—The K1 and K2 bytes are located in the first STS-1 of an STS-*n*. The corresponding bytes locations in the second through *n*th STS-1s are currently undefined. These bytes are used mainly for automatic protection switching (APS) between LTE for switching on line errors (for instance, in systems using linear APS, or in bidirectional line-switched rings [BLSRs]). The K2 byte is also used for detecting alarm indication signals (AIS-L) at the line level and remote defect indication (RDI-L) signals. The receiving side monitors these bytes to decide when to switch to an alternate physical path.

- **D4–D12**—The D4 through D12 bytes are located in the first STS-1 of and STS-*n* and are used for line data communications. The corresponding byte locations in the second through *n*th STS-1s are currently undefined. These 9 bytes form a 576-kbps channel that can also be used for OAM&P between line level equipment. A few vendors use these bytes for differing purposes, but most OAM&P is done in the section D1 through D3 bytes. Cisco uses these bytes in the 15454 and 15327 products to tunnel other vendors' D1 through D3 bytes through their network.

- **S1/Z1**—Synchronization status bytes are used to allow equipment to choose the best clocking source if several are available. Bits 1 though 4 are undefined, and bits 5 through 8 are used to convey the synchronization status of the network element. This byte helps to avoid timing loops. The S1 byte is only valid in the first STS in an STS-*N* signal. In STS-2*N*, the byte is referred to as the Z1 byte and is reserved for future growth.

- **M0/M1/Z2**—This byte is called M0 if the STS is nonconcatenated (individual STS-1s populate the STS frame) and is used to convey the B2 error count back to the source referred to as *remote error indication* (REI-L, formerly known as FEBE). The byte is called M1 if the frame is concatenated (a single frame not maintained as individual

STSs) and performs similar functions to M0, but it is not in every line overhead section. If the bytes not used by M0 and M1, the byte is referred to as the *Z1 byte* and is defined for future growth.

- **E2 orderwire**—Express orderwire. The E2 byte is located in the first STS-1 of an STS-*n* and is a 64-kbps voice channel used by craft personnel. The corresponding byte locations in the second through *n*th STS-1s are currently undefined.

Path Overhead

The path overhead (POH) is considered part of the payload and not part of the TOH as described in the preceding section. It consists of a single byte that leads off each new row of the SONET payload. This provides for 9 path overhead bytes (9 rows × 1 byte each) that form a column that leads off the SPE. If an STS frame consists of multiple STS-1s (nonconcatenated), there will be separate path overhead bytes for each STS-1. If the frame is concatenated, only one path overhead exists. Figure 3-28 shows the POH bytes.

Figure 3-28 *Path Overhead*

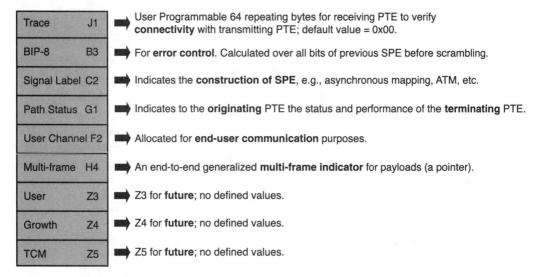

Trace	J1	→	User Programmable 64 repeating bytes for receiving PTE to verify **connectivity** with transmitting PTE; default value = 0x00.
BIP-8	B3	→	For **error control**. Calculated over all bits of previous SPE before scrambling.
Signal Label	C2	→	Indicates the **construction of SPE**, e.g., asynchronous mapping, ATM, etc.
Path Status	G1	→	Indicates to the **originating** PTE the status and performance of the **terminating** PTE.
User Channel	F2	→	Allocated for **end-user communication** purposes.
Multi-frame	H4	→	An end-to-end generalized **multi-frame indicator** for payloads (a pointer).
User	Z3	→	Z3 for **future**; no defined values.
Growth	Z4	→	Z4 for **future**; no defined values.
TCM	Z5	→	Z5 for **future**; no defined values.

The following list describes the POH bytes in greater detail:

- **J1 trace byte**—The first byte in the STS SPE. Its location is indicated by the pointer (H1, H2 bytes). This byte provides an STS path trail trace identifier. This byte is a user-programmable field allowing for a 64-byte repeating sequence to be transmitted. Its purpose is to allow the end pieces of equipment to positively identify the far-end system to ensure an appropriate continued connection. This information could be any kind of identifier (IP address, E.164 address, and so on). It supports end-to-end monitoring of an STS path.

- **B3**—This byte is similar to the B1 and B2 bytes in SOH and LOH. The STS path BIP-8 is calculated over all bits (783 bytes for an STS-1 SPE or $n \times 783$ bytes for an STS-Nc, regardless of any pointer adjustments). It is calculated against the previous SPE only (no TOH is included) before the SPE is scrambled.

- **C2 signal label**—This byte is used to tell the receiving equipment the type of payload that is contained in the SPE. It tells the receiving equipment whether the SPE contains VTs, a DS3, ATM cells, HDLC-framed IP traffic, and so on, so the receiving equipment will know how to process the incoming SPE. Table 3-3 shows the C2 byte mapping.

Table 3-3   *STS Path Signal Level Assignment*

Code (Hex)	Contents of STS SPE
00	Unequipped
01	Equipped—Nonspecific payload
02	VT-structured STS-1 SPE
03	Locked VT mode
04	Asynchronous mapping for DS-3
12	Asynchronous mapping for DS-4NA
13	Mapping for ATM
14	Mapping for DQDB
15	Asynchronous mapping for FDDI
16	HDLC over SONET mapping
FE	0.181 test signal (TSS1 to TSS3)

- **G1 path status byte**—This byte is actually divided into two parts. The first part is found in bits 1 through 4 where the B3 count information is carried back to the originating equipment. This is called the path remote error indicator (REI-P, formally called FEBE). Bits 5 through 7 carry the STS path remote defect indicator (RDI-P). Bit 8 is undefined. This byte allows for the bidirectional end-to-end transport of both path performance and path terminating status information.

- **F2 user channel**—This byte is used for communications between PTE. It is up to the equipment vendor to define its use.

- **H4 multiframe indicator**—This byte is used based on the type of payload of the SPE. It is normally used when the SPE is broken into multiple VT units or groups and acts as a pointer at the VT level like the H bytes worked at the SPE level.

- **Z3/Z4/Z5**—Like the other Z bytes, these bytes are reserved for future growth. Although Z5 is officially a growth byte, ANSI has defined some specific functions that it should be used for. Z5 is allocated for tandem connection maintenance and the path data channel on ANSI specifications.

The section "SONET Framing" mentioned that an SPE could be broken into distinct sections called virtual tributary groups and that each VTG was 108 bytes. Inside each VTG could either be four VT1.5s, three VT2s, two VT3s, or one VT6. Each VT group has 12 bytes of overhead. Although it is beyond the scope of this book to review VTGs in detail, note that overhead bytes exist at the VTG level as well. These bytes are used for such things as parity checking, LOP-V, AIS-V, RDI-V and REI-V, and tracing, with a couple of bytes left over for growth. (See Figure 3-29.)

Figure 3-29 *VTGs*

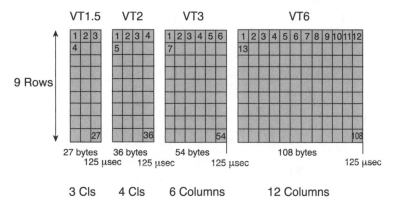

With an understanding of the SONET layers and the overhead bytes that support each layer, you can now look at the equipment used by each layer. The layout of this equipment depends on the network design, as described in the "SONET Topologies" section later in this chapter.

SONET Network Elements

As traffic traverses a SONET network, it can pass through any of a number of different pieces of equipment, or network elements, designed for a specific function and intended to support one or more of the SONET layers. Figure 3-30 shows the pieces of equipment normally found in a SONET network and where they might be found.

Figure 3-30 *Equipment Used in a SONET Network*

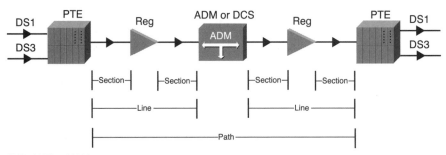

ADM = Add/Drop Multiplexer
DCS = Digital Cross-Connect System
PTE = Path Terminating Equipment
REG = Regenerator

As traffic enters a SONET network, the first piece of equipment it encounters is the PTE. The PTE is responsible for the end-to-end passage of the traffic to its destination PTE device. The PTE can be any network element, such as a terminal multiplexer, that processes up to the path layer. (See Figure 3-31.)

Figure 3-31 *Network Elements*

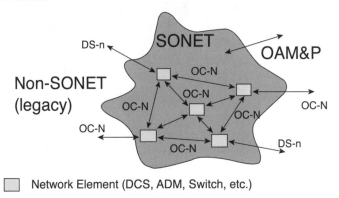

Network Element (DCS, ADM, Switch, etc.)

Terminal Multiplexer

The terminal multiplexer (TM) acts as a concentrator for traffic coming in on a T1, T3 or an E1 line, and then packages it into the appropriate payload envelope and STS frame. (See Figure 3-32.)

Figure 3-32 *Terminal Multiplexer*

A TM is normally associated with the processing of DS signals but can also include ATM or Ethernet traffic. Equipment that processes ATM and Ethernet traffic is sometimes referred to as *multiplexing gateways*. In this text, the term *terminal multiplexer* is used to include ATM and Ethernet traffic. TMs are considered edge devices, normally configured in a point-to-point relationship with a SONET ring, and are usually located on customer sites.

Add/Drop Multiplexers

Add/drop multiplexers (ADMs) can perform many different functions. An ADM's main function is to add or drop individual STS signals from the higher-rate traffic stream without having to fully multiplex or demultiplex that traffic. The extracted traffic would be passed to the appropriate equipment and new traffic from connected equipment could then be added into the traffic stream without disturbing the existing signals. (See Figure 3-33.)

Figure 3-33 *ADMs*

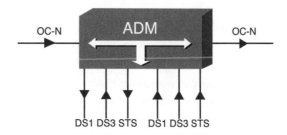

An ADM normally gets its traffic from a TM, but can play the role of a TM itself and directly interface with non-SONET equipment. An ADM can also act as a pass-through node by just performing regeneration functions. This allows SONET traffic from other ADMs to transit this device on its way to another device down the fiber. Whereas TMs usually have one set of fiber that connects to the ADMs (unless APS protection has been configured), ADMs have at least three sets of optical interfaces: two for the interfaces to the ring, and one or more to interface to the TMs or other devices.

Regenerators or Amplifiers

Regenerators or amplifiers are used when the distance between SONET devices is too great for the signal to arrive at its destination in an identifiable form. The exact distance at which the optical signal becomes too weak (attenuates) for the receiver varies based on the type of fiber being used and the type and strength of the laser producing the optical signal. (See Figure 3-34.)

Figure 3-34 *Regenerators*

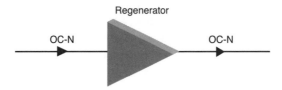

Regenerators take the incoming optical signal, convert it back into its electrical equivalent, and then generate a new optical signal. This results in delay and extra processing, but the new signal is returned to its original state free from distortion. In addition to re-creating the signal, the SOH bytes have to be processed, which means regeneration also performs the following tasks:

- **Framing**—Must obtain frame synchronization with the adjacent device by identifying the frame pattern contained in the A1 and A2 bytes of the section overhead.

- **Performance monitoring**—Includes physical layer and section layer performance monitoring, by checking for transmission errors using the B1 byte of the section overhead.

- **Passing of the embedded operation channel**—Bytes D1, D2, and D3 of the SOH can be used to establish a communication channel with the adjacent device for exchanging provisioning, management, and maintenance information.

- **Maintaining a craft interface**—Used by maintenance personnel as a voice channel for communication.

- **Alarms**—A regenerator must be able to recognize the local (red) alarm condition (loss of signal or loss of frame) and convey these conditions to its operation support system (OSS)/NMS (network management systems).

Amplifiers take the incoming signal and amplify it. They do not change any information in any of the layers, not even the section layer. Distortion in the signal that exists at the time of amplification is considered as part of the signal and is amplified. Optical signals can only be amplified a certain number of times before the amplified distortion interferes with the integrity of the original signal. The number of times amplification without regeneration can occur depends on the amplifier. With the Cisco ONS 15216 EDFA amplifier, this number is around four times. A regenerator or an ADM could perform this amplification.

Digital Cross-Connects

In the early days of phone systems as individual phone lines needed to be joined into higher-speed lines, a person physically configured the wiring to create patch panels. A major step forward was made with the development of digital cross-connects (DCS). With DCS devices, wiring could be brought in to one location, and through equipment a signal could be switched from one DS1 to another DS1 without manual rewiring. These DCS devices can aggregate traffic and feed that traffic into an ADM. (See Figure 3-35.)

Figure 3-35 *DCS*

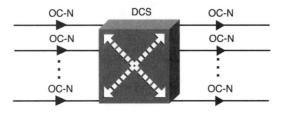

SONET DCS

SONET DCSs are advanced ADMs and are used to manage the traffic passing through the various optical fibers. An ADM allows traffic from STSs to be extracted and new traffic re-inserted, but does not allow traffic from one STS to be moved to another STS for transport. This is called *cross-connecting*. Cross-connects used to be accomplished by fully demultiplexing the traffic and sending it across a cable to a patch panel, where lots of cabling would be used to manually interconnect the signals. The traffic would then be sent out another cable to be multiplexed into a new traffic stream. The SONET DCS can take traffic from one stream and multiplex that traffic into another stream. DCS 1 has traffic coming in on fiber 1. Some of the traffic needs to go onto fiber number 2, and some of the traffic needs to go onto fiber number 3. The level at which the cross-connect can operate is based on the type of cross-connect equipment being used. There are basically two types:

- **Broadband DCSs**—Can make two-way cross-connections at the DS3, STS-1, and STS-*N*c levels. It is best used as a SONET hub, where it can be used for STS-1 grooming, signal regeneration, or traffic switching. Switching as it is used here is under administrative control, and is *not* switching that is based on information derived from the traffic stream. (See Figure 3-36.)

Figure 3-36 *Broadband DCSs*

* **Wideband DCSs**—Can make two-way cross-connections down to the VT (DS1) level. This gives you greater granularity of service. Individual DS1s can be routed independently through the SONET network. (See Figure 3-37.)

Figure 3-37 *Wideband DCSs*

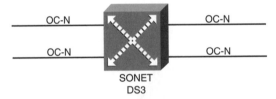

The ONS 15454 SONET Transport Network Element has now dramatically revolutionaries the ownership and operational cost structures of building and integrating the SONET transport infrastructure with the rest of the network. Many of the network elements based on the traditional specification reference model have now been collapsed into a single platform with a significantly smaller form factor and cost a fraction of the traditional elements.

Depending on the cards used in the 15454, they can be either a broadband DCS or a wideband DCS.

A couple of other pieces of equipment that might be found in a SONET network, but not necessarily in every network, include the following:

* Digital loop carriers

* Drop and continue

The next two sections discuss this equipment in greater detail.

Digital Loop Carriers

SONET digital loop carriers (DLCs) are similar to the TM but are used to aggregate even lower-speed traffic. The DLC itself is actually a system of multiplexers and switches designed to perform concentration from remote terminals to the community dial office and from there to the central office DLCs, which take individual DS0 or 56-kbps DDS or even DSL traffic and aggregate it into the higher speeds needed to send them across a SONET network. The DLC does not provide a local switching function between input channels, only aggregation. DLCs do not reside at customer locations but are maintained at the central offices.

Drop and Continue

SONET drop and continue nodes are ADMs with additional capabilities used in multipoint applications. An ADM either extracts the desired traffic, delivering it to the appropriate equipment, *or* it passes the traffic through to the next node. The drop and continue node extracts the desired traffic delivering it to the proper equipment *and* passes it through to the next node. This feature is used in videoconferencing, TV broadcast, and so on.

SONET Topologies

Before beginning, note that this is not a networking design book. This section introduces you to structures, which you can combine to create any number of network designs. As in all designs, there are always tradeoffs between cost and "best" design. Considered these factors when creating your final network configuration.

In any SONET network, a number of designs may be in use, and in fact many different types of designs can be intermixed. When torn apart, SONET topologies generally break down into multiple categories: point to point, point to multipoint/linear, mesh, and ring. (See Figure 3-38.)

Figure 3-38 *SONET Topologies*

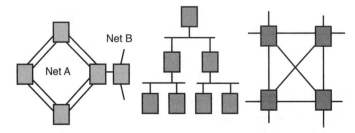

Point-to-Point Topology

The simplest SONET/SDH topology is called *point-to-point*. It involves a direct fiber link between TMs shown generically in Figure 3-39 as PTE. The fiber span might include regenerators, depending on distance, but there will be no other intermediate network elements. This configuration might be used between two customer locations over dark fiber or it could be used to connect two nodes within a carrier network. As discussed earlier, the PTE equipment is performing a multiplex function so that, for example, multiple DS1 signals can be transported between the two locations over the point-to-point link. If protection is required for this link, a linear protection scheme such as 1+1 or 1:1 is used. Protection is discussed in more detail later in this section.

Figure 3-39 *Point-to-Point Topology*

In today's market, equipment that supports point-to-point technology is more prevalent than equipment that supports rings. In a point-to-point environment, the PTE can be routers, switches, or any device capable of supporting OC-*n* optical interfaces. Vendor interoperability is not normally an issue because the SONET standards implementation is rather straightforward.

Traffic can pass across a point-to-point topology in either unprotected or protected mode. Protected mode offers some options that may be vendor or equipment specific that allow for card redundancy, line redundancy, or card and line redundancy. Redundancy can support revertive options (the option to automatically switch back to the original line once repaired) as well. Many routers may have an optical interface compatible with OC-*n* standards for passing traffic but implement protection through the routing protocol and not through SONET.

Point-to-point topologies can be interconnected to take on more complex forms where linear networks can be created. ADM can create networks that function like rivers, with traffic from tributaries entering and leaving the network at different locations.

Point-to-Multipoint/Linear Topology

The linear bus topology has one or more intermediate points for dropping or adding traffic along the path. The intermediate nodes are add/drop multiplexers. This topology is sometimes called *point-to-multipoint*, although you should not confuse multipoint with multicast. The benefit to this topology is that there are now multiple nodes, as shown in Figure 3-40 (four in this specific example) that can communicate over the same fiber link. In the figure, for instance, the PTE on the left could establish one or more SONET paths to each of the other three nodes in the diagram. The ADMs would be administered to drop the appropriate traffic at each of the intermediate nodes. The paths could be at the VT level or at the STS level.

The operation at the ADMs could be drop and insert or drop and continue. Drop and insert implies that the dropped traffic channel is terminated at that location, and a new traffic channel can be inserted into the vacated time slot to communicate with a downstream node. Drop and continue means that the signal is dropped, but it is also transported to the downstream nodes. In the drop and continue case, a multicast function is performed.

Figure 3-40 *Point-to-Multipoint/Linear Topology*

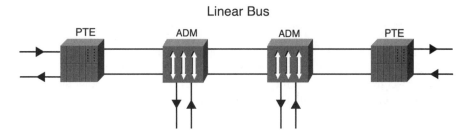

Mesh Topology

A mesh topology is a more complex topology most likely found in carrier backbones or in very large enterprise networks. It provides a much broader range of connectivity than the previous topologies and is appropriate when there is a need to support a broad range of many-to-many connectivity. At intermediate hub locations where multiple fiber routes intersect, cross-connect equipment is used to move traffic from one route to another. Individual channels might or might not be terminated at the cross-connect locations, depending on other functions performed at that node.

Protection and restoration schemes are still evolving for mesh networks. Today, protection tends to be provided over individual point-to-point links or rings within the mesh using techniques discussed later in this chapter. The Cisco Path Protected Mesh Network (PPMN) solution takes this one step further and provides ring-like protection over a mesh network. This is possible because the ADMs and cross-connects that comprise the mesh are being made "topology aware" through the use of autodiscovery protocols borrowed from the IP world. Future networks may take this scenario one step further, and dedicated protection resources might not be established. (See Figure 3-41.)

Figure 3-41 *Mesh Topology*

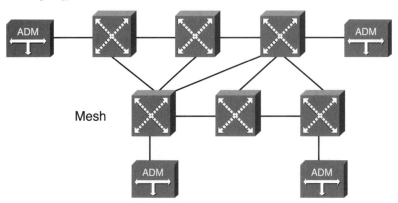

When a failure occurs, the affected nodes identify a new path across the network and use signaling to establish the replacement path. This fast restoration approach should result in far fewer resources being dedicated for protection. At issue is whether the new path can be established quickly enough to satisfy performance requirements.

Ring Topology

The ring topology is, by far, the most common in today's service provider networks. It is common because it is the most resilient. Rings are based on two or four fibers. Transmission is in one direction on one half of the fibers and in the opposite direction on the other half. Half the bandwidth can be reserved for protection. Quick recovery from a fiber cut anywhere on the ring can be accomplished by switching to the signal being transmitted in the opposite direction. Ring topologies have been so successful at providing reliable transport that even long-haul carriers often use multiple, very large circumference rings in their nationwide networks.

Add/drop multiplexers are used at nodes on the ring for traffic origination or termination. It's not unusual for rings to be connected to other rings—in that case, cross-connects provide the interconnection function.

Two dominant types of rings are used in SONET networks: unidirectional path-switched rings (UPSRs) and bidirectional line-switched rings (BLSRs). Both rings types are discussed in detail later in this chapter. (See Figure 3-42.)

Rings require all segments to operate at the same speed; so if one segment needs OC-48 speeds, the entire ring must consist of OC-48 segments. For long-haul applications, this might not be cost effective.

Figure 3-42 *Ring Topology*

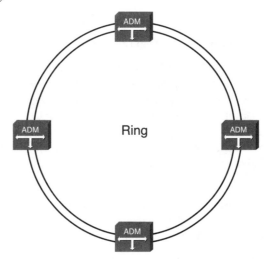

Rings come in several varieties. Rings can be line switched or path switched. They can be unidirectional or bidirectional. They can also be two fiber or four fiber. Extrapolating the different types of rings, you could theoretically build eight different types of rings.

Automatic Protection Switching

A discussion of network types does require that you understand one of SONET's most important features: its automatic protection switching (APS). The next few sections walk you through APS and the various SONET topologies:

- Automatic protection system (APS)
  - 1+1 protection
  - 1:1 and 1:*n* protection
  - Unidirectional switching
  - Bidirectional switching
- Point-to-point networks
  - Point-to-point architecture
  - Linear APS

- Ring networks

  — Unidirectional path-switched ring (UPSR)

  — Bidirectional line-switched ring (BLSR)

There is often confusion between the meaning of protection and restoration. Here, we provide a high-level definition of protection. As you'll see in the review of protection options, protection implies that a backup resource has been established for recovery if the primary fails. Because backup is pre-established, the switchover can be very rapid, usually in less than 50 milliseconds. Because failures are often localized and quick recovery is required, most protection schemes are localized based on a linear recovery of a single link or recovery of a ring within the network topology.

It is much less common for protection to be applied end to end across a complex network. Because of their proven performance, protection strategies dominate today's service provider networks even though they require as much as 50 percent of the network capacity to be reserved for the protection function.

Here, we have a high-level definition of restoration. The key distinction relative to protection is that restoration does not have a predefined dedicated path reserved for switchover in the event of a failure. Restoration implies re-establishing the end-to-end path based on resources available after the failure.

Historically, restoration involved actions such as rerouting a customer's traffic so that it continued to follow a path that was protected on every link. It even sometimes included restoring the route by cable splicing or laying new fiber. Because restoration involved at least reprovisioning the route and perhaps even new construction, it was not considered to be an alternative to protection, rather it was an action that took place after protection to ensure that the service continued to be reliable. More recently, as IP networking protocols for topology awareness and path establishment are designed into SONET network nodes, the use of "fast restoration" as an alternative to protection is being evaluated.

To fully understand SONET topology, you need to understand some of its basic terms that center on protection. Terms such as *1:1* and *1+1* and *unidirectional* and *bidirectional switching* are used in just about every conversation that deals with configuring SONET. This section introduces you to these terms and their meanings, and later you learn how APS is applied when the actual topology is discussed.

APS is a standard devised to provide for link recovery in the case of failure. Link recovery is made possible by having SONET devices with two sets of fiber. One set (transmit and receive) is used for working traffic, and the other set (also a transmit and receive) is used for protection. The working and protect fibers take different physical paths and must be in different conduits to be effective. The fibers used for protection may or may not carry a copy of the working traffic depending on how protection has been configured (1:n or 1+1).

APS protection switching has always been known for its recovery speed. SONET defines a maximum switch time of 50 milliseconds for a BLSR ring with no extra traffic and less than 1200 km of fiber. The specifications actually state 60 milliseconds, with 10 ms for discovery of the problem and 50 ms to perform the switch, but most people refer to this as just 50 ms. Most SONET networks are much smaller and simpler than this example and switch even faster. Compare this to the 250 milliseconds needed in the more commonly used Y cable switchover. This rapid recovery minimizes the impact to the traffic when there is a problem. A problem does not have to be defined as a catastrophic failure such as a loss of signal (LOS), loss of frame (LOF), line, and so on for a switchover to occur; the problem could be a degraded signal, such as BER exceeding a configured limit. Protection switching can also be initialized manually to allow for routine maintenance or testing without having to take the circuits out of service.

When configuring protection, you have many options. Protection can be for linear or ring topologies. Each type of topology has specific choices of 1:1, 1:n, or 1+1 protection. These can further be configured into using either unidirectional or bidirectional switching mechanisms, as discussed in the following sections.

1+1 Protection

One of the most fundamental protection schemes is 1+1 protection. In the 1+1 architecture, the head-end signal is continuously bridged (at the electrical level) to working and protection equipment so that the same payloads are transmitted identically to the tail-end working and protection equipment. At the tail end, the working and protection signals are monitored independently and identically for failures. The receiving-end equipment chooses either the working or the protection signals. Because of the continuous head-end bridge, the 1+1 architecture does not allow an unprotected extra traffic channel to be provided. See Figure 3-43, which depicts 1+1 protection.

Figure 3-43 *1+1 Protection*

The purpose of this graphic is to explain 1+1 protection. Walk through the configuration. Be sure to understand that the signal is bridged on both paths and that the receiver can choose between the two based on local criteria.

As depicted in the diagram, the signal is bridged onto two different paths and sent to the receiver. The receiver can choose either of the two signals. Switchover to the alternative signal is a local decision that does not require coordination with any other nodes. Because the signal is always available on both paths, the time to switch to the protected path is very short. The receiver detects failure and immediately switches. Recovery times of much less than 50 milliseconds are possible.

Note that as with all protection schemes, for 1+1 protection to be effective the two paths must be physically diverse.

1:1 and 1:n Protection

In 1:1 and 1:*n* protection, traffic is carried on the working line and not on the protection line until a failure occurs. (See Figure 3-44.)

Figure 3-44 *1:1 Protection*

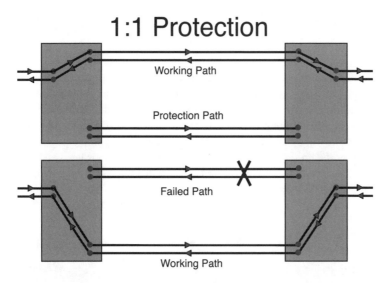

Figure 3-45 shows an example of 1:*n* protection. This is a more general case of 1:1 protection. As the name implies, one protection path is established as a backup for *n* working paths. If a failure occurs, an APS protocol is again used to switchover to the protection facility. In the figure, the lowest of the multiple working paths has failed, so its signal is bridged to the reserved protection path. Note that any one of the multiple working paths could be bridged to the protection path, but only one path can be protected at a time. If a failure that affects multiple working lines occurs, a priority must be established to identify which line gets protection.

Figure 3-45 *1:n Protection*

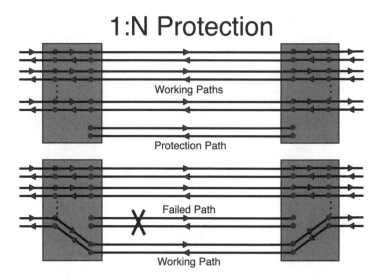

1:*n* protection supports a revertive option, which allows the traffic to automatically switch back to the original working fibers after a repair and a configured wait time called *Wait To Restore* (WTR).

In 1:1, there must be one protect line for every working line. Because the protect line does not actually carry traffic when it is not in use, it is possible, if the vendor equipment allows it, for the protect line to carry other nonpriority traffic called *extra traffic*. The extra traffic is dropped if the protection feature kicks in. In 1:*n* protection, a single line can protect one or more working lines.

Unidirectional Switching (UPSR)

If protection is configured as unidirectional and a single fiber (transmit or receive) fails (initial state in Figure 3-46), only the failed fiber switches to the protect path (final state in Figure 3-46).

Figure 3-46 shows an example of unidirectional protection switching. It is a situation that might apply when only one direction of transmission fails. In the diagram on the left, only the west-to-east direction of transmission has failed. This might be due to a component failure in the equipment or the cut of an individual fiber. In the diagram on the right, the protection path is now used for the failed path, but the opposite direction of transmission remains in its original state because that path is still working satisfactorily. This switching of only the failed direction of transmission is called *unidirectional switching*.

Figure 3-46 *Unidirectional Protection Switching*

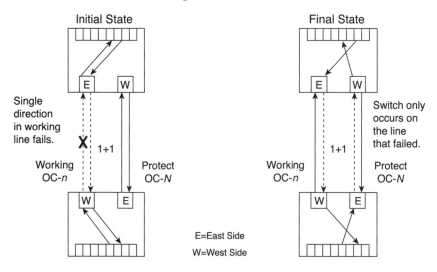

Bidirectional Switching (BLSR)

If protection is configured to be bidirectional, and a single fiber fails (initial state in Figure 3-47), both the transmit and receive working lines switch to the protect fibers.

Figure 3-47 *Bidirectional Protection Switching—Initial State*

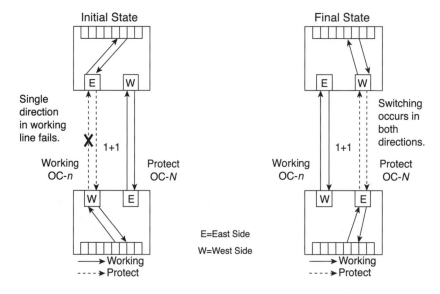

Figure 3-47 shows an example of bidirectional switching. The failure scenario on the left is identical to the previous graphic. The difference here is that protection switching takes place for both directions of transmission even though only one direction has failed. This is called *bidirectional switching*. (See Figure 3-48.)

Figure 3-48 *Bidirectional Protection Switching*

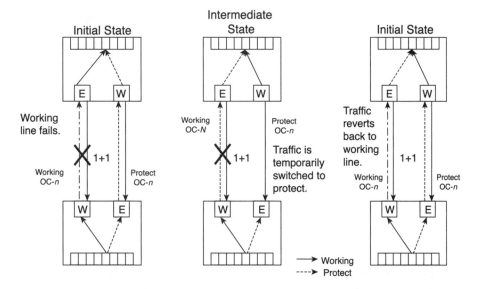

Nonrevertive

The purpose of Figure 3-49 is to explain the term *nonrevertive protection* and prepare for contrast with revertive protection. The key point is that we continue to use the "protection" path even after the original path has been restored.

Figure 3-49 explains the term *nonrevertive protection*. On the left, you see a failure of the working line. On the right, a switchover to the protection line has occurred. Even after the original working line has been restored to its proper operating condition, the system will not revert to use the original line as the working path. Instead, it continues to use what was originally the protection line as the working line. In other words, it does not revert. Nonrevertive protection is most common in 1+1 protection schemes.

Figure 3-49 *Nonrevertive*

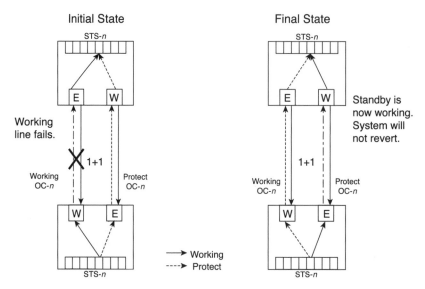

Revertive

Refer back to Figure 3-48. The purpose of this graphic is to explain the term *revertive protection* and contrast it with nonrevertive. Walk through the three diagrams starting on the left. The key point is that after the original line is restored, you switch back to it.

Figure 3-49 shows an example of revertive protection. In this case, you have a failure of the working line on the left. In the center, you have switched over to the protection line. On the right, you have switched traffic back to the original working line after it has been restored. In other words, you have reverted to the original working line. Revertive protection is most common with 1:*n* protection schemes. 1:1 protection may use revertive or nonrevertive.

For bidirectional switching to work, both sides must communicate to perform the switch. This is accomplished through the use of an APS channel. If protection is configured to be unidirectional, the receiving side makes the decision to switch and does not tell the transmitting side. Unidirectional switching does not require an APS channel.

Some vendors have other combinations, but these are the ones that are discussed in this book.

Unidirectional Path-Switched Ring (UPSR)

UPSR rings are the simplest types of ring to implement. Two fibers interconnect each network element to form two counterrotating rings. One ring provides flow for the working traffic. By convention, working traffic flows clockwise through the ring. The other ring flows in a counterclockwise direction and is used for protect traffic. Figure 3-50 shows how working and backup traffic flows around a UPSR ring.

Figure 3-50 *A UPSR Concept*

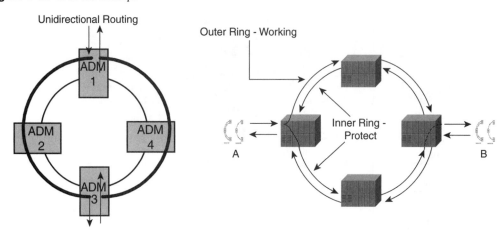

Next, apply what you learned about network topologies and protection to review the operation of a unidirectional path-switched ring.

The greater the communications bandwidth carried by optical fibers, the greater the cost advantages of ring structures as compared with linear structures. A ring is the simplest and most cost-effective way of linking a number of network elements. It offers the highest availability. Various protection mechanisms are commercially available for this type of network architecture, only some of which have been standardized in ANSI Recommendation T1.105.1. A basic distinction is made between ring structures with unidirectional and bidirectional connections.

In a UPSR, traffic is transmitted simultaneously over both the working line and the protection line, one in a clockwise direction and the other in a counterclockwise direction. The better signal is always selected by the receiver to be the working line. Refer back to Figure 3-50.

If an interruption occurs between network elements A and B, the receiver switches to the protection path and immediately takes up the connection. This is the same as 1+1 protection. So, the K1-K2 protocol is not needed and the switchover is very fast because it is done locally only. Only one of the two rings is used in case of a failure. The UPSR is typically used in access networks.

The UPSR is optimized for a carrier to provide services to a selection of customers and to aggregate their traffic into the central office for transmission over the network. So, the UPSR is most often found in access or metropolitan-area networks. (See Figure 3-51.)

Figure 3-51 *Access/Metro Area*

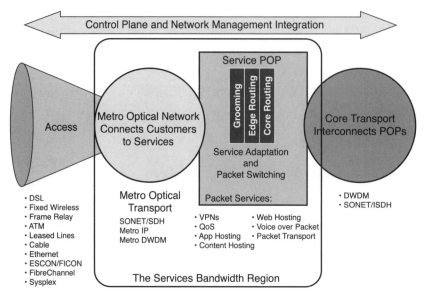

The UPSR is optimized for a carrier to provide services to a selection of customers and to aggregate their traffic into the central office for transmission over the network. So, the UPSR is most often found in access or metropolitan-area networks.

In this case, the ring is mainly a wire-saving issue, and the actual bandwidth need per end node is relatively low; one wire or fiber can carry the whole aggregated bandwidth of a certain access area.

The advantage of the UPSR architecture is the simplicity of operation. No extra signaling protocol is needed to perform protection switching in a case of failure. The protection switching mechanism is working autonomously in each node, always selecting the signal with the better quality. In the case of a failure, UPSR provides fast switchover independent of the rest of the network. (See Figure 3-52.)

Figure 3-52 shows six SONET network devices connected to a UPSR consisting of a two-fiber topology using 1+1 protection. The traffic pattern on the two fibers is indicated by the arrows. One fiber is used for the working facility, and the other fiber is used for the protection facility. Each fiber is carrying three working channels or three protection channels. The working traffic flows clockwise around the ring, whereas the protection traffic flows counterclockwise around the ring.

This implementation uses 1+1 protection, and the maximum capacity for the total of all user traffic on the ring equals the line rate between the connected devices. (See Figure 3-53.)

Figure 3-52 *A UPSR Traffic Flow*

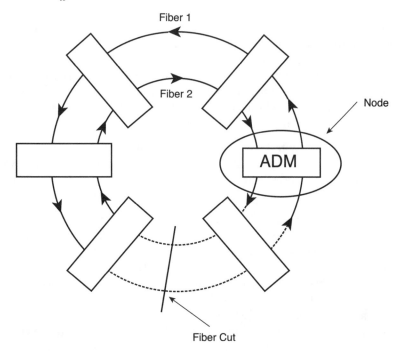

Figure 3-53 *A UPSR Traffic Flow*

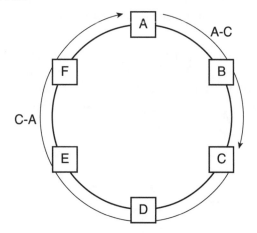

In Figure 3-53, when node A wants to talk to node C, traffic flows clockwise from node A through node B to get to node C. When node C wants to respond back or talk to node A, node C must take the long way around the ring going through nodes D, E, and F. The working traffic always flows in one direction, making it unidirectional. This also makes traffic flow asymmetrically, meaning it takes longer for node C to talk to node A than for node A to talk to node C. In small networks (either in the circumference of the ring or number of nodes — four or less nodes), this is not a big factor; if a UPSR is used in a large ring (more than four nodes), however, this could create some problems with applications that are trying to implement flow control.

To overcome flow-control issues, a UPSR should be limited to fewer nodes and kept in closer geographic proximity. UPSRs can be used effectively in campus networks, small metro rings, and as access rings that are used to feed larger metro or regional BLSRs. (BLSRs are discuss in the section "Two-Fiber Bidirectional Line-Switched Ring [BLSR].")

Because any traffic placed on the ring must travel the entire length of the ring, the ring cannot carry more traffic than that available across a single span of the ring. For example, in an OC-12 point-to-point network, each node can send up to 12 STS-1s between each other, effectively using an OC-24's worth of bandwidth. In an OC-12 ring with six nodes, the total available bandwidth of all six nodes is OC-12. If node A puts three STS-1s on the ring, only nine STS-1s worth of bandwidth is left for any node to use on the ring. If node B wants an STS-3 to talk to node C and node C needs an STS-3 to talk to node D, you can quickly run out of bandwidth. (In Figure 3-54, only working traffic shown.)

This bandwidth issue tends to make UPSR rings more efficient in a hub-and-spoke environment where most of the traffic is destined for a common node. This node might be providing access to a corporate headquarters or may be providing Internet connectivity. UPSR rings are normally found on the edges of large metro and regional rings where traffic is being aggregated for distribution to other segments in the core or to be transported through the core.

Because of the simplicity of UPSR rings, many vendors have implemented them pretty close to the published standards. So, it is possible for one vendor's UPSR-compatible piece of equipment to coexist in a ring with a different vendor's UPSR-compatible piece of equipment. (Exceptions always apply, so do not take this as an absolute.)

Figure 3-54 *Traffic Flow in a UPSR Ring*

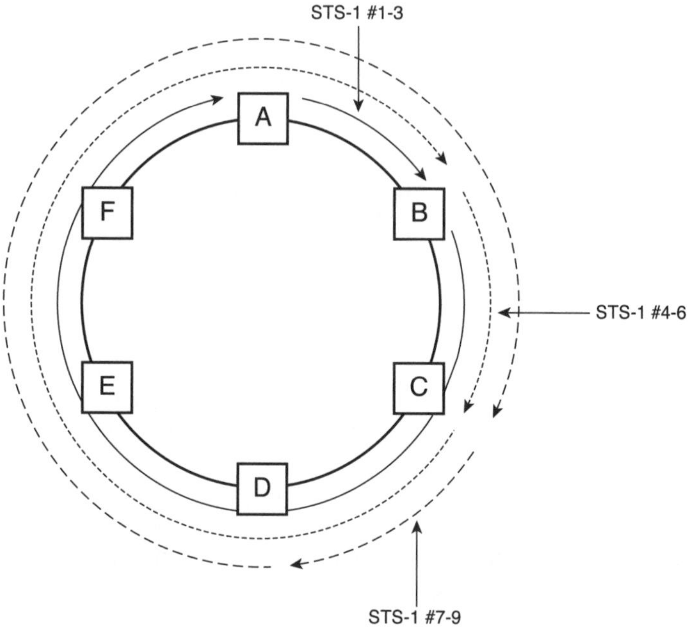

Protection in a UPSR Ring

UPSR rings are built around 1+1-type of APS protection. As discussed in the "1+1 Protection" section earlier, the main feature of 1+1 protection is that traffic is permanently bridged onto both the working and protect fibers. That means the receiving side always monitors both traffic streams and decides the best traffic to use.

A UPSR ring is path switched. As such, it looks at the bit error rates (BERs) and alarm indicators in the path and VT overhead to determine when to switch. The selection of traffic can be at the SPE or VT levels. This type of switching does not require that all traffic coming across the working fiber be switched. Only the traffic failing to meet specified criteria is switched. In this case, the K1 and K2 bytes found in the line overhead are not used. There is no need for switch requests, channel numbers, or other information found in these bytes to be communicated with the originator of the traffic. Path switching is extremely fast and results in very little loss of traffic. Figure 3-55 shows a fiber cut on a UPSR ring and how traffic flow occurs.

Next, examine the operation of a UPSR ring when there is a fiber cut. The figure shows working traffic between nodes A and C. Next, the fiber between network device A and network device B is cut and is interrupting the flow of the working traffic from A to C. The receiver in network device C detects the failure of the ring.

Figure 3-55 *UPSR Ring with a Fiber Cut*

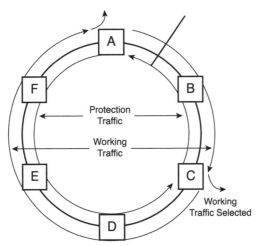

Traffic from node A to node C is unable to use the original path, so the protect path is selected. Node C would still use the working path to communicate to node A.

In a UPSR ring, the protect lines are dedicated to passing protect traffic. If a node failure occurs, all traffic to or from that node is lost. Other nodal traffic is selected by the receiving node from either the working or protect lines as appropriate. (See Figure 3-56.)

Figure 3-56 *Fiber Cut Recovery Steps*

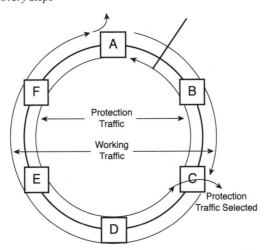

When the network node C detects the signal loss from a network node A, network node C switches to the protection path. Because node C is located in the downstream data path with respect to node A, only the receiving node C is aware of the problem. For network node A, the situation remains unchanged, so no actions need to be undertaken from node A.

The switchover to the protection path between nodes A and C can be done within a few SONET frame times (one frame is 125 microseconds), so little traffic is lost. Some applications (for example, voice) probably won't even detect that there has been a failure. (See Figure 3-57.)

Figure 3-57 *Node Failure*

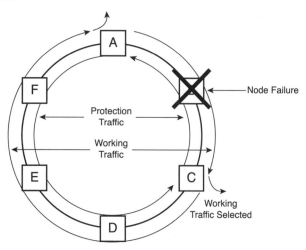

Figure 3-58 shows the same connections over a UPSR ring as the previous graphic. There is working traffic between nodes A and C. In this example, the network node B between node A and node C breaks down. This failure interrupts the working traffic flow. Because node C is located in the downstream data path with respect to node A, the receiver in network node C detects the signal loss from node A.

When node C detects the signal loss, node C automatically switches over to receive the traffic from node A on the protection path. The operation is essentially the same as the previous example when we had a fiber cut. Only node C is aware of the protection switching while the situation for node A remains unchanged.

This and the previous example demonstrate that a UPSR ring can quickly recover from a single fiber cut of node failure. The operation is essentially the same in both cases. Note, however, that in the event of a node failure traffic to and from the failed node is lost.

Figure 3-58 *Node Failure*

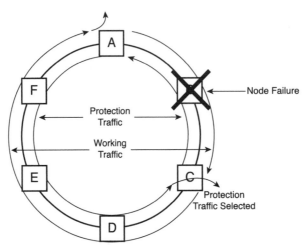

To facilitate vendor interoperability, standards for path-switched rings have been developed. Even though standards have been defined for both unidirectional and bidirectional rings, only the unidirectional ring has been developed and deployed. That's why all the previous examples have focused on UPSR. The BPSR is specified by the ANSI Standard T1.105. It was never accepted by the equipment manufacturers, and so no products according to this specification are available on the market. The UPSR standards, developed by Bellcore, have resulted in products manufactured by multiple vendors.

Two-Fiber Bidirectional Line-Switched Ring (BLSR)

Next, examine the basic operation of a BLSR. In a BLSR, the protection switching is based on the line, and a 1:1 or 1:*n* protection scheme can be used. We define both two-fiber and four-fiber BLSRs. For a bidirectional ring using a two-fiber connection, normal routing of the working traffic is such that both directions of a bidirectional connection travel along the ring through the same ring nodes but in two opposite directions. If two adjacent notes such as A and B are exchanging data streams, for example, the data stream from A to B flows clockwise using the first fiber while the data stream from B to A flows counterclockwise using the second fiber. (See Figure 3-59.)

In a two-fiber ring with a shared protection configuration, only half of the capacity of a fiber is used for the working traffic, and the other half is used for the protection traffic of the working traffic flowing in the opposite direction.

Figure 3-59 *BLSR Concepts*

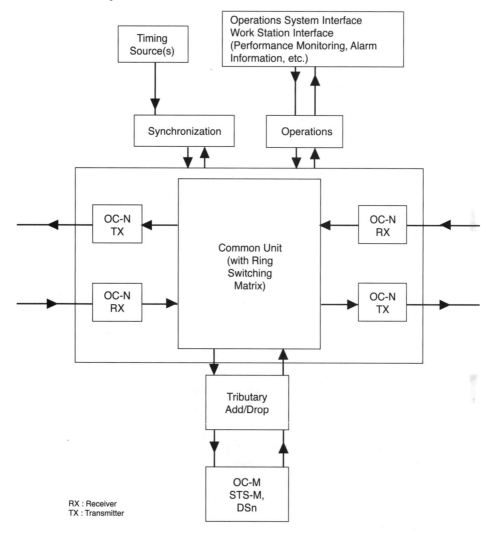

A two-fiber BLSR includes a set of nodes interconnected by a pair of fibers, possibly including regenerators and optical amplifiers between nodes. Figure 3-60 shows a simplified illustration of an NE (a node) in a two-fiber BLSR. To provide the maximum restoration (that is, 100-percent restoration of restorable traffic) for single failures, it is necessary to reserve 50 percent of the ring's capacity for protection. Therefore, a two-fiber optical carrier level (OC-n) ring effectively has a span capacity of OC-($n/2$).

Figure 3-60 *BLSR Concepts*

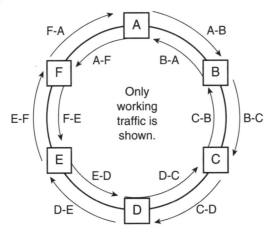

In a four-fiber ring with dedicated protection configuration, two fibers are used for bidirectional routing of the working traffic, and the other two fibers are used to carry the protection traffic. So each direction has a working and a protect fiber.

Protection in a BLSR requires a node to wrap traffic back into the protection path and away from the failure. You learn more about both fiber and node failures later in this chapter.

In both BLSR types, a key attribute is the capability to support spatial reuse of the ring bandwidth. Spatial reuse is the capability to reuse the capacity of the ring on different ring segments for nonoverlapping traffic. Spatial reuse is also discussed more fully later in this chapter.

BLSR rings were designed to overcome some of the limitations of the UPSR ring. A BLSR ring's traffic is passed symmetrically rather than asymmetrically, allowing BLSR rings to perform well in larger multinode environments. In a BLSR ring when node A wants to talk to node C, the working traffic by default is passed via the shortest route to node C. When node C needs to respond to node A, the traffic returns via the shortest route, which is in the reverse direction. So the working traffic between nodes is bidirectional. Figure 3-61 shows a typical two-fiber BLSR ring topology.

The BLSR is more complex to handle than the UPSR ring. To perform proper protection switching in the case of failure, coordination is required between the ADMs so that the switchover to the protection circuit is harmonious at both ends.

A major advantage of this technology is that low-priority traffic can be transported on the protection channels, whereas this feature is impossible in the UPSR technology. This lower-order traffic is discarded in cases where the protection line is needed.

Figure 3-61 *BLSR Ring Topology*

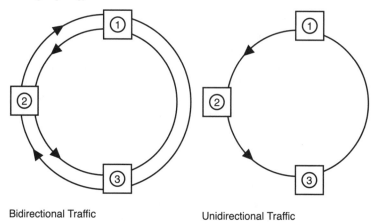

Bidirectional Traffic   Unidirectional Traffic

Another advantage is that bandwidth can be reused on nonoverlapping segments of the ring. When traffic is removed from the ring at its drop point, the STS-*n* used to carry the traffic can be provisioned to carry traffic between other nodes on downstream segments of the ring.

These advantages mean that the BLSR is very efficient in terms of the use of the network infrastructure, especially if there is a lot of traffic between adjacent nodes on the ring, as shown in Figure 3-62.

Figure 3-62 *BLSR Ring Topology*

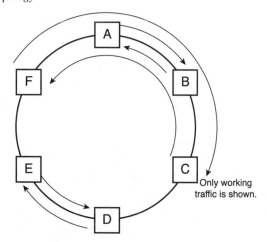

Only working
traffic is shown.

Protection switching is typically not as fast as UPSR; however, protection switching in less than 50 milliseconds is standard.

The maximum number of nodes on a BLSR is 16. This is partly driven by the 50-millisecond performance requirement and partly due to the protection switching coordination as defined by the K1 and K2 bytes only being able to address up to 16 nodes. Four bits of the K1 byte are used for ring APS messages, whereas the other four bits are used to carry the destination node ID. Similarly, four bits of the K2 byte are used for ring APS messages, whereas the other four bits are used to carry the source node ID. (See Figure 3-63.)

Figure 3-63 *Maximum Bandwidth Capacity*

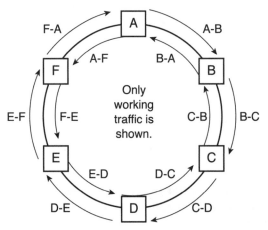

Figure 3-63 shows how the working channels could be provisioned across the ring. Traffic is routed over the shortest path around the ring if at all possible. Duplex traffic between two nodes is carried in opposite directions over the same ring links. Multiple channels can be carried over the same ring segments up to the capacity of the ring. Through the capability to support spatial reuse on nonoverlapping links, the maximum ring capacity occurs when all traffic flows between adjacent nodes. In this case, the total capacity of the ring from all ring nodes is equal to the number of nodes times one half the line rate ($n \times$ line rate/2).

The minimum capacity occurs when all traffic from all nodes is being terminated at the same node. In this case, no spatial reuse is possible and the capacity is equal to the line rate, the same as for UPSR. In a real-world scenario, the actual capacity of the ring is somewhere between these two extremes.

For a two-fiber BLSR ring, the line rate must always be an even multiple of STS-3s, because in this case each fiber is used for working and protection traffic.

Figure 3-64 describes how the maximum capacity of a BLSR is derived. It's primarily a reference. In this figure, a BLSR ring consisting of two fibers is shown. The traffic flow in the two fibers travels in opposite directions. The traffic from node A toward node F is flowing counterclockwise through the ring, whereas the traffic from F toward A is flowing clockwise through the ring.

If two OC-12 channels are used, one for each direction, $6 \times$ STS-1s per fiber are used to carry the working traffic, whereas the remaining $6 \times$ STS-1s are used to carry the protection traffic. Therefore, the maximum bandwidth that can be used for user traffic under normal circumstances is half the maximum bandwidth capacity of a fiber, multiplied by the numbers of nodes connected to the ring.

Figure 3-64 *Extra Traffic*

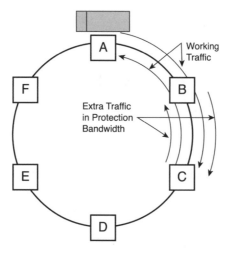

It is possible in BLSR technology to use the bandwidth (half of the line rate), which is reserved for protection traffic to feed some extra, normally low-priority traffic through the ring. This extra traffic is not protected and is lost if a failure in the ring occurs.

Feeding extra traffic through the network is not possible with UPSR technology. Extra traffic handling adds complexity, and in UPSR simplicity is important; therefore, typical needs do not include extra traffic.

If traffic in a BLSR ring is destined for a single node, that is, node A, a BLSR ring will not have any bandwidth advantage over a UPSR ring. If the traffic is distributed around the ring such that there are equal requirements for node A to talk to node B, and node D talks a lot to node E and node F talks a lot to node A, a BLSR ring can provide more bandwidth across a single ring. (See Figure 3-65.)

Figure 3-65 *Bandwidth Usage in a Two-Fiber BLSR Ring*

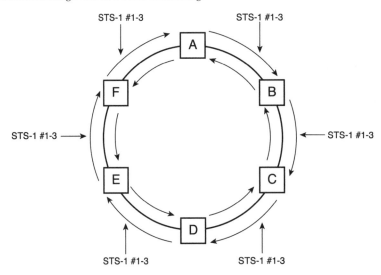

The maximum potential bandwidth a BLSR ring can have is calculated with the following formula:

(Bandwidth of a span on the ring/2) × (Number of nodes in the ring)

This means an OC-48 BLSR ring with six nodes has a maximum potential bandwidth of OC-144.

$(48/2) \times 6 = 144$

The reason the value represents the maximum potential bandwidth is because this value can only be achieved if all nodal communication occurs between directly connected nodes. Most of the time, traffic needs to pass through nodes to reach its destination, which limits the maximum bandwidth.

Protection in a Two-Fiber BLSR Ring

BLSR rings are based on 1:1 protection. As such, traffic is only placed on the working line, and the protect lines are unused until needed for protection. Because the protect lines are unused, it is possible to send extra traffic across the line. This traffic should be low-priority bandwidth because it could be dropped in the event of a working line failure.

BLSR, as its name implies, switches based on the information based in the line overhead. When a switch occurs, all STSs on that link are switched to the protect path. (See Figure 3-66.)

Figure 3-66 *Fiber Cut*

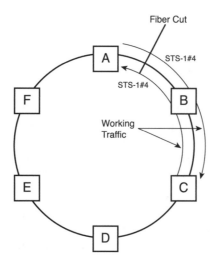

Figure 3-67 shows a BLSR ring with the nodes on the ring interconnected by two fibers. Both fibers carrying working traffic and one half of the capacity of each fiber are reserved for protection. In this example, we have working traffic between nodes A and C that travels on the short path between the nodes. The fiber is cut between nodes A and B. Both nodes detect the failure.

Figure 3-67 *Fiber Cut*

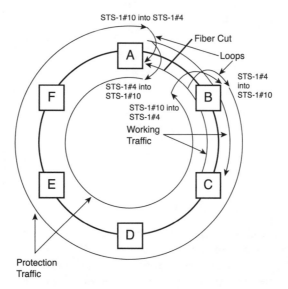

In Figure 3-67, both nodes, A and B, detect the failure. Node A and node B notify all the nodes in the ring about the error condition. Then nodes A and B loop back their fiber connection.

After completing this loopback function, the original two rings appear as one single ring. This ring now carries the traffic for both directions, using the bandwidth reserved for the protection traffic. No traffic is lost.

Because of the wraparound of the ring, the long path instead of the short path is used for communication. So, you need to take into account that the delay is increasing when a protection switch is performed. This could be critical, especially when the BLSR ring is used for voice transport. During the design phase of the ring, you need to take into account that even in case of a ring wrap the delay is still acceptable.

To achieve a proper delay, the number of nodes connected to a ring can be limited or the nodes on the ring can be arranged in a way to achieve a proper balance between short path and long path. But then the ring reuse capability is lost, so a tradeoff in the design needs to be made between the amount of reuse capability and the delay. (See Figure 3-68.)

Figure 3-68 *K1, K2 Bytes*

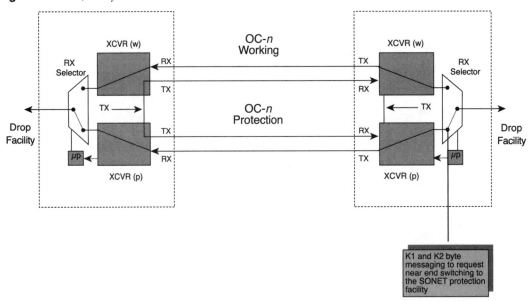

Protection switching in a BLSR requires coordination between the affected nodes. That coordination is provided through use of the K1, K2 bytes that appear in the line overhead of every SONET frame. The next several graphics shows how the K1, K2 bytes would have been used to recover from the fiber cut of the previous example. (See Figure 3-69.)

Figure 3-69 *K1, K2 Bytes*

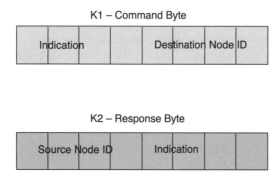

K1 – Command Byte

| Indication | | | Destination Node | ID |

K2 – Response Byte

| Source Node ID | | Indication | | |

Figure 3-69 shows an overview of the coding of the K1 and K2 bytes. K1 is called the *command byte*. It sends indications to a destination node that range from "no request" to "forced switch." K2 is called the *response byte*. It is used to indicate that line alarms have been generated or that protection switching has occurred. Both K1 and K2 have a 4-bit address field. The K1 byte inserts the address of the node to which the command is being sent, the destination node. The K2 byte inserts the address of the node providing the response, the source node. Note that with 4 bits only, 16 nodes can be uniquely addressed. This is the protocol limit that forces BLSRs to have a maximum of 16 nodes. (See Figure 3-70.)

Figure 3-70 *Fiber Cut with K1, K2 Bytes*

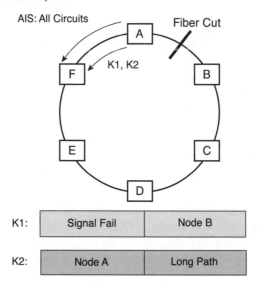

K1: | Signal Fail | Node B |

K2: | Node A | Long Path |

Here, the same fiber cut situation experienced in the earlier example. Our focus here is on the use of the K1 and K2 bytes to facilitate the switchover. Node A detects the failure. It sends the AIS on all circuits so that any downstream termination is aware of the problem. Node A then sets the K1 byte in the line overhead to indicate that a signal failure has occurred. It addresses the indication to node B because it knows that B is the adjacent node in the failed direction. Node A uses the K2 byte to provide its address so that downstream nodes are aware of the source and indicates that the message is being sent on the long path around the ring. Note that even though the K1 byte is addressed to node B, all intermediate nodes can observe the byte and will be aware of the failure. (See Figure 3-71.)

Figure 3-71 *Fiber Cut with K1, K2 Bytes*

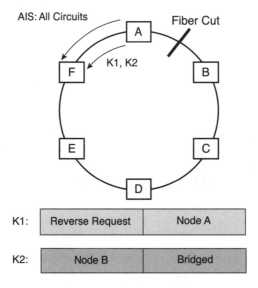

Pick up the example after node B has received the K1, K2 bytes from the previous figure, when node B received the indication of a failure, it bridged the affected circuits onto the protection path in the reverse direction. The looped circuits in the figure depict the bridge created by node B. Node B uses the K1 byte to command node A to loop traffic at node A onto the protection path as well. Node B uses the K2 byte to let node A (and intermediate nodes who observe the byte) know that it has bridged the traffic as requested. (See Figure 3-72.)

When node A receives the indications from node B, it bridges the traffic on its end to complete the protection switching. It sets the K2 byte to indicate to downstream nodes that protection switching has occurred. The K1 byte indicates that there is no further request, as shown in Figure 3-72. (See Figure 3-73.)

Figure 3-72 *Fiber Cut with K1, K2 Bytes*

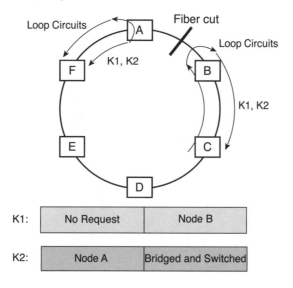

Figure 3-73 *Node Failure*

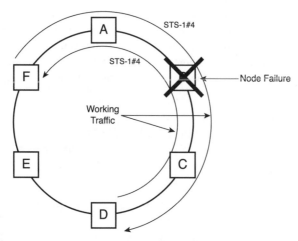

Next is a brief example of BLSR operation on a two-fiber ring when a node fails. Figure 3-74 illustrates traffic flow between nodes D and F when a failure occurs at node B. The failure is detected at nodes A and C.

Figure 3-74 *Node Failure*

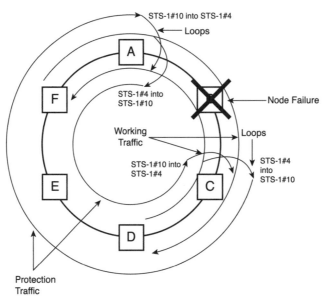

Both nodes A and C detect the failure because the connection toward node B is interrupted.
Nodes A and C notify all other nodes on the ring about the error condition through use of
the K1 and K2 bytes. In a manner similar to the recovery from a fiber cut, nodes A and C
loop traffic onto the protection path and essentially form a single ring. The bandwidth
previously used to carry the protection traffic is now used for the working traffic. The traffic
to and from node B is lost. Any low-priority traffic carried on the protection path is now
also dropped.

BLSR uses a protection scheme called *shared protection*. Shared protection is required
because of the construction of the BLSR ring and the reuse of STSs around the ring. This
creates a situation in which the STSs on a protection fiber cannot be guaranteed to protect
traffic from a specific working STS. This is another reason why BLSR rings are more
complex and require more planning.

Shared protection, which provides BLSR its capability to reuse bandwidth, brings with
it additional problems when a node failure occurs in a BLSR ring. Part A of Figure 3-75
shows an OC-12 BLSR ring carrying traffic between nodes F and B and between nodes B
and E. At this point, everything is fine. Following the traffic from node F to node B in part
B of Figure 3-75, you see that traffic STS-1#1 leaves node F destined for node B. When the
traffic arrives at node A, the STS-1#1 is placed on the protect line as STS-1#7 and sent out
for node B. When the traffic for node B arrives at node C for node B, node C cannot deliver
the traffic, so it places the traffic on the working line in STS-1#1. STS-1#1 has a connection
to node E. The traffic is delivered to the wrong node. This is called a *misconnection*.

Figure 3-75 *Two-Fiber BLSR Ring Passing Traffic Subject to Node Failure*

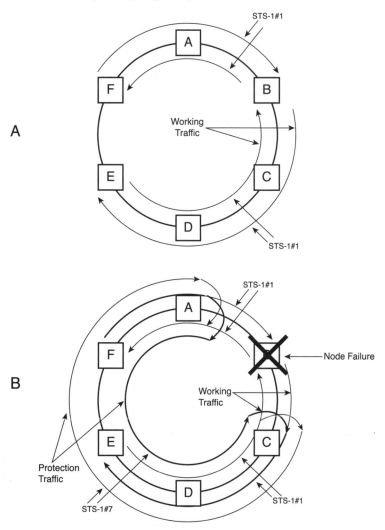

The misconnection is solved with the use of a *squelch table*. This table contains a list of inaccessible nodes. Any traffic received by a node for the inaccessible node is never placed on the fiber and is removed if discovered.

In some situations, it's possible that bridging traffic after a node failure could lead to a misconnection. Figure 3-76 shows a specific example of this. Squelching is a technique that has been defined to avoid the misconnection problem. Squelching involves sending the AIS in all channels that normally terminated in the failed node. (See Figure 3-76.)

Figure 3-76 *Squelching*

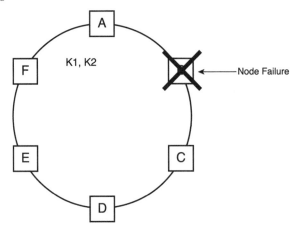

In this BLSR scenario, node B has failed. Both nodes E and F have working traffic being sent to node B, as shown in Figure 3-77. Recall that BLSRs support spatial reuse. Because the communication is on different links of the ring, both the F to B traffic and the E to B traffic happen to be using STS-1#1.

Figure 3-77 *Squelching*

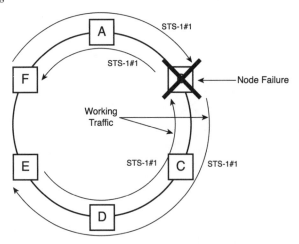

After node B fails, the ring is looped back by nodes A and C. This action provides protection for most traffic, but it also causes a misconnection between nodes E and F because both had used the STS-1#1 channel to communicate with the failed node, node B. (See Figure 3-78.)

Figure 3-78 *Squelching Misconnection*

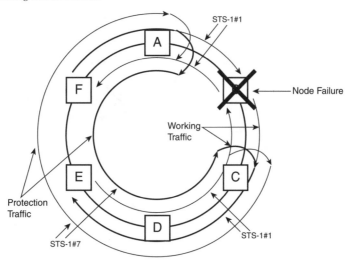

The misconnection is avoided by the insertion of an AIS path by nodes A and C into channel STS-1#1. In an AIS path, all the bits belonging to that path are set to 1 so that the information carried in that channel is invalidated. Now nodes E and F are informed about the error condition of the ring, and a misconnection is prevented.

Misconnection can only occur in BLSR technology when a node is cut off and traffic happens to be terminating on that node from both directions on the same channel.

In some implementations, the path trace might also be used to avoid this problem. If nodes E and F monitor the path trace byte, they will recognize that it has changed after the misconnection. This change should be sufficient indication that a fault has occurred, and the traffic should not be terminated. (See Figure 3-79.)

To summarize, squelching is used to avoid misconnection in the event of node failures. Misconnection can occur if the failed node happens to be terminating traffic on the same STS-n in both directions of the ring. When this occurs, the two nodes that had originally been connected to the failed node could end up misconnected to each other. Squelching is used to avoid this situation

Here is a brief summary of the standardization status for line-switched rings. The ULSR is specified by the ANSI Standard T1.105-1. Equipment manufacturers never accepted this standard, and so no products according to this specification are available on the market.

This section has covered the BLSR developed by Bellcore (now Telcordia). Products for BLSRs have been manufactured by many different vendors.

Figure 3-79 *Squelching—Path AIS Insertion*

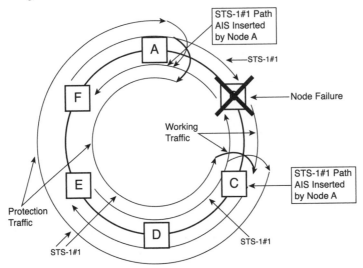

Now that you understand more about rings, look at Figure 3-80. This figure provides a high-level summary and comparison or UPSRs and BLSRs. Briefly stated, UPSRs are relatively simple and provide fast protection switching. UPSRs make most economic sense in access networks where all traffic tends to be focused on a common node, the central office. If all traffic is terminating on a common node, the lack of spatial reuse capability on UPSR is not a limitation.

BLSRs are more complex and require coordination via the K1, K2 bytes for protection switching. However, BLSRs can support spatial reuse and low-priority traffic on the protection bandwidth. BLSRs are more efficient if there is a lot of meshed internodal traffic on the ring. They make most economic sense in network backbones where the more meshed traffic patterns can take advantage of the BLSR efficiencies.

The discussion so far has focused on individual SONET rings, and you should by now be convinced that they can provide very reliable transmission over a individual ring. However, do not lose sight of the fact that an end-to-end connection often traverses many different physical paths. If protection of the end-to-end communication is required, each segment of the path must be protected.

Figure 3-80 *UPSR/BLSR Comparison*

	UPSR	BLSR
Number of Fibers	Two	Two or Four
Protection Switching	Fast and simple. Requires no coordination between nodes.	Slower and more complex. Requires coordination via K1, K2 bytes.
Efficiency	No spatial reuse. Total capacity for all customer traffic equals line rate.	Efficient reuse of capacity if a high degree of internodal traffic exists.
Areas of Application	Predominantly access networks where most or all traffic terminates at the central office.	Predominantly network backbone for interoffice traffic.

Figure 3-81 shows a connection between two routers via the public network. In the core of the network, most carriers use BLSRs, and all traffic is inherently protected. In the access networks, UPSRs are typically used to provide path-level protection. In many cases, UPSR protection is not an inherent part of the transport. It may be positioned as a service option that must be purchased; so if protection is required, you must buy the option. At the edges of the network, a SONET ring may not be available on the last segment to the customer location. If end-to-end protection is required, a linear protection scheme such as 1:1 or 1:*n* must be established for these segments. In some cases, this may be the most difficult segment to protect because diverse routing of the physical link is not available.

Figure 3-81 *Protection*

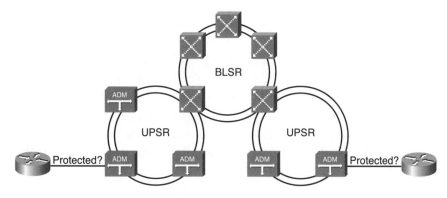

BLSR rings can also have problems dealing with timing issues because the looping of the traffic can also create timing loops.

Four-Fiber Bidirectional Line-Switched Ring (BLSR)

In a four-fiber BLSR ring, there are two full fiber strands dedicated to working and two full fiber strands for protection. (See Figure 3-82.) The two fibers defined as working pass traffic in opposite directions and have access to all of their bandwidth. The protect fibers have all their bandwidth reserved for protection. This gives a four-fiber BLSR ring some advantages over a two-fiber BLSR ring. The main reasons for the increase in cost is the extra fiber, and every network element needs four line cards for west, west-protect, east, and east-protect.

Figure 3-82 *Four-Fiber BLSR*

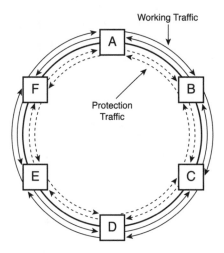

BLSR rings such as the four-fiber version can also provide greater survivability in the case of fiber cuts or failures. These features don't come without a price. BLSR rings are generally more complex and expensive to set up and manage.

When two nodes want to communicate, the system by default wants to use the shortest path to the node. In a BLSR ring, nodal information has to be exchanged between the nodes so that a shortest path can be determined.

This nodal information exchange is one reason that most vendor equipment will not interoperate in the confines of a BLSR ring. Standards are good at specifying what must be done but do not necessarily describe how. This leaves much up to interpretation by the vendor, which can reduce vendor interoperability.

BLSR rings are best used in longer distance—multimode or distributed access environments. Distributed access means that the communications between the nodes are relatively balanced. This ring type is well suited for the metro area where multiple businesses or offices are interconnecting and exchanging information on a fairly consistent basis throughout the ring.

BLSR topologies are found primarily in core and backbone networks, but are starting to approach the access and metro networks, too, under some situations

Protection in a Four-Fiber BLSR Ring

When extra fibers are introduced into a BLSR ring, additional protection options are available. These extra protection options are what make a four-fiber BLSR ring so attractive to service providers.

A four-fiber BLSR ring has the additional capacity to do either a span switch or a ring switch. A *span switch* is used when a failure occurs between two nodes; the protect path can carry the traffic between the nodes, and the ring does not need to be looped. This conserves ring resources by allowing the protect path between nodes to carry the traffic instead of having to take the long path around the ring. (See Figure 3-83.) All the information about span switch or ring switch is carried in the K1 byte.

Figure 3-83 *Four-Fiber BLSR Ring with a Working Line Failure*

SPAN SWITCH

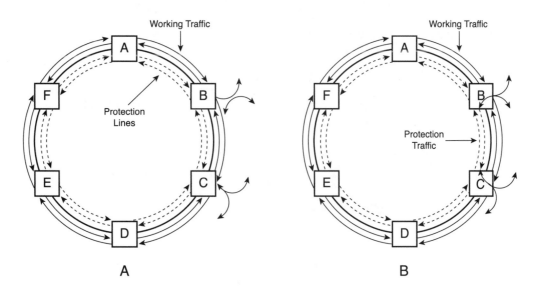

Example A is showing a four-fiber BLSR ring working under no line failures, whereas B shows a line failure and shows how the traffic would switch.

It would take a cut of both the protect and working fibers to produce a ring switch. The four-fiber BLSR ring is capable of surviving multiple failures, making it desirable in the large-scale ring implementations.

Path-Protected Mesh Network

Path-protected mesh networks (PPMN) is a solution that applies UPSR protection to a mesh or semi-mesh network. A key advantage to PPMN is that it can provide ring-like protection without requiring construction of a ring. PPMN networks also support easy network provisioning.

PPMN uses extensions of routing protocols to allow all the nodes in a mesh to be topology aware. When a path is required between two nodes, the originating node calculates the shortest path to the destination and then uses a signaling protocol to automatically establish the path between source and destination. If protection of the new path is requested, the originating node computes the "second shortest" path to the destination, making sure that it does not share any links with the first path. The second path is then established using the signaling protocol.

After both paths have been established, they can be considered to be a virtual ring that connects the source and destination nodes. At this point, the source bridges traffic onto both paths, and the destination picks whichever signal it prefers. This is standard 1+1 protection as used in a UPSR now applied to the virtual ring. (See Figure 3-84.)

Figure 3-84 *Path-Protected Mesh Networks (PPMN)*

Path Established from Node 1 to Node 2

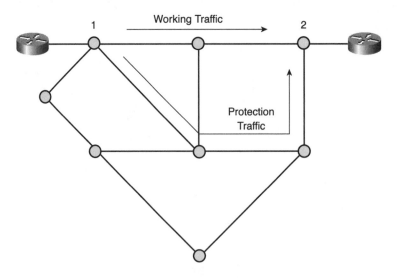

In this example of a PPMN network, the autodiscovery and signaling protocols establishes two paths between nodes 1 and 2. When both paths are established, nodes 1 and 2 are part of a virtual ring. For the direction of node 1 to node 2, node 1 bridges the traffic onto both paths, and node 2 receives two copies of the traffic. (See Figure 3-85.)

Figure 3-85 *PPMN Example*

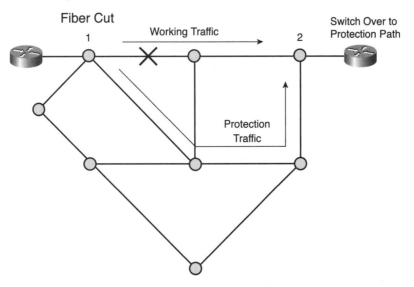

Next, there is a fiber failure in the example network. Node 2 recovers from the failure just by switching to the protection path. Protection switching is fast and does not require coordination between the nodes. Operation is essentially the same as UPSR. (See Figure 3-86.)

Here, you see protection in the event of a node failure. As in the case of the fiber failure, node failure recovery is accomplished just by switching over to the protection path. Recovery is fast, and no coordination with the other nodes is required. This is a fairly straightforward application of 1+1 protection over the virtual ring.

Cisco BLSR Ring Enhancements

Although the four-ring BLSR is not beyond the standards, it is not implemented by most vendors due to the complexity of controlling both span and ring switches. Instead, the vendors have expanded the number of rings that can exist on a BLSR ring. The standard calls for BLSR to support 16 nodes. This limitation is due to the K1 and K2 bytes and the 4 bits reserved for node identification. Cisco has enhanced the capabilities so that a BLSR ring will support up to 25 nodes.

Figure 3-86 *PPMN Example*

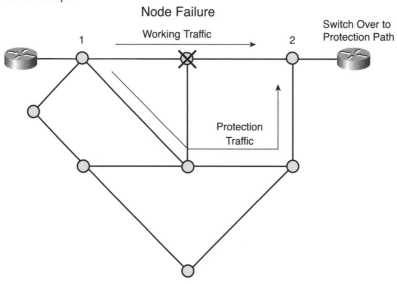

SONET Alarms

Whenever SONET is used in a network, whether it is a point-to-point circuit, UPSR or
BLSR ring, or other technologies such as packet over SONET or ATM over SONET,
SONET provides alarm reporting, which is important for monitoring and troubleshooting.
As with all SONET functionality, this information is carried in the headers used. Details of
the individual bytes were discussed earlier in the section "SONET Framing." In this section,
you learn how some of the bytes in those headers are used to capture alarm information and
pass that information through the network.

Understanding the alarms and recovery mechanisms found in SONET will enable you to
manage your network more effectively. This section will help you to do the following:

- Describe the events associated with the initiation of automatic protection switching

- Describe the events associated with a loss of signal in the section layer

Error-Causing Events

Earlier in this chapter (Figure 3-68), the K1 and K2 bytes were discussed (with regard to
their role in protection switchover). At the line level, the following items can activate the
switchover process:

- **Loss of signal (LOS)**—Loss of signal is declared when an all-0 pattern is detected
  on the incoming OC-*n* data stream before descrambling. When LOS is detected, a
  downstream multiplex line alarm indication signal (AIS) must be transmitted.

- **Loss of frame (LOF)** — Loss of frame is declared when all incoming STS-n signals have an out-of-frame (OOF) indication for a specific time period. The OOF could be triggered by a transmitter failure or a high BER condition. The high BER condition could be caused by problems such as too high or too low received optical power levels. When an LOF is detected, a downstream multiplex line AIS must be transmitted.

- **Line BIP-8 errors** — One byte, B2, is allocated in each STS-1 frame for line error monitoring. The BIP-8 is a bit interleaved parity 8 code using an even parity checksum. The line BIP-8 is calculated over all bits of the LOH and the STS-1 envelope capacity of the previous STS-1 frame before scrambling.

At the path level, the following conditions can also activate the protection switching process:

- **Loss of pointer (LOP)** — When a pointer processor cannot obtain a valid pointer condition, an LOP state is declared and a downstream VT path AIS or STS-N path AIS must be transmitted.

- **Path BIP-8 errors** — One byte, B3, is allocated for a path error monitoring function. This function must be a bit interleaved parity 8 code using even parity. The path BIP-8 is calculated over all bits of the previous STS synchronous payload envelope (SPE) before scrambling.

- **VT path BIP-2 errors** — The BIP-2 is a bit interleaved parity 8 code using even parity and is calculated over the VT path overhead bytes (V5, J2, N2, K4).

After an error has occurred, that information has to be propagated through the network. Figure 3-87 shows the messages used.

Figure 3-87 *Loss of Signal Messages*

As Figure 3-87 illustrates, the upstream node detects an LOS. This would occur at the section level. An AIS would be propagated to the downstream node and a remote defect indication (RDI) would be sent to the next upstream node.

Error Event Propagation

The relationship between the different alarms that are reported by a SONET network is critical to understanding the exact nature of a failure or degradation of service conditions when managing a PoS network.

Figure 3-88 shows a global overview of the signal propagation. Depending on the problem, different levels may detect it and pass it on to the other levels.

Figure 3-88 *Event Propagation in SONET*

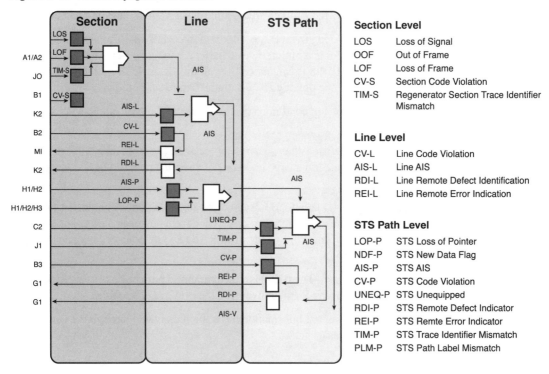

Section Level

LOS	Loss of Signal
OOF	Out of Frame
LOF	Loss of Frame
CV-S	Section Code Violation
TIM-S	Regenerator Section Trace Identifier Mismatch

Line Level

CV-L	Line Code Violation
AIS-L	Line AIS
RDI-L	Line Remote Defect Identification
REI-L	Line Remote Error Indication

STS Path Level

LOP-P	STS Loss of Pointer
NDF-P	STS New Data Flag
AIS-P	STS AIS
CV-P	STS Code Violation
UNEQ-P	STS Unequipped
RDI-P	STS Remote Defect Indicator
REI-P	STS Remte Error Indicator
TIM-P	STS Trace Identifier Mismatch
PLM-P	STS Path Label Mismatch

Following the flows in Figure 3-88, you can see that LOF or LOS in the section/regenerator section area leads to an AIS that is also handed over to the line layer, as well as to the STS path layer.

The alarm hierarchy shown in Figure 3-88 is important because it allows root-cause analyses that shows where the failure actually occurred. This chart gives you the understanding of what causes an alarm and what you need to look at.

An LOS is raised when the synchronous transport signal (STS-n) level drops below the threshold at which a BER of 1 in 10^3 is predicted. An LOS could result from a cut cable, excessive attenuation of the signal, or equipment fault. LOS clears when two consecutive framing patterns are received, and no new LOS condition is detected. (See Figure 3-89.)

Figure 3-89 *Example of a SONET Event Propagation*

— Loss of Signal (LOS)
    • Section LOS (SLOS)
— Loss of Frame (LOF)
    • Section LOF (SLOF)
— Alarm Indication Signal (AIS)
    • Line AIS (LAIS)
    • Path AIS (PAIS)
—Remote Defect Indicator (RDI)
    • Line RDI (LRDI)
    • Path RDI (PRDI)
— Loss of Pointer (LOP)
    • Path LOP (PLOP)

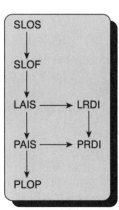

As shown in the previous figure, an LOS causes a section loss of frame (SLOF) at the section layer, which results in a line alarm indication signal (LAIS) for the line/multiplex section on the equipment where the problem occurred.

The LAIS is an all-1 characteristic or adapted information signal. The LAIS is generated to replace the normal traffic signal when it contains a defect condition to prevent consequential downstream failures being declared or alarms being raised.

The error condition is propagated to the adjacent line/multiplex layer equipment. The remote defect indication (LRDI) signal for the line/multiplex layer is used by the neighbor equipment to signal the error condition back to the equipment where the error was detected. (The LRDI was previously known as a far-end receive failure [FERF].)

Further on, an alarm indication signal (PAIS) for the path will be generated with its path remote defect indicator (PRDI) signal.

In the end, the pointer is lost, which leads to a path loss of pointer (PLOP) signal. A loss of pointer (LOP) defect is declared when either a valid pointer is not detected in eight consecutive frames, or when eight consecutive frames are detected with the new data flag (NDF) set to 1001 without a valid concatenation indicator

Performance Event Collection

Errors on lines can come and go, sometimes so quickly it may not actually cause a major disruption of service. These errors might not be significant enough for immediate action but should be noted because they might be a precursor to future problems. SONET network elements collect performance information such as the following:

- Parity errors in the section BIP byte B1

- Parity errors in the line BIP byte B2

- Parity errors in the path BIP byte B3

- Received remote error indication (RDI) signals

- Pointer events

- VT path BIP, bits 1 and 2 of the V5 byte

The network elements must store the different types of performance information with associated threshold register settings. When the threshold register count exceeds the associated threshold register settings, a threshold cross alert (TCA) is sent out to the appropriate operating system (OS) to inform the user of the error condition.

These error conditions, as well as degradation in signal quality indicated by signal degrade (SD) and signal fail (SF), may lead to an automatic protection switching (APS) process.

Performance statistics for the regenerator section, multiplex/line section, and the path section are kept separately by the network in what are called *current 15-minute registers* and *current day registers*.

A current 15-minute register contains the error count for the last 15 minutes and is reset to 0 each time the 15-minute time limit is reached. The content of the register is saved to an optional previous 15-minute register or just thrown away. The same applies to the current day register. It is also possible to manually reset the registers to 0 at any time. The content of the register is read out for diagnostic purposes.

Errored seconds (ES) and severely errored seconds (SES) rate the amount of errors in a specific time interval. The following definitions of ES and SES are taken from part of RFC 1595, *Definitions of Managed Objects for the SONET Interface Type*:

- **Errored seconds**—At each layer, an ES is a second with one or more coding violations at that layer, or when one or more incoming defects (for example, SEF, LOS, AIS, LOP) at that layer have occurred.

- **Severely errored seconds**—At each layer, an SES is a second with x or more code violations at that layer, or a second during which at least one or more incoming defects at that layer has occurred. Values of x vary depending on the line rate and the bit error rate.

Now that you have a basic understanding of SONET and its structure, topology, and alarms, the next major aspect of SONET is its mechanism to pass management traffic through the network. Management traffic is generally referred to as operations, administration, management, and provisioning (OAM&P).

SONET Management

You can manage SONET devices remotely through the use of in-band management channels. Management traffic embeds the SOH and LOH bytes, resulting in the section data communication channel (SDCC) and line data communication channel (LDCC) as depicted in Figure 3-90.

Figure 3-90 *SONET Device Management via the SDCC and LDCC*

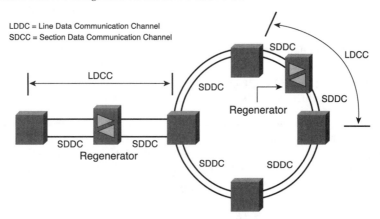

These channels can transport operations and management messages. These types of messages for service providers will comply with Telecommunication Management Network (TMN) specification. This is dictated by the standards; however, implementation of these bytes is left up to the vendor. Although SONET is a standard, there is little interoperability between vendors in the use of these bytes.

The section layer DCC (SDCC) bytes are the D1 through D3 bytes. These 3 bytes provide a 192-kbps communications channel. The line layer DCC (LDCC) bytes are bytes D4 through D12. These 9 bytes provide a 576-kbps communications channel. Most SONET systems use SDCC bytes for management purposes and leave the LDCC bytes alone.

Cisco Transport Manager (CTM) enables customers to manage SONET and optical transport products collectively under one management system. Cisco sells a craft tool and element management system (EMS) for comprehensive SONET and optical transport management called *CTM*.

Cisco SONET Enhancements

Standards have provided a means for equipment from different vendors to interoperate, giving the customer choice. Companies developing to the standards are given access to a larger market; however, it is still necessary to differentiate yourself from the crowd. Cisco has done so by providing enhancements to both their SONET management technique and to their SONET ring capabilities.

SONET Network Element Management

Cisco, in its 15454 and 15327 pieces of equipment, has opted to develop its management application based on the TCP/IP stack. This IP feature coupled with an OSPF-based topology-discovery mechanism makes these the leading SONET transport network elements with IP intelligence.

Although this might make their management applications incompatible with others, it also allows them to use generic tools such as browsers to manage the node. This feature leverages the power of the Internet with its ease of use and universal accessibility to provide robust management systems.

Interoperability of management systems is often a problem; however, Cisco has provided another feature not commonly found in SONET systems to enhance interoperability. Because the LDCC bytes are not normally used, Cisco tunnels its SDCC bytes through the Cisco 15454 or 15327 SONET network. (See Figure 3-91.) If a vendor is using the LDCC bytes, Cisco enhancement will not apply.

Figure 3-91 *A 15454 SONET Ring Tunneling Management Traffic*

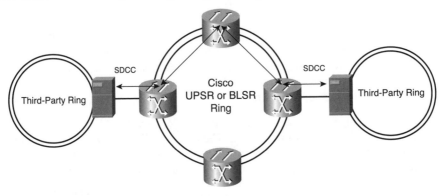

Cisco tunnels the bytes by copying the SDCC bytes into three of the LDCC bytes and then restoring them as the traffic leave the network. This gives Cisco the capability to transport traffic from three different SONET networks simultaneously across any given span.

Cisco UPSR Ring Enhancements

Another major feature found in Cisco SONET networks is based on UPSR rings. When a network grows, it cannot always be true to its original design. Growth tends to be where it is most needed at the time, and this can result in networks that are more pasted together than taking advantage of proper design. Figure 3-92 shows one such example.

Figure 3-92 *A Random Growth Network*

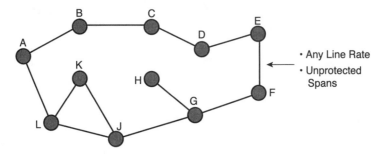

In this environment, original ring protection would not be available for all spans. In this situation, you might look for linear protection, but that would have to be implemented on a span-by-span basis. Cisco has developed a system called *path-protected mesh network* to address this protection problem: UPSR-style protection, but at the circuit level. At the time you create a circuit, you can automatically or manually select two paths for traffic: a working path and a protect path. (See Figure 3-93.)

Figure 3-93 *PPMN Circuit*

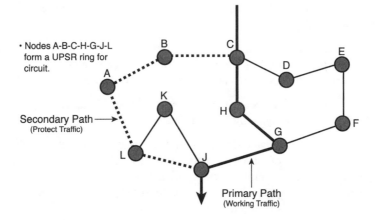

If there is a failure, the receive end detects the failure and automatically switches to the protect path. (See Figure 3-94.)

Figure 3-94   *PPMN Circuit Using the Protect Path*

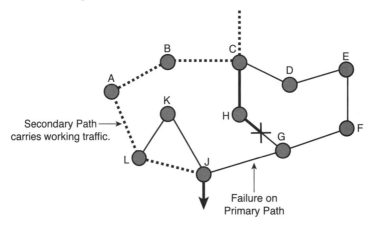

Summary

In this chapter, you have learned that SONET/SDH define the standards for transmission of higher-rate signals in public communications networks.

You learned that starting with 155 Mbps, SONET and SDH support the same rate hierarchy, which increases by a factor of four in the current hierarchy.

You learned that SONET transmission segments are broken into sections, lines, and paths for management purposes.

This chapter also explained that SONET/SDH uses a time-division frame that is repeated 8000 times per second. That frame size is rate-dependent, and the base frame is 810 bytes. Higher-order frames contain an integer multiple of 810 bytes.

You also learned that the management overhead and payload are time-division multiplexed in the SONET/SDH signal by byte interleaving.

As you learned in this chapter, SONET/SDH can transport subrate signals such as DS1s and E1s through the use of virtual tributaries/tributary units. You learned that squelching is required to ensure that misconnections are not made. This chapter also explained that squelching is also required for extra traffic because the extra traffic might be dropped when a protection switch is required.

You learned that SONET/SDH can be deployed in point-to-point, point-to-multipoint, mesh, and ring topologies; rings are most common.

For ring topologies, UPSRs are most common in access networks, and BLSRs are most common in interoffice networks. (See Figure 3-95.)

Figure 3-95 *UPSR/BLSR Ring Locations*

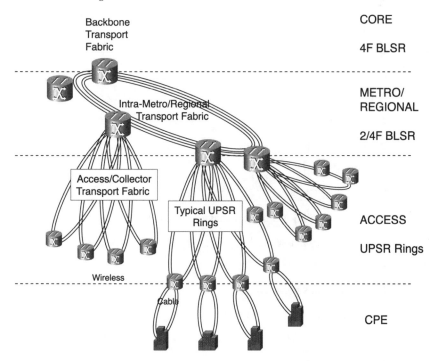

SONET protection can be based on 1+1, 1:1, or 1:*n* schemes; protection switching can be at the line level or the path level.

SONET/SDH define the fundamental transmission infrastructure for telecommunications networks worldwide, and a family of rates that range from 51.84 Mbps to 39.808 Gbps is currently defined.

Review Questions

1 SONET electrical signals are represented by

   A OC-N

   B STM-N

   C STS-N

   D SDH-N

2 What is the line rate for an OC-12?

   A 51.84 Mbps

   B 155.52 Mbps

   C 622.08 Mbps

   D 2.488.32 Gbps

3 An OC-48 SONET frame is repeated how many times per second?

   A 4000 times per second

   B 8000 times per second

   C 16000 times per second

   D 48000 times per second

4 What do SONET TOH bytes carry? (Choose two.)

   A Payload data

   B Path overhead

   C Line overhead

   D Section overhead

   E SPE data

5 Which of the following layers is responsible for framing and scrambling?

   A Path layer

   B Section layer

   C Line layer

   D Photonic layer

6 How many rows are in a SONET frame?

   A 1

   B 9

   C 90

   D 270

7 How many VTGs are in an STS-1 SPE?

   A 1

   B 2

   C 3

   D 4

   E 7

8 A VT1.5 encapsulates what type of traffic?

   A DS0

   B DS1

   C E1

   D DS3

9 Match these SONET layers to their network elements

   A Section   _____End to end

   B Line   _____Regenerator to regenerator

   C Path   _____Multiplexer to multiplexer

10 What is a SONET regenerator's primary role?

   A 1r

   B 2rs

   C 3rs

   D 4rs

11 Which of the following is true of a terminal multiplexer?

   A Exists at the edge of a SONET network

   B Exists in the core of a SONET network

   C Exists on the fiber span of a SONET link

12   A SONET add/drop multiplexer has at least how many ring and add/drop ports?

   A   One ring port and two add/drop ports

   B   Three ring ports and two add/drop ports

   C   Two ring ports and one add/drop port

13   Which digital cross-connect system is capable of switching at the DS1 level between SONET OC-ns?

   A   Wideband DCS

   B   Broadband DCS

   C   Narrowband DCS

   D   Low-band DCS

14   A SONET broadband DCS is capable of switching which of the following?

   A   Minimum of DS1 levels and up

   B   Minimum level of DS3 levels and up

   C   Minimum level of STS-3 levels and up

   D   Minimum level of STS-1 levels and up

15   Which of the following devices is used to multiplex low-speed (56-kbps, voice) services onto a high-speed SONET network?

   A   Terminal multiplexer

   B   Add/drop multiplexer

   C   Digital loop carrier

   D   Digital cross-connect

16   Bidirectional automatic protection switching uses the _____ byte(s) in the line overhead.

   A   H1 and H2

   B   D1 and D2

   C   K1 and K2

   D   H3

17 Which of the following SONET ring topologies duplicates data on both the working and protect rings all the time?

A   UPSR

B   BLSR

C   ULSR

D   BPSR

18 Extra traffic is possible in a _____ ring.

A   BLSR

B   UPSR

C   ULSR

D   BPSR

19 What is the maximum traffic-carrying capacity of an OC-12 UPSR ring with four nodes?

A   OC-3

B   OC-6

C   OC-12

D   OC-24

E   OC-48

20 What is the maximum traffic carrying capacity of an OC-12 BLSR ring with four nodes?

A   OC-3

B   OC-6

C   OC-12

D   OC-24

E   OC-48

21 Which OC levels can support a BLSR ring?

   A OC-1

   B OC-3

   C OC-12

   D OC-48

   E OC-192

22 Which type of ring can perform either a span switch or a ring switch?

   A 2-fiber UPSR

   B 4-fiber UPSR

   C 2-fiber BLSR

   D 4-fiber BLSR

23 An STS-12 frame is how many bytes?

   A 9700

   B 9720

   C 9740

   D 9780

24 In which section(s) can you find the data communication channel(s)?

   A Photonic

   B Section

   C Line

   D Path

   E Virtual tributary

25 How many DS1s are in a DS3?

   A 21

   B 24

   C 28

   D 32

This chapter covers the following topics:

- Choosing a Cisco Metro-Area Network Platform
- Cisco ONS 15454 Optical Platform
- Cisco ONS 15327 Optical Platform
- Cisco Transport Controller (CTC)

CHAPTER 4

ONS 15454 and ONS 15327 Optical Platforms

Cisco Systems offers a variety of optical transport platforms for the metropolitan-area network (MAN). This book concentrates on the ONS 15454 and 15327 add / drop multiplexers (ADMs), the 15216 optical component family, Packet over SONET (PoS), and Dynamic Packet Transport (DPT). None of these technologies are mutually exclusive. They work hand in hand for migration or integration, or can be used standalone.

The ONS 15454 and 15327 combination supports legacy voice circuits and is primarily electrical ADMs. These two shelves also provide migration to dense wavelength division multiplexing (DWDM) architectures and metropolitan Ethernet services. Speeds and feeds supported in these two platforms start at DS1 (1.544 Mbps) through OC-192 (10 Gbps). The 15216 family of products represents the Cisco all-optical suite of products. The 15216 family operates with ONS 15454 DWDM capabilities or can be used with other compliant DWDM equipment. Packet over SONET and DPT represent the Cisco high-end router and switch customer premises equipment (CPE) optical interfaces, which interoperate with existing SONET technology because of the framing employed. Other Cisco optical platforms are mentioned in the previous chapters when necessary, but they are not focused on.

This chapter covers the architecture, features, and functionality of the ONS 15454 and ONS 15327 platforms. Upon completing this chapter, you will be able to do the following:

- Position the Cisco ONS 15454 and ONS 15327 platforms in a SONET MAN network.
- Identify the Cisco ONS 15454 and ONS 15327 hardware and software components required to create a SONET MAN network.
- List the features provided by the Cisco Transport Controller.

Choosing a Cisco Metro-Area Network Platform

Although optical networking is primarily used by service providers who need to transport enormous amounts of traffic, it no longer corners the market of optical usage. Bandwidth, distance, and security needs are making optical networking a mandate at many medium to large enterprises. Price points have come down to levels that are affordable to the enterprise market segment. Enterprises normally do not own their own fiber facilities, so they lease fiber, wavelengths, and services from service providers using optical networking. The

Cisco ONS platforms provide aggregation and transport to support both service providers and enterprises. If an enterprise owns or leases its own fiber, it can choose to control its own service needs.

The ONS 15454 has much more capacity and scalability than the ONS 15327 but costs more. The Cisco ONS 15327 is the platform of choice in areas where size or service requirements are small and fixed. The ONS 15327 features a small form factor for areas where real estate is a premium and supports DS1 through OC-48 connectivity.

The ONS 15454 provides a much higher quantity of interfaces and capabilities. Although the device is much larger and versatile than the ONS 15327, up to three ONS 15454s are supported in a standard 19-inch rack. This platform is selected when the ONS 15327 will not support the amount of interfaces needed, when scalability is necessary, or when DWDM services are needed.

The ONS 15454 platform is positioned as an access or metro ring platform. Access rings are in geographical areas of a metropolitan environment and feed into metropolitan rings that link rings throughout the entire metropolitan geography. Metropolitan rings are sometimes aggregated onto backbone rings for long-haul services between metropolitan areas.

Figure 4-1 displays the multiservice capabilities of the two mentioned optical platforms. Different service types are received at various locations throughout the rings and transported via SONET at speeds from OC-3 to as high as OC-192.

Figure 4-1 *Cisco Metro-Area Network Transport Solutions*

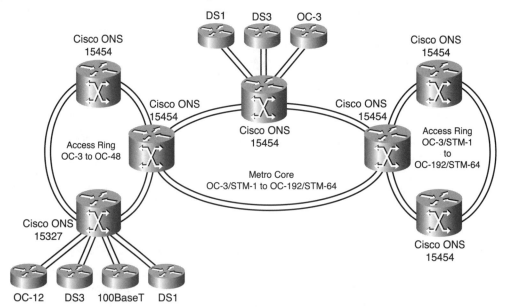

The ONS 15454 is capable of supporting up to five rings. The ring configurations supported are as follows:

- Five UPSR rings
- One two-fiber BLSR ring and four UPSR rings
- Two two-fiber BLSR rings and three UPSR rings
- One four-fiber BLSR ring, one two-fiber BLSR ring, and two UPSR rings
- One four-Fiber BLSR ring and three UPSR rings

Actual capabilities can change depending on CTC version capabilities. You can aggregate multiple service types onto a single fiber because each traffic type is encapsulated into its own container before being multiplexed onto the fiber for transport.

Cisco ONS 15454 Optical Platform

The Cisco ONS 15454 was designed to enable users to get new services provisioned quickly through a browser-based graphical interface. If a browser is not available, an ONS 15454 node could be accessed via a nine-pin RS-232 port. The RS-232 port supports VT100 emulations such that TL1 commands might be entered directly without the need of a browser. The platform supports up to 12 cards in which services can be provisioned. The platform can fill the need for high-density services or allow a migration path for new services. The Cisco ONS 15454 architecture is optimized specifically to provide the following features:

- **SONET transport**—Integrates the capabilities of OC-3, OC-12, OC-48, and OC-192 SONET add / drop multiplexing in one platform.

- **Integrated optical networking**—Provides scalable SONET transport, International Telecommunications Union (ITU) grid wavelengths, DWDM optics, and optical services.

- **Multiservice interfaces**—Supports time-division multiplexing (TDM), Ethernet, framed ATM, data, and video services, which allow service providers to sell new features and support their existing features.

- **Integrated M1:3 multiplexing using a digital cross-connect system (DCS)**—The M1:3 functionality, which historically involved another piece of equipment in service provider ADM environments, has been built in to the ONS 15454 shelf. In addition to the DCS, the broadband cross-connect system allows the shelf to integrate legacy electrical circuits with high-speed optical interfaces up to 10 Gbps in speed. The DCS functionality operates in the electrical realm and is broken down into the wideband digital cross-connect system (WDCS) and broadband digital cross-connect system (BDCS). The WDCS is used to interconnect a large number of DS1s, whereas the BDCS is used to cross-connect signals at DS3 and higher speeds. The M1:3 multiplexing capabilities allow DS1 channels to be aggregated into a 44.76 Mbps DS3 signal. A channeled DS3 service can be broken down into 28 composite DS1 signals.

The Cisco ONS 15454 provides subtending ring functionality through its capability to support up to five rings in one platform. For pure backbone ring aggregation, the 15600 platform is a better choice.

In addition, the Cisco ONS 15454 platform supports high-speed data interfaces, which allow the service provider to aggregate lower-speed traffic onto the same fiber pairs. The aggregation capabilities of the Cisco ONS 15454 save capital expenditures (CapEx) on the cost of cards in the switch by providing four high-speed slots into the core without the requirement for separate low-speed interfaces for each access device needing to connect into the network.

Figure 4-2 contains an architectural breakdown diagram of the ONS 15454. The following information is based on this diagram.

Figure 4-2   *Architecture of the ONS 15454 Shelf*

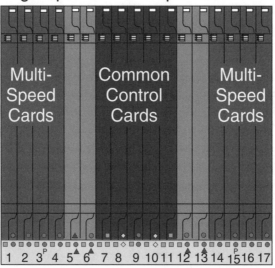

Three types of cards are supported on the ONS 15454. Each slot is card ready so that when a new card is plugged in, the card automatically boots up and becomes ready for service. Starting from the left side of Figure 4-2 is slot 1. Slots 1 through 4 are multispeed slots. Multispeed slots accept DS1 (1.544 Mbps) through Gigabit Ethernet (1 Gbps) cards. Slots 5 and 6 are high-speed cards. The high-speed cards accept OC-48 (2.5 Gbps) speed cards. Slots 5 and 6 can accept any of the cards that are supported in the multispeed slots but will not use the full switching matrix capabilities of the device. Slots 7 through 11 are used for common control cards. Slot 7 and 11 are reserved for the timing and control card (TCC), and slots 8 and 10 are reserved for cross-connect cards (XC). Slot 9 is used for an alarm interface card (AIC). Slots 12 and 13 are used for high-speed cards, which can also accept multispeed cards, whereas slots 14 through 17 are used for multispeed cards only.

If the ONS 15454 were logically cut down the middle of slot 9, the left side of the chassis would be a mirror reflection of the right side. The features and functionality of both sides are identical.

The ONS 15454 shelf is available in two versions. The legacy NEBS-3 shelf supports up to OC-48 speeds, whereas the ANSI shelf (also NEBS compliant) supports up to OC-192 (10 Gbps) speeds. This ANSI version of the shelf is required for the OC-192, G-series Gigabit Ethernet, and OC-48 any-slot cards. The new ANSI shelf assembly follows the same rules as the NEBS-3 shelf assembly but allows the incorporation of the new line cards with use of new XC-10G cross-connect cards. The OC-48 any-slot cards allows OC-48 interfaces in the multispeed slots.

Some cards that run on the NEBS-3 shelf assembly cannot be used in the 10 Gbps-capable ANSI shelf assembly. Table 4-1 contains a compatibility list.

Table 4-1   *Card Compatibility*

Card	2.5-Gbps Shelf Assembly	10-Gbps Shelf Assembly
Control Cards		
TCC+	Yes	Yes
XC	Yes	No
XC-VT	Yes	No
XC-10G	No	Yes
AIC	Yes	Yes
Electrical Cards		
EC1–12	Yes	Yes
DS1–14, DS1n–14	Yes	Yes
DS3–12, DS3n–12	Yes	Yes
DS3XM–6	Yes	Yes
Optical Cards		
OC-3 cards	Yes	Yes
OC-12 cards	Yes	Yes
OC-48 cards	Yes, only slots 5, 6, 12, 13	Yes, only slots 5, 6, 12, 13
OC-48AS	No	Yes
OC-192	No	Yes, only slots 5, 6, 12, 13
Ethernet Cards		
E-100T-12	Yes	No

continues

Table 4-1   *Card Compatibility (Continued)*

Card	2.5-Gbps Shelf Assembly	10-Gbps Shelf Assembly
E-100T-12-G	Yes	Yes
E-1000-2	Yes	No
E-1000-2-G	Yes	Yes
G-1000-4	No	Yes

Upon physical inspection, the two shelf assemblies appear to be identical. The differences are underlying in the architecture and cannot be easily verified by physical inspection. Luckily, both shelves have a product sticker on the lower inside left side when viewing the front of the cards. This sticker contains the product name. The old shelf name includes NEBS-3, whereas the new shelf name includes ANSI.

Control Cards

Control cards are used for shelf operations, administration, management, and provisioning (OAM&P). Table 4-2 contains a feature comparison of the various control cards.

Table 4-2   *The Control Cards*

Card	Purpose	Node Specification	Card Slots
Timing and Control Card Plus (TCC+)	Management access	10BASE-T LAN and RS-232 craft interface for management access.	7 or 11
XC	Cross-connect ability at STS level	E-series Gigabit Ethernet is maximum speed supported.	8 or 10
XC-VT	Cross-connect capability at STS or VT level	E-series Gigabit Ethernet is maximum speed supported.	8 or 10
XC-10G	Cross-connecting STS/STM and VT/AU	OC-192 maximum ring/circuit speed supported.	8 or 10
Alarm interface controller card (AIC)	External alarm reporting		9

The following sections describe each control card in detail.

TCC+

The TCC card is currently in end-of-life (EOL) status. The TCC+ card is the TCC card replacement, required in CTC versions later than release 2.2. The TCC+ card has all the features of the TCC but more processing and memory capabilities required for some of the newer line cards. This is the main control card that houses the central intelligence of the

ONS 15454 shelf and must be installed to be able to set up and configure the system. The card provides LAN connectivity through an autosensing 10/100-Mbps RJ-45 Ethernet port and craft port access via an RS-232 craft interface. The LAN port might be used for CTC connectivity, whereas the craft interface might be used for TL1 commands. The RS-232 port supports VT100 emulation.

The TCC+ card holds the system database, allowing the shelf to activate and cross-connect the other cards. Provisioning, alarm reporting, maintenance, diagnostics, IP networking, SONET data communications channel (DCC) termination, and system fault detection all occur through this card. Although the shelf can operate with one TCC+, redundant TCC+ cards are included with all Cisco configurations.

XC and XC-VT

The XC and XC-VT cross-connect cards are not compatible with the ANSI shelf assembly and the OC-192, G1000-4, or OC-48IR/OC-48LR any-slot cards. The XC (cross-connect) and the XC-VT (cross-connect virtual tributary) are supported in the NEBS-3 shelf.

The XC card is used when the only circuits to cross-connect are at the synchronous transport signal (STS) level. The XC card can support a switching matrix of 288×288 cross-connect ports, which equates to 144 bidirectional STS-1 circuits because each circuit has a connection to and from the fabric. This cross-connect card has enough capacity to ensure all STSs can cross-connect without any blocking taking place in the switching fabric. The 288-x-288 switching matrix exemplifies a fully loaded shelf configuration. An example of a fully loaded shelf is one in which each multispeed slot is populated by an OC-12 card and each high-speed slot is populated with an OC-48 card. Although Gigabit Ethernet represents a much faster line speed than OC-12, the largest provisioning of an E-series Gigabit Ethernet card is STS-12. This is discussed further in Chapter 7, "Configuring Metro Ethernet." There are eight multispeed slots and four high-speed slots. The following math dictates the need for 288-x-288 STS for a nonblocking switch fabric:

OC-12: $12 \times 8 = 96$ STS (8 multispeed slots)
OC-48: $48 \times 4 = 192$ STS (4 high-speed slots)

$96 + 192 = 288$ STS

If the service provided needs to be cross-connected at the virtual tributary (VT) level, the XC-VT card is needed. The XC card can be used as an intermediary switching point but will carry the VT1.5 in an STS-1 and have no ability to add/drop, rearrange, or manage the VT1.5. This concept is known as a *VT tunnel* (VTT) connection. The XC-VT card has all the switching capacity of the XC card plus the capability to cross-connect up to 672 VT1.5s depending on the ring type used. For more information on this topic, refer to the "Understanding the 15454 XC and XC-VT Switching Matrix" document at Cisco.com. A VT1.5 is a 1.722-Mbps SONET container with the capability to carry a DS1 at 1.544 Mbps and the associated VT overhead. A DS1 card on the ONS 15454 supports 14 ports, which fills one-half of an STS-1. The shelf could support a maximum of 168 DS1s if all slots were used for DS1 connectivity with no uplinks. This is not a realistic design because most shelves aggregate services and connect via a high-speed interface.

XC-10G

The XC-10G is the cross-connect card required to support the OC-192, G-series Gigabit Ethernet, and OC-48AS (any-slot) cards. The XC-10G card must be installed in the ANSI shelf assembly. To keep the nonblocking nature of the STS cross-connects, the XC-10G supports 1152 STS-1 ports or 576 bidirectional STS-1 circuits. The number of VT cross-connects supported is the same as the XCVT at 672. The OC-192 and legacy OC-48 cards operate in the high-speed slots, whereas the OC-48 any-slot cards work in any of the high-speed or multispeed slots. Why the need for 1152 cross-connects? If each of the multispeed slots are now capable of supporting 48 STS with the OC-48 any-slot card and the high-speed slots are capable of 192 STS with the OC-192 cards, the needs are as follows:

OC-48: $48 \times 8 = 384$ STS (8 multispeed slots)
OC-192: $192 \times 4 = 768$ STS (4 high-speed slots)

$384 + 768 = 1152$ STS

AIC

The alarm indicator card (AIC) is an optional card supported only via slot 9. The card provides for up to four external alarm contacts and for external controls using the backplane wire-wrap connectors. The AIC also gives the customer access to the orderwire bytes in the SONET frames. The orderwire bytes allow a SONET frame to transport a 64-kbps voice channel between ONS 15454 nodes. There are two different orderwire bytes, one carried in the section header of the SONET frame called the local orderwire and one carried in the line overhead of the SONET frame called the express orderwire. Two RJ-11 connectors on the front of the AIC card give access to these channels. Refer to Chapter 3, "SONET Overview," for more information about section and line overhead bytes.

Electrical Cards

Electrical cards support technologies that use copper cable and electrical signals to transmit information. The electrical interfaces are predominantly associated with DS1, DS3, and DS3XM-6 interfaces. There is, however, a version of SONET that is capable of being transmitted over copper wire. This version is supported by the EC1 (electrical carrier) card that carries an STS-1.

Table 4-3 describes the electrical cards that the Cisco ONS 15454 supports.

Table 4-3   *ONS 15454 Electrical Cards*

Card	Telcordia Interface Compliance	Speed	Cabling/Fiber
EC1–12	12 GR-253 STS-1 interfaces	51.84 Mbps	Single 75-ohm 728A or equivalent coaxial span
DS1–14 DS1n–14	14 GR-499 DS-1 interfaces	1.544 Mbps	100-ohm twisted-pair copper cable
DS3–12, DS3–12E DS3n–12, DS3n–12E	12 GR-499 DS-3 interfaces	44.736 Mbps	Single 75-ohm 728A or equivalent coaxial span
DS3XM-6	6 GR-499-CORE M13 multiplexing functions	Converts 6 framed DS3 network connections to 28×6 or 168 VT-1.5 signals	

The electrical cards do not contain any connectors. All connectors are located on the back of the ONS 15454 shelf. Four different types of back panels can be ordered for the 15454 depending on needs. Table 4-4 describes the four different connector panels.

Table 4-4   *Electrical Interface Assembly (EIA) Configurations*

EIA Type	Cards Supported	A Side Columns Map To	A Side Product Number	B Side Columns Map To	B Side Product Number
BNC	DS3 DS3XM-6 EC1	Slot 2	15454-EIA-BNC-A24	Slot 14	15454-EIA-BNC-B24
		Slot 4		Slot 16	
High- density BNC	DS3 DS3XM-6 EC1	Slot 1	15454-EIA-BNC-A48	Slot 13	15454-EIA-BNC-B48
		Slot 2		Slot 14	
		Slot 4		Slot 16	
		Slot 5		Slot 17	
SMB	DS1	Slot 1	15454-EIA-SMB-A84	Slot 12	15454-EIA-SMB-B84
	DS3	Slot 2		Slot 13	
	EC1	Slot 3		Slot 14	
	DS3XM-6	Slot 4		Slot 15	
		Slot 5		Slot 16	

continues

Table 4-4 *Electrical Interface Assembly (EIA) Configurations (Continued)*

EIA Type	Cards Supported	A Side Columns Map To	A Side Product Number	B Side Columns Map To	B Side Product Number
		Slot 6		Slot 17	
AMP champ	DS1	Slot 1	15454-EIA-AMP-A84	Slot 12	15454-EIA-AMP-B84
		Slot 2		Slot 13	
		Slot 3		Slot 14	
		Slot 4		Slot 15	
		Slot 5		Slot 16	
		Slot 6		Slot 17	

The connector panels are divided into an A side and a B side, which match the slots on the front of the shelf. (A = slots 1 through 6, and B = slots 12 through 17.) Figure 4-3 contains a graphic of the backplane connector architecture. The types of cards used and density needs of the user determine the appropriate EIA to use. The reason the electrical cards do not have interfaces is for protection purposes. The ONS 15454 might be providing card-level redundancy, but only one physical interface is coming from the customer. The ONS 15454 handles this operation internally according to the administrator's protection configuration.

Figure 4-3 *Cisco ONS 15454 Backplane Connectors*

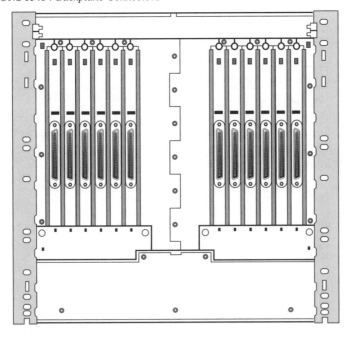

Optical Cards

The ONS 15454 supports a number of optical cards geared for the type of service and distances a MAN system would need. All the cards listed in Table 4-5 are supported via single-mode fiber using SC connectors and are compliant with the Telcordia GR-253 recommendation.

Table 4-5   *ONS 15454 Optical Interfaces*

Card	Ports	Wavelength
OC-3-IR-4 1310	4	1310 nm
OC-12-IR 1310	1	1310 nm
OC-12-LR 1310	1	1310 nm
OC-12-LR1550	1	1550 nm
OC-48-IR 1310	1	1310 nm
OC-48-LR 1550	1	1550 nm
OC-48-IR-AS-1310	1	1310 nm
OC-48-LR-AS-1550	1	1550 nm
OC-192-LR-1550	1	1550 nm
OC-48-ELR ITU 100-GHz DWDM cards	1	32 orderable wavelengths
OC-48-ELR ITU 200-Ghz DWDM Cards	1	18 orderable wavelengths

The card-naming nomenclature dictates four different card types: SR, IR, LR, and ELR. The acronyms stand for the span that the optical signal can travel, as follows:

- Short reach (SR)
- Intermediate reach (IR)
- Long reach (LR)
- Extended long reach (ELR)
- Although many factors affect the distances that an optical signal can travel, a 1550-nanometer (nm) laser normally attenuates less per kilometer than a 1310-nm laser. The ELR cards operate in the 1520–1560-nm range and receive similar optical attenuation as the 1550-nm lasers but use more powerful ("hotter") lasers and use more sensitive photodiode receivers. This allows for a larger optical span budget (hence the ELR characterization). ELR cards and optical span budgets are discussed further in Chapter 8, "Implementing DWDM in Metropolitan-Area Networks."

Table 4-6 contains the approximate distances that might be traveled given optimal fiber. The fiber characterization testing results should be cross-referenced against the documented span budgets before deploying an optical system.

Table 4-6 *Signal Distances of Optical Cards*

Range	Approximate Distance
Short range (SR)	2 km
Intermediate range (IR)	15–20 km
Long range (LR)	40–80 km
Extended long range (ELR)	65–80 km

Distances can be extended through the use of optical amplifiers or regenerators. Chapter 3 covers regenerators, and Chapter 8 covers optical amplifiers.

Ethernet Cards

The ONS 15454 supports several Ethernet cards supporting 10/100 Mbps and 1000 Mbps. The 10/100 cards support autosensing 10- or 100-Mbps Ethernet connectivity, whereas the Gigabit Ethernet cards support optical interfaces through the use of gigabit interface controllers (GBICs). The Ethernet cards are described in Table 4-7. The Ethernet cards are discussed in detail in Chapter 8.

Table 4-7 *Ethernet Cards*

Card	Speed	Interface
E100T-12 or E100T-12-G Ethernet card	10/100 Mbps	Category 5 UTP; RJ-45 connector
E1000-2 or E1000-2-G Gigabit Ethernet card	1 Gbps	SX (short range) or LX (long range) GBIC
G1000-4 Gigabit Ethernet card	1 Gbps	SX, LX, or ZX GBIC

E100T-12 and E100T-G

The Ethernet E100-T E-series cards support twelve 10/100 ports. The card can be placed into any of the multispeed slots and supports a maximum of 12 STSs for transport. If the Ethernet port is used as a 10-Mbps Ethernet interface, it can easily fit into an STS-1 transport, but most of the STS-1 capacity would be wasted. Multiple Ethernet ports cannot be combined onto a single STS to maximize use of an STS.

The G version of the card is the version required when the XC-10G cross-connect card is used due to electrical requirements on the backplane.

E1000-2 and E1000-2-G

These cards support Gigabit interfaces via GBIC adapters. The maximum bandwidth available to the E1000-2 card is 12 STS (622 Mbps) because it is compatible with the XC and XC-VT cross-connect cards. The E1000-2-G card is compatible with the XC-10G and the ANSI shelf. These cards are covered further in Chapter 7, "Configuring Metro Ethernet."

G1000-4

This card is available only with the XC-10G due to the 48 STS capacity it uses. The card can support wire-speed Gigabit Ethernet on up to two interfaces. The 48 STS can be broken down to more granular levels to support different customers. This card is discussed in more detail in Chapter 8.

The ONS 15327 platform is the companion to the ONS 15454 platform, providing edge access into the core for those environments that do not require the port density provided by the ONS 15454. The following section describes the ONS 15327 platform.

Cisco ONS 15327 Platform

The ONS 15327 has been touted for its small form factor and its capability to be easily installed as edge equipment. Up to twelve 15327s can fit in a 7-foot rack. Release 3.3 of the CTC software merges the release schedules of the ONS 15327 and ONS 15454 platforms. The unified software release provides consistent features and functionality, as well as a consistent look and feel for configuration and provisioning. The ONS 15327 has fully achieved the Industrial Temperature (I-Temp) rating and is NEBS-3 compliant.

The ONS 15327 platform is DC powered but can easily be coupled with a rectifier that provides AC power conversion.

The ONS 15327 has eight slots. Four slots are used for the system cards, which provide for management access as well as DS1 and DS3 support. The remaining four slots are high-speed slots supporting OC-3 through OC-48 connectivity, including Ethernet. All cable connections on the ONS 15327 are via the front. There is no backplane connector panel on the back of the device like there is on the ONS 15454. Figure 4-4 is an architectural breakdown of the slots on the ONS 15327.

Figure 4-4 *Card Locations Within the 15327*

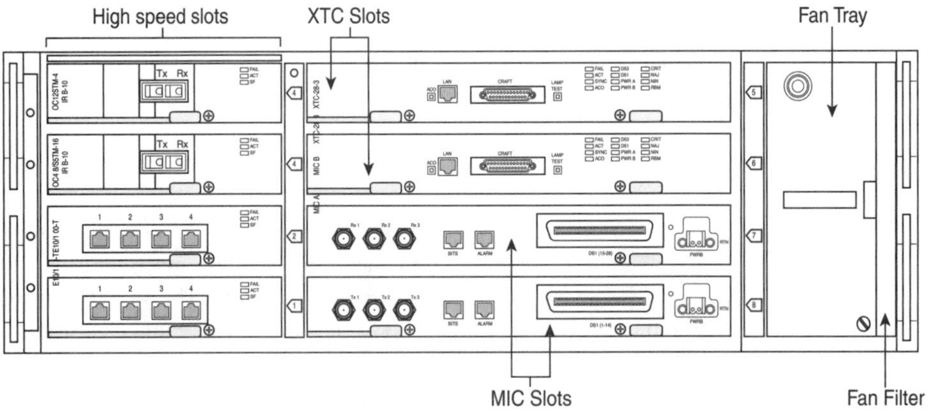

Common Control and Electrical Cards

The control cards are mounted horizontally in the upper center of the shelf. The main control card on the 15327 is called the *XTC* because it combines the TCC and XCVT cards used on the 15454 into a single card. The XTC card provides for LAN and craft port accessibility for management of the node.

The following two versions of the XTC card are available:

- **XTC-14**—Provides 14 DS1 interfaces via a 64-pin champ connector on the mechanical interface card (MIC).

- **XTC-28-3**—Provides 28 DS1 interfaces via two 64-pin champ connectors on the MIC cards. This card also supports three DS3 interfaces on the MIC cards.

Two MICs are available on the ONS 15327, providing DS1 and DS3 physical interfaces and redundant power connection points, external building-integrated timing supply (BITS) timing, and alarm interfaces. The MIC cards are required in pairs. There are two orderable parts: A side and B side. Figure 4-4 displays the MICs. Notice that each card contains three coaxial connectors, one BITS interface, one alarm interface, and one AMP champ connector. If the XTC-14 cross-connect card is used, only the top champ connector is active and the coaxial connectors are not operable. If the XTC-28-3 card is used, all 28 DS1 interfaces via the champ connectors are activated, as are the 3 DS3 interfaces. Notice that the A side (bottom) card contains the DS3 transmit (Tx) coaxial interface, whereas the B side (top) coaxial interface supports the receive (Rx) coaxial interface.

Optical and Ethernet Cards

The four slots on the left side of the shelf are high-speed slots providing STS-48 connectivity to the XTC. The cards supported in the high-speed slots are as follows:

- 4-port OC-3 IR 1310
- 1-port OC-12 IR 1310
- 1-port OC-12 LR 1550
- 1-port OC-48 IR 1310
- 1-port OC-48 LR 1550
- 4-port 10/100BASE-T Ethernet
- 2-port Gigabit Ethernet (SFP optics)

All the optical cards support SC interfaces with the exception of the two-port Gigabit Ethernet interfaces, which use small form-factor pluggable optics (SFP), and the four-port OC-3 card, which uses LC connectors because of their small size. Release 3.3 of CTC on the ONS 15327 introduced the four-port Gigabit Ethernet card, whereas release 3.4 introduced the two-port Gigabit Ethernet card. The ONS 15327 can participate in point-to-point (terminal mode), linear add / drop connections and UPSR rings from OC-3 through OC-48. The ONS 15327 can also support participation in up to two two-fiber BLSR rings.

Cisco Transport Controller

When either the ONS 15454 or ONS 15327 is installed and powered, the node needs to be configured. Access to the node is achieved through either the LAN port or the craft interface on the XTC control card. The most common way to manage the ONS products is to use a graphical application called the Cisco Transport Controller (CTC). The CTC software resides on the ONS node. CTC is a Java-based applet in which Java Beans are downloaded to the administrator's directory as long as the proper Java Runtime Environment (JRE) and policy key are stored on the administrator's laptop. Access is provided via a standard web browser in which the administrator types in the IP address of the node.

The versions of the JRE that are compatible with releases as of the publication of this book are JRE 1.2.2 or JRE 1.3. This software can be found on the CD-ROM that contains the node software or from www.sun.com. You can manually create the policy key via instructions available at Cisco.com.

After you install the software on your computer, you can access the network or node by starting up your browser software and typing in the IP address of one of the nodes in the browser's URL address line. The default IP address for all new nodes is 192.1.0.2. The current IP address displays on the LED panel on the front of the fan tray for the ONS 15454.

When accessing the CTC, you can work from the following three different views:

- **Network**—Displays the entire networks from which you can access an individual node, alarms, and circuit creation, review, and add new users and other system-level options for the entire ring. The Network view enables you to view and manage both ONS 15454 and ONS 15327 that have DCC connections to the node that you are logged in to.

- **Node**—Provides access to the node configuration options, alarms, circuit creation, adding new users and other node-level options, including the Card view of that node only. This is the initial view mode after login.

- **Card**—Displays the card and its associated alarms, options, and circuit creation, and other card-level options. From this view, you can activate and configure the individual ports for that card only.

After you enter the node's IP address, the Java Beans are downloaded to the client machine (first time login only). After the appropriate files have been downloaded, a CTC logon screen displays, and you are asked to enter your login name and password. This name and password must exist on the node to which you are trying to connect. The default login for a new installation is CISCO15 (case-sensitive) with no password. This username is a permanent part of the ONS product. The name cannot be changed, and the account cannot be deleted. The first thing you should do is replace the null password with a complex password to protect this account. Configuration details are discussed in Chapter 5, "Configuring the ONS 15454 and ONS 15327." After a valid username / password combination has been successfully entered, more Java Beans are downloaded to the administrator's desktop. The administrator will see the tccp – Cisco Transport Controller window.

Node configuration and management is available from a series of views, tabs, and subtabs associated with particular tabs. The tabs and their defined features are as follows:

- **Alarms**—Lists current alarms. The ONS products use standard Telcordia categories to characterize levels of issues. Alarms signify a network issue classified by severity level. The different severity levels are represented by different colors. The available levels are as follows: critical (red), major (orange), or minor (yellow). If event-level information is turned on, these events display as blue.

- **Conditions**—Lists the current conditions on the node. This information can include alarms and fault conditions. Conditions notify the user of an event that does not require action, such as a switch to a secondary timing reference or a user-initiated manual reset.

- **History**—Lists the history of node alarms, including date, type, and severity. The following are the two subtabs on this tab:

    — **Session**—Only alarms and events that have occurred on this CTC session.

    — **Node**—Alarms and events stored in a fixed-size log.

- **Circuits**—Enables you to create, delete, edit, and map circuits.
- **Provisioning**—Used to initially configure the node, create rings, set up security, and so on. The Provisioning tab has 10 subtabs, as follows:
  - **General**—Basic node information such as name, location, and so forth.
  - **Ether Bridge**—Metro Ethernet card configuration information.
  - **Network**—IP address, subnet mask, and default gateway configuration.
  - **Protection**—Card protection-level configuration. Protection levels include 1:1, 1:*n*, or 1+1 card.
  - **Ring**—BLSR ring configuration and management.
  - **Security**—User accounts management.
  - **SNMP**—Management reporting information.
  - **SONET DCC**—Configuration and management of section DCC and tunnel DCC information.
  - **Timing**—System clock reference configuration and management.
  - **Alarming**—Alarm profile configuration.

  The Provisioning tab is covered extensively in Chapter 5, "Configuring ONS 15454 and ONS 15327.

- **Inventory**—Node card inventory and management information. Information includes part number, serial number, and CLEI codes for the cards. Cards can be deleted and reset from this tab.
- **Maintenance**—Enables you to perform maintenance tasks. The Maintenance tab has 11 subtabs:
  - **Database**—Used to back up and restore the node's database configuration information.
  - **Ether Bridge**—Used to view MAC address table information and trunk utilization performance management.
  - **Protection**—Initiate protection group switching for maintenance windows / hardware upgrade.
  - **Ring**—Configure ring switching. Utilized when adding nodes to a ring.
  - **Software**—Used to download, activate, or revert system software.
  - **XC Cards**—Enables the administrator to switch active and standby XC cards. Used during hardware upgrades.
  - **Diagnostic**—Test all the system LEDs.
  - **Routing Table**—Displays the IP routing table. In addition to configuring the node-level information, CTC configures individual card options and the associated ports on the card.

To configure a card, double-click the card in the graphic area of the top pane. The display changes to reflect the configuration options of that particular line card.

To return to the Node view, click the up arrow or the home icon on the toolbar. Figure 4-5 shows the Network view that results from this selection. This Network view reflects a ring configuration with five devices.

Figure 4-5   *Network View*

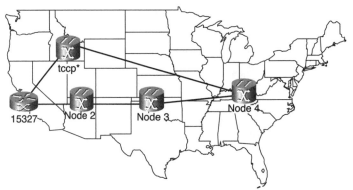

From Network view, all the nodes in the ring display, as does their alarm status. Double-clicking the node takes you into Node view of that device. When configuration is complete, close your browser window. There is no logout process; just exit the Java application. New attempts to access the node with the browser require a new login.

Summary

This chapter introduced the slots, cards, and interface of the Cisco ONS 15454 and ONS 15327 optical platforms. The capabilities of the two platforms were presented and the cards that would operate within them were reviewed. The Cisco Transport Controller (CTC) is used to configure and manage the devices.

Chapter 5 details the configuration and setup of the various ring architectures used in the metropolitan area.

Review Questions

1 The largest ring size that the ONS 15327 can be a part of is

   A   OC-3

   B   OC-12

   C   OC-48

   D   OC-192

2 For 1:*n* protection, which slots on the ONS 15454 must be used for the protection cards?

   A   1,17

   B   2,16

   C   3,15

   D   6,10

3 How many DS3s are supported when the ONS 15327 is configured with the XTC-14 and MIC-28-3?

   A   0

   B   1

   C   2

   D   3

4 How many DS1s are supported when the ONS 15327 is configured with the XTC-14 and MIC-A/MIC-B cards?

   A   3

   B   14

   C   28

   D   672

5 How many STS cross-connect ports does the XC-10G support?

   A   144

   B   288

   C   336

   D   572

   E   1152

6 The ONS 15327 supports (Select all the correct answers.)

   A   OC-3

   B   Gigabit Ethernet

   C   OC-192

   D   OC-12

   E   OC-48

   F   10/100 Ethernet

7 Which of the following fulfill the requirements of running the Cisco Transport Controller application?

   A CTC does not require any special installation. Point any web browser to the IP address of the ONS device.

   B Requires that it runs on a Sun workstation.

   C Can run from a JRE-compliant web browser but requires installation of the JRE and policy file.

   D Does not require a browser. CTC is a standalone application.

8 The three view levels of the CTC are

   A Network view

   B Node view

   C Provisioning view

   D Card view

This chapter covers the following topics:

- Initially Configuring the Cisco ONS Nodes
- Configuring SONET Timing
- Configuring the Cisco ONS Optical Platform

Configuring ONS 15454 and ONS 15327

This chapter covers the methods used to set up and configure the ONS 15454 and ONS 15327 products. The configuration and provisioning of both products are identical. The only differences are the displays of the Node and Card views, which is attributable to the hardware differences in each platform. The configuration windows might vary slightly with the version of Cisco Transport Controller (CTC) used, but the concepts remain the same.

All networks start somewhere. That somewhere is usually basic node configuration. These nodes are then interconnected to create the network. This chapter reviews ONS 15454 and 15327 node setup procedures, and then examines the various ways the devices can be joined together to form a network. The ONS 15454 and ONS 15327 support preprovisioning of cards and configurations to allow for flexible deployment.

The ONS 15216 devices do not require any configuration. Recall that most of these devices are passive and do not receive any AC or DC power. The only exceptions to this rule are the 15216 erbium-doped fiber amplifier (EDFA) and 15216 optical performance monitor (OPM), which are both active devices. The 15216 EDFA version 2 (v2) can also be managed and monitored. The only configuration necessary for the 15216 EDFA v2 is management and alarm information.

Initial Configuration of the Cisco ONS 15454 and 15327

The CTC software on the ONS 15454 and ONS 15327 is configurable via a standard web browser with Java Runtime Environment (JRE) support. The JRE needs to be installed for access to either of the ONS shelves. The appropriate JRE is shipped with the ONS 15454 and ONS 15327, but the requirements might change depending on the version of CTC being run. At the time of this writing, JRE 1.2.2_05 operates with any shipping version of CTC software. If the CD is lost, the JRE can be downloaded from www.sun.com. The other software requirement is that of the Java Policy File. The Java Policy File ships with the CTC CD but can be manually created from a text file. You can find instructions on creating the Java Policy File at Cisco.com. Table 5-1 lists the end-user computer requirements for running the CTC software.

Table 5-1 *Cisco Transport Controller PC Hardware Requirements*

Computer Requirements for CTC Area	Requirements	Notes
Processor	Pentium II 300 MHz, UltraSPARC, or equivalent	300 MHz is the minimum recommended processor speed. You can use computers with less processor speed; however, you might experience longer response times and slower performance.
RAM	128 MB	
Hard drive	2 GB	CTC application files are downloaded from the XTC card to your computer's Temp directory. These files occupy 3 to 5 MB of hard drive space.
Operating system	PC: Windows 95, Windows 98, Windows NT 4.0, or Windows 2000 Workstation: Solaris 2.6 or 2.7	
Web browser	PC: Netscape Navigator 4.51 or higher, or Netscape Communicator 4.61 or higher, or Internet Explorer 4.0 (Service Pack 2) or higher Workstation: Netscape Navigator 4.73 or higher	Either Netscape Communicator 4.73 (Windows) or 4.76 (Solaris) are installed by the CTC Setup Wizard included on the Cisco Software and documentation CDs.
Java Runtime Environment	JRE 1.2.2_05 with Java plug-in 1.2.2 minimum JRE 1.3.1_02 (PC) recommended JRE 1.3.0_01 (Solaris) recommended	Use JRE 1.2.2_05 if you connect to ONS 15454s running CTC Release 2.2.1 or earlier. (The earliest available ONS 15327 software is CTC Release 2.3.) Use JRE 1.3.1_02 if all ONS 15454s that you connect to are running CTC Release 2.2.2 or later. JRE 1.3.1_02 is installed by the CTC Setup Wizard included on the Cisco ONS 15327 Software and documentation CDs.
Java Policy File	A Java Policy File modified for CTC must be installed	A modified Java Policy File is installed by the CTC Setup Wizard included on the Cisco ONS 15327 Software and documentation CDs.
Cable	User-supplied Category 5 straight-through cable with RJ-45 connectors on each end to connect the computer to the ONS 15327 directly or though a LAN	

After the appropriate software has been installed, verify connectivity to the ONS device with a standard ICMP ping. The default IP address of the ONS 15327 and ONS 15454 is 192.1.0.2. Ensure that your PC is set up with an appropriate IP address and subnet mask to establish connectivity with the device. If ping replies are received, proceed to launch the web browser to the IP address of the device in which you want to configure. The first time a user connects to either the ONS 15327 or ONS 15454 device, the device sends down Java Beans to the client. These files are large (few megabytes), and it is not advisable to perform this operation over a low-speed link.

After the appropriate Java Beans have been installed, a login prompt appears. The initial username of CISCO15 (case-sensitive) with no password is activated by default on the box. Use this username or password combination the first time you log in to the device. It is advisable (from a security perspective) to create a new Superuser access account or to change the password of the default CISCO15 account. Click the **Login** button. The Cisco Transport Controller Login prompt displays.

After a successful username and password have been entered, the CTC software begins caching more Java Beans. After caching is completed, the Node view of the device appears.

Troubleshooting CTC

If CTC will not load, the first step in troubleshooting is to determine the compatibility of the installed JRE version and CTC software. You can find matrices of this compatibility at Cisco.com.

If multiple versions of the JRE environment are installed on the machine, statically set the version of the JRE to the appropriate version. The Java Plug-In control panel will appear in Control Panel or as a program group depending on the version of OS in which it was installed (Windows platforms only). Launch the Plug-In control panel and click the **Advanced** tab. The default is to autosense the JRE version, which does not always work. Statically set this to the appropriate version and click the **Apply** button. Figure 5-1 shows this process.

If the CTC software will still not load, it is advisable to delete the CTC cache. This appears as a button in the web browser window.

If the CTC software will still not load, delete the entire web browser offline cache. It is not necessary to delete cookies. A different version of the JRE might be necessary depending on the version of CTC used. Check Cisco.com to see which version of the JRE is used for the CTC version used.

Figure 5-1   *Java Console Configuration*

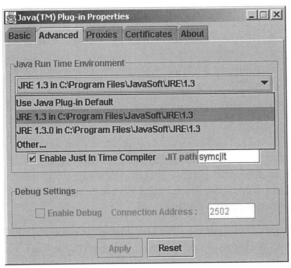

Entering Basic Node Information

Basic node information includes those items that allow the system to be remotely managed and make the ONS 15454 and 15327 ready to accept network and user traffic. This information includes the following:

- Node Name, location, and contact
- IP address, subnet mask, and gateway address
- Securing access

This information should be part of an overall network design, and this data is needed at the time you are ready to start configuring the node.

When you are logged in, a display known as Node view appears. You can view the entire node from here.

The tabs in all views run from left to right. On the left side of the bottom pane, you will notice subtabs. The Provisioning tab is where most initial configuration is performed, so this tab is inspected first. Notice the different subtabs available from the Provisioning tab. This chapter discusses the General, Network, and Security tabs in detail:

- General
- Ether Bridge
- Network
- Protection

- Ring
- Security
- SNMP
- SONET DCC
- Timing

General Subtab

A picture of the General subtab appears in Figure 5-2 below the General subtab parameters. The General subtab contains various fields that are used to enter generic information about the node, such as the following:

- **Node Name**—Type a name for the node. For Transaction Language 1 (TL1) compliance, names must begin with an alpha character and have no more than 20 alphanumeric characters. TL1 is a service provider command language that can be used for provisioning. ONS 15454 and ONS 15327 TL1 command references can be found at Cisco.com. This name must be unique within the network for proper nodal identification. The node name will be seen in Network view and should describe the device for accurate network management.

- **Contact**—Type the name of the node and phone number of the contact. Although not required, this information gives a person troubleshooting a problem someone to contact in case of issues with the node.

- **Location**—Type the node location (for example, a city name or specific office location). This is also an optional field but can prove useful in troubleshooting and reporting.

- **Latitude and Longitude**—This optional information is used to position the node on the Network view map. This information is based on the default U.S. map and will not be accurate for other map GIFs that might be used.

NOTE   Positioning of network nodes on the Network view map can also be accomplished by pressing the **Ctrl** key and dragging the node icon to a new location.

- **SNTP Server**—CTC can uses a Simple Network Time Protocol (SNTP) server to set the date and time of the node. When selected, you will also need to enter the IP address of the SNTP server in the adjacent field.

  If you do not use an SNTP server, complete the Date Time and Time Zone fields. The ONS 15454 uses these fields for alarm dates and times, and other SNMP-based network management.

NOTE   CTC displays all alarms in the login node's time zone for cross-network consistency.

- **Date**—Type the current date.
- **Time**—Type the current time.
- **Time Zone**—Select the time zone.

Figure 5-2   *CTC General Subtab Configuration*

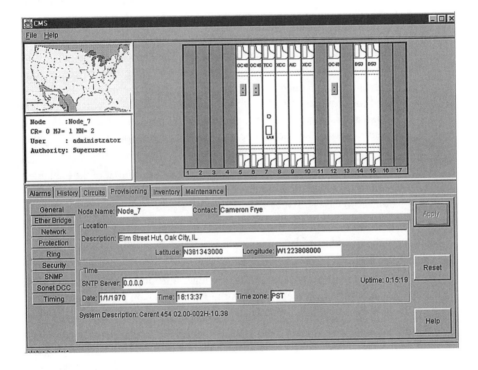

Click the **Apply** button to save the entered information.

Network Subtab

The Network subtab has three subtabs:

- General
- Static Routing
- OSPF

General

The General subtab is used to configure the node's management port IP address, subnet mask, and default gateway. Beware that changes to this information require a system reboot. This information should be configured and the timing control card (TCC) reloaded prior to any service provisioning.

The Network subtab enables you to manage the node through your corporate or maintenance network without having to have a cable plugged directly from your workstation into the node. To gain initial access to the node, you need the IP address, subnet mask, and possibly a default gateway that will be used. This information should be part of the network design plan and should be obtained from the appropriate person. Remember that the default node IP address is 192.1.0.2, and your workstation must be on the same network to access the node. When the IP address information is changed, a reload of the TCC is required for the changes to take effect. During this process, you lose connectivity and need to change the IP address of your workstation again to regain connectivity.

Figure 5-3 shows the Network subtab.

Figure 5-3   *Network Subtab*

You can enter the following information from the General subtab of the Network subtab:

- **IP Address**—Type the IP address assigned to the node.
- **Prevent LCD IP Config**—This prevents the Cisco ONS 15454 IP address from being changed using the fan tray module when selected. This option is valid only for the 15454. The 15327 does not have an LCD display that can be used to change the address.

  If the IP address is changed from the LCD panel, the default gateway must be set. Figure 5-3 displays the Fan Tray Module LCD screen. Notice that alarm conditions are visual from this module. The IP address information is configurable via the fan tray module as Slot-0. The Slot, Status, and Port buttons are used to perform this operation. Documentation on this procedure is available at Cisco.com.

Figure 5-4   *ONS 15454 Fan Tray Module*

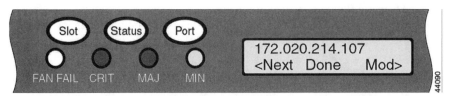

- **Default Router**—For the node to communicate with a device on a network to which it is not directly connected, the IP address of a default router needs to be entered. The node then forwards all packets not destined to the local network to the default router. If the management console is part of the same network as the node, this field can be left blank. Recall that this information is required for fan tray module configuration.

- **Subnet Mask Length**—Type the subnet mask length. This value is entered as a decimal number representing the subnet mask length in bits. The subnet mask length needs to be the same for all nodes in the same subnet.

    The MAC address is for display purposes and cannot be changed through this subtab. This information is useful for troubleshooting Layer 2 issues if they were to arise.

- **Forward DHCP Request To**—When this box is checked, the node forwards Dynamic Host Configuration Protocol (DHCP) requests to the IP address entered in the Request To field. DHCP is a TCP/IP protocol that enables your CTC computer to get temporary IP addresses from a DHCP server. If you enable DHCP, CTC computers that are on the same network as the node can obtain temporary IP addresses from the DHCP server.

- **TCC CORBA (IIOP) Listener Port**—This option is needed when management traffic between the CTC and the node will be traveling through a firewall. This option sets a listener port to allow communications. Without this option set, the CTC software uses dynamic ports.

Clicking the **Apply** button or selecting any other tab or subtab causes a confirmation box to appear. When you select **Yes**, both 15454 TCC+ cards reboot. As a reminder, you must change the IP address of your computer and relog in to continue configuring the node.

Static Routing

The Static Routing subtab on the Network subtab allows the configuration of static routes in the network. This is used for environments where there is not a dynamic routing protocol running and the management port will need connectivity to remote networks (or vice versa). The Create button is used to create a new static route. When a static route exists, it can be selected and the Edit or Delete buttons can be used to edit or remove the static route. Figure 5-5 shows the configuration parameters for creating a static route.

Figure 5-5   *Static Route Configuration*

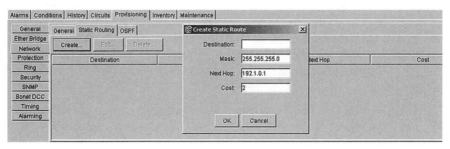

OSPF

The OSPF subtab on the Network subtab allows the configuration of the OSPF dynamic routing protocol to run on the TCC card:

- **DCC OSPF Area ID**—This parameter cannot be changed at this time. It must be set to area 0 in dotted-decimal notation (that is, 0.0.0.0). This is the OSPF process that will run on the trunk side of the ONS 15454 and ONS 15327. This allows these devices to automatically route circuits over spans (including PPMN networks).

- **DCC Metric**—This parameter is the same as the OSPF cost configurable in a Cisco router.

- **OSPF Active on LAN**—This check box activates OSPF routing on the Ethernet management port on the ONS 15454 or ONS 15327 device. After the check box is selected, the LAN port area ID must be entered. This area ID must not be area 0. This device must be an OSPF area border router (ABR). If this device needs to be connected to an existing OSPF area 0, a virtual link must be configured.

- **Authentication**—OSPF can be set up with authentication types including clear mode and MD5 authentication. If authentication is turned on, an authentication key must be used.

Some other OSPF configuration parameters include the following:

- OSPF Priorities and Intervals Configuration
- OSPF Area Range Tables
- OSPF Virtual Link Tables

Security Subtab

The Security subtab enables you to control access to the CTC program. The CISCO15 default username cannot be changed or deleted. It is important to protect this account with a complex password (eight or more characters with mixed case and at least one special character). There is another default Superuser account: cerent454. This account exists for backward-compatibility with older versions of the CTC software. Up to 500 users can be added to a single ONS device. They can be added either at the node level or at a network level. If you add users at the node level, users are created only at that node. If you add users after the node is part of the network, you can add users at the network level. This enables you to selectively add the users to as many nodes in the network as you want with a single command.

At this point in your initial configuration, you are not part of a network and adding many users now will create extra work later. It is a good idea to protect the CISCO15 user and add a different Superuser account at this time.

To create a new user, follow these steps:

Step 1   Select the **Security** subtab. Users can be assigned one of the following security levels:

- **Superusers**—This user can perform all functions on the node. To protect unintended access, a Superuser account times out after 15 minutes of inactivity.

- **Provisioning**—This user can access provisioning and maintenance options. This account is designed for users who will be creating circuits. Provisioning accounts time out after 30 minutes of inactivity.

- **Maintenance**—This user can access only the maintenance options. This account is designed to be used by troubleshooters and administrators who need to perform maintenance activities on the node. Maintenance accounts time out after 60 minutes of inactivity.

- **Retrieve**—This user can retrieve and view CTC information but cannot set or modify any parameters. Retrieve accounts do not time out after connection.

Step 2   Click the **Create** button to add an account. The following information is needed:

- **Name**—Type the username.

- **Password**—Type the user password. The password must be a minimum of 6 and a maximum of 10 alphanumeric (a–z, A–Z, 0–9) and special characters (+, #,%), where at least 2 characters are nonalphabetic and at least 1 character is a special character.

- **Confirm Password**—Type the password again to confirm it.

- **Security Level**—Select the user's security level from the four levels just discussed.

Step 3   When finished, Click **OK**.

Protection Subtab

Protection schemes are a function of the level of redundancy necessary for the network. Although card protection can be configured any time, it is good practice to design the network for the desired redundancy level.

To set up your card protection groups, select the **Protection** subtab.

Protection groups are available in both electrical or optical **protection** schemes. Protection groups can be defined as one of following three types:

- **1:1**—This is an electrical protection scheme used for the following types of cards: DS1, DS3, EC1–12, and DS3XM. This type of protection requires the working card to be in an even slot while the protect card is in the adjacent odd slot. The even and odd groupings change based on the side of the node the card is on. On the left side of the chassis, slot 1 protects slot 2, 3 protects 4, and 5 protects 6; whereas on the right side of the chassis, slot 17 protects slot 16, 15 protects 14, and 13 protects 12. On the 15327, the DS1 and DS3 circuitry is on the XTC cards. When two cards are installed, you automatically have 1:1 protection configured.

- **1:n**—Can be used for protecting multiple DS1 or DS3 cards with a single protect card. Special versions of the DS1 or DS3 cards are required for this type of protection and are identified by the n at the end of the card designation: DS1n or DS3n. The n cards are required for only the protect card and must be placed in either slot 3 or slot 15. An n card protects only cards placed on the same side of the chassis. A DS1n placed in slot 3 can be used to protect only DS1 cards in slots 1 through 6. A DS3n card placed in slot 15 can protect only DS3 cards in slots 12 through 17. An n card can be used in any slot and works like the associated DS card, but a DS card cannot work as an n card to provide 1:n protection. Planning is necessary to ensure that all cards requiring protection can receive protection. A chassis can support 1:1 and 1:n on the same side of the chassis. This would enable three DS1 cards in slots 1, 2, and 4 protected by an DS1n in slot 3 and also enable you to have a DS1 card in slot 6 having 1:1 protection with the DS1 protect card in slot 5. The maximum level of 1:n protection is 1:5; 1:n protection is not available on the 15327.

- **1+1**—This is an optical protection scheme. This scheme uses SONET Automatic Protection Switching (APS) as its protection mechanism. With this protection in place, cards do not have to be in neighboring slots, and there is no special version of the card to provide protection. (Refer to Chapter 3, "SONET Overview," for a more detailed explanation of APS protection.)

To configure a protection group, follow this procedure:

With the **Protection** subtab selected, click **Create**. Figure 5-6 displays the different fields for the creation of a protection group.

Figure 5-6   *Protection Group Creation Screen*

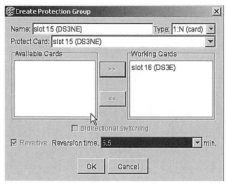

In the Create Protection Group dialog box, enter the following:

- **Name**—Type a name for the protection group. The name can have up to 32 alphanumeric characters.

- **Type**—Choose the protection type (1:1, 1:n, or 1+1) from the drop-down menu. The protection selected determines the cards that are available to serve as protect and working cards. If you choose 1:n protection, for example, only DS1n and DS3n cards display.

- **Protect Card or Port**—Choose the protect card (if on the 15454 and using 1:1 or 1:n) or protect port (if using 1+1) from the drop-down menu.

  Based on these selections, you see a list of available working cards or ports displayed under Available Cards or Available Ports.

  From the Available Cards or Ports list, choose the card or port that you want to be the working card or port. These are the cards or ports that will be protected by the card or port previously selected in Protect Cards or Ports.

  Click the top arrow button to move each card and port to the Working Cards or Working Ports list.

- **Bidirectional switching** (optical cards only)—Click if you want both the transmit and the receive channels to switch at the same time if a one-fiber failure occurs.

- **Revertive**—When selected, the ONS device reverts traffic to the working card or port after failure conditions stay corrected for the amount of time entered in Reversion time. This alleviates flapping of the circuit if the card is continuously rebooting or has other issues.

- **Reversion Time**—If Revertive is checked, enter the amount of time following failure condition correction that the ONS device should switch back to the working card or port.

This completes the discussion on basic node configuration. Now, you need to set up the timing for the node. Synchronous timing cannot be overemphasized when setting up your SONET network. Accurate transmission of traffic requires precise detection of individual bits that are arriving at picosecond intervals. The next section explains how the nodes handle timing issues and how you should configure timing.

Configuring SONET Timing

Many problems in SONET networks can be tied back to improperly configured timing. As speeds increase, the time available for the network to determine whether a bit is a 1 or a 0 has decreased dramatically. To provide a high degree of accuracy, network clocking has been designed in a hierarchy with the highest level being a Stratum 1 level clock and the lowest being a Stratum 4 level clock. The various stratum levels define the accuracy that the clock is able to maintain. The minimum level of slips required by SONET is 20 parts per million (ppm). The Stratum level clock required to maintain this level is a Stratum 3 level clock, which is sufficient but not recommended. Stratum 3 slightly exceeds the requirement of 20 ppm. Both the ONS 15454 and the ONS 15327 products have an internal Stratum 3 clock. Routers, desktop PCs, and some wristwatches currently have internal Stratum 4 clocks. Most high-speed interfaces of DS3 speed or above have a built-in Stratum level 3 oscillator.

Timing in Linear Networks

The ONS products can define their timing as coming from one of three sources: external, line, or internal, as described in the following list. Figure 5-7 depicts how these various timing methods work.

Figure 5-7   *SONET Timing Options*

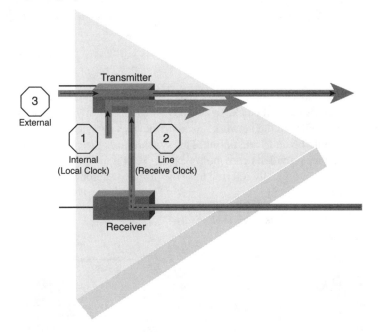

1   **Internal clock**—This clock is located on the control card and is rated at a Stratum 3 clock. It is not recommended for ongoing SONET operations.

    2 **Line clock**—This timing option synchronizes the transmitter clock with the clock derived from the receive signal. This is the preferred clock for interconnection to a service provider network when the clocking used to transmit the signal is better than the Stratum 3 of the internal clock. Most service provider networks are timed to a Stratum 1 timing source.

    3 **External clock**—This timing option synchronizes the transmitter clock with an external source. The external clock is usually rated as a Stratum 1. This could be used by a service provider in a central office to provide clocking to a network.

Figure 5-8 depicts a timing scenario where the network element (NE) on the left side of the link (West) is receiving timing from an external timing reference and distributing that timing through the network. The mid-span regenerator is passing the timing (through timing) to the NE on the right side of the link (East). The East side NE is receiving the timing signal from the received signal. This is an example of line timing.

Figure 5-8   *Timing of Lines in a Private Network*

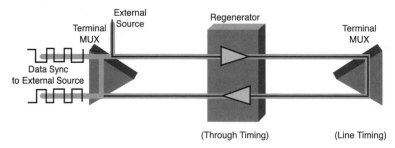

Figure 5-9   *Timing Your Network from a Service Provider*

A service provider's digital network typically provides a higher-quality clock source than the equipment's internal clock. To take advantage of this high-quality clock and reduce the percentage of slips in the network, customer devices connected to a service provider network should be set to line timing. This synchronizes the customer premise multiplexers with the service provider's core multiplexers. (See Figure 5-9.) The master reference clock in the network is typically referred to as the primary reference clock (PRC) or the primary reference source (PRS).

Figure 5-9   *Timing Your Network from a Service Provider*

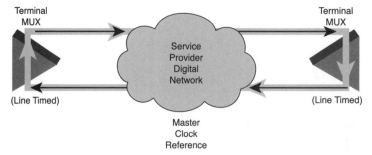

Timing in Metro Rings

When working with a metro optical network, the timing might be derived from the service provider core network and distributed into the metro optical network. Figure 5-10 depicts a scenario in which the gateway network element (GNE) is distributing Stratum 1 timing via the S1 byte in the line overhead in both the clockwise and counterclockwise directions. ST1 is the S1 byte abbreviation for Stratum 1 timing. Clocks drift after time, and the accuracy of the source decreases with every node that is traversed. A great deal of caution should be taken when timing from a service provider network. To ensure accurate timing from a service provider, this should be built in to a Service Level Agreement (SLA). Some service providers do not provide timing to their customers.

Figure 5-10   *Metro Ring Timed from a Service Provider*

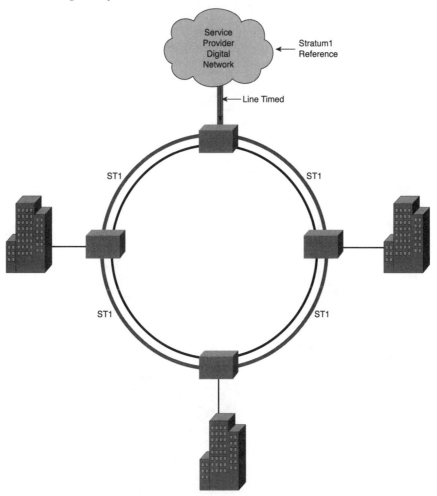

Stratum 1 clocks are created from the rare earth element Cesium or Rubidium. The clocking is derived from the vibrations of the atomic nuclei. This clock used to be so prohibitively expensive that there were only a couple in the entire United States. These clocks can be purchased now for less than $10,000. Stratum 1 clocks distributed via satellite are even less expensive. A global positioning satellite (GPS) constellation can provide Stratum 1 timing to almost anywhere in the world. To add a level of redundancy in the network, two clocking sources can be obtained on the ONS 15454 and ONS 15327. One is used as the primary clock, and the other as the secondary clock. All external references are obtained via pins on the back of the devices. These pins are wire wrapped to a building-integrated timing supply (BITS) source that uses an unframed T1 for delivery of timing. Figure 5-12 depicts a network in which node 1 is connected to two BITS timing sources that provide redundant timing in the ring. Nodes 2 through 4 are line timing from node 1. Node 3 is also distributing timing via the BITS out ports on the back of the ONS device. This timing can be used to time another ring.

Line timing should always flow in the same direction. In Figure 5-11, nodes 2 through 4 would define slot 5 as their primary reference, slot 6 as their secondary reference, and their internal clock as their tertiary reference. This would ensure timing accuracy during times of network problems.

Figure 5-11 *Metro Ring Network with 2 Clock Sources*

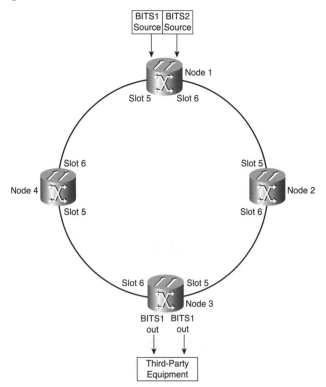

In addition to obtaining an accurate clock signal for use throughout your network, you must also maintain that clocking in the event of some network component failure. Proper planning regarding the providing of timing during network failures is essential to maintain traffic flow with acceptable bit error rates (BERs).

Ring Synchronization

To understand how clocking works, examine a typical SONET ring with six nodes. Figure 5-12 depicts a SONET ring in which timing at node A is supplied by an external clock source. Nodes B through F are deployed at branch offices without any external clock. All nodes in the ring are ONS add/drop multiplexers (ADMs). Nodes B through F are line timing in the same direction from the received SONET signal sourced at node A. Timing is unidirectional, and all nodes should be timed in the same direction. Multiple timing references provide resiliency to line-timed nodes on the ring. Although all the nodes in the ring have a primary timing reference of the slot that is receiving the signal in the clockwise direction, the secondary timing reference can be set to the slot that is receiving the signal in the counterclockwise direction.

Figure 5-12   *SONET Ring with Six Nodes*

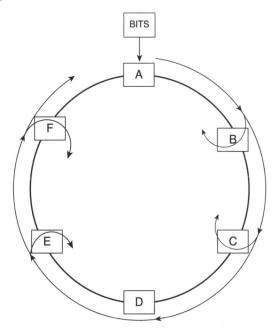

In Figure 5-13, the fiber between nodes B and C has been pulled out of the ground by a backhoe at a construction site. The outage is immediately detected by the loss of signal (LOS) on the receivers at nodes B and C. Node C was originally receiving its primary timing from node A. Because node C is configured with a secondary timing reference on the protect ring that points to node D, it instantly reverts to timing from node D. Node D does not know about this failure yet because the ring has not converged. Because node D is not aware of the failure, it continues clocking the signal from C. The less-accurate clock of C is sending traffic out and is being propagated between nodes B and C. These two nodes are now clocking themselves independently from the rest of the ring, and both are using line timing. The clocks between nodes C and D shift apart from the rest of the ring, leading to an increased BER. Eventually, it could result in complete loss of communications if the clocks drift far enough apart.

Figure 5-13 *Normal Ring Timing*

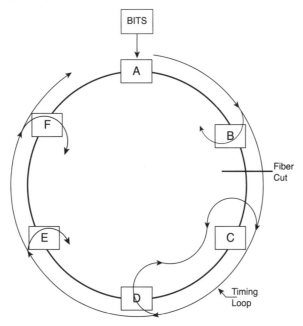

The solution to the timing loop issue is called synchronization protection switching (SPS). SPS is a system whereby quality of clocking is controlled by synchronization status messaging (SSM) conveyed in bits 5 through 8 of the S1 byte in the SONET line overhead. SPS is considered a line-switching function but is used for both bidirectional line-switched rings (BLSRs) and unidirectional path-switched rings (UPSRs).

The S1 byte is used to trace the source of a reference-timing signal. The S1 byte can be set to the level of the stratum clock used as a source, such as Stratum 1 traceable (S1); do not use (DUS), or holdover (HO). Figure 5-14 displays normal ring timing operation. Notice that each node is receiving timing from the primary reference but sending a DUS message the opposite direction so that none of the other nodes use that ring for timing. The S1 byte set to Stratum 1 traceable is forwarded clockwise around the ring to inform the nodes about the timing source. In the counterclockwise direction, the S1 byte is set to DUS to prevent other nodes from using this signal for timing regeneration. Node A is sending S1 SSM on both rings. Configuration defines which S1 byte the line-timed nodes listen to. The S1 byte is set to Stratum 1 traceable when it is sent out by node A to inform the other nodes on the ring that the Stratum 1 timing source is available.

Figure 5-14   *Synchronization Status Messaging*

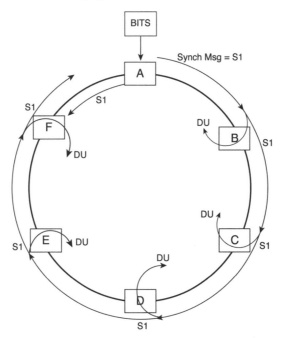

When a node loses its timing source, the node enters an HO state where it waits for the configured timing Wait To Restore (WTR) timer to expire. Figure 5-15 shows the initial scenario with a fiber cut between nodes B and C. Node C has lost its timing source because of the wraparound of the ring. Node C now receives the S1 byte set to DU via the transmission line coming from node D. Therefore, node C has no other possible clock source to regenerate its timing from and is switching into HO state using the internal clock for its transmission lines.

Figure 5-15 *Fiber Cut Stage One*

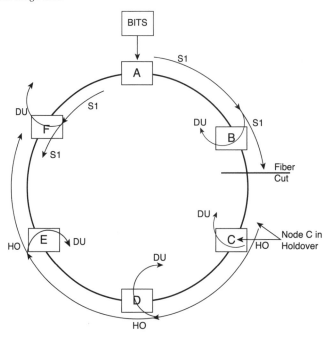

Figure 5-16 *SPS Recovering from Fiber Cut*

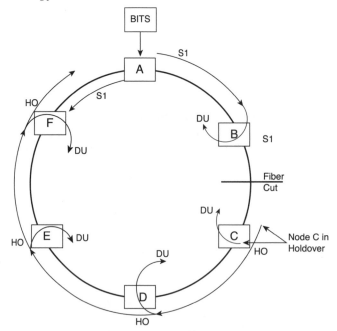

Node F's WTR timer expires and starts looking for its secondary reference element. The S1 byte traveling counterclockwise around the ring from node A never stopped transmitting; Node F is now ready to listen to it. Node F selects the incoming transmission line from node A as the new timing source from which to regenerate the timing. Figure 5-16 shows this operation.

The process of each node's WTR timer expiration continues until all nodes regain the Stratum 1 traceable clock source from node A. The ring is once again synchronized (converged) to the Stratum 1 traceable BITS timing reference. The time that is needed to perform SPS is in the sub-50-millisecond range.

Now that you understand the workings of SONET timing and protection, you are ready to configure ring timing.

Timing Subtab

To configure timing, select the **Timing** subtab from the Provisioning tab in Node view. From this tab, you can configure the timing parameters needed. To view the bottom portion of the screen, you must scroll down. This portion of the screen displays the Reference Lists section.

The Timing subtab is divided into multiple sections:

- **General Timing**—Defines the main timing method for the node
- **BITS Facilities**—Provides information about any building-integrated timing supply (BITS) clocks used
- **Reference Lists**—Shows information about the nodes primary, secondary, and tertiary clocking sources

In the General Timing section, you provide the following information:

- **Timing Mode**—Set to External if the device derives its timing from a BITS source; set to Line if timing is derived from an OC-n card that is optically connected to the timing node. A third option, Mixed, enables you to set both external and line-timing references. Mixed might be used in networks where BITS clocks reside on different nodes. Internally timing the node can be quite confusing for new users because the only two options are External and Line. When internally timing a node, this field should be set to External, and the Reference Lists section should all be set to Internal.

CAUTION   Because mixed timing might cause timing loops, Cisco does not recommend its use. Use Mixed mode with care.

- **SSM Message Set**—Synchronization status messages maintain network clocking synchronization when a failure has occurred. The format of the messages is either Generation 1 or Generation 2. Choose the message set level supported by your network. Details of the appropriate message set are defined by your clocking source. Generation 1 is widely deployed but older than Generation 2. The Generation 1 message set lacks the Stratum 3e (Enhanced Stratum 3) and transit node clock (TNC) messages. Stratum3e was invented by Telcordia (Bellcore) and is not acknowledged by all standards bodies. Table 5-2 shows both message sets.

Table 5-2   *SSM Generation 1 and Generation 2 Message Set Messages*

SSM Generation 1 Message Set Message	Quality	Description
PRS	1	Primary reference source (Stratum 1).
STU	2	Sync traceability unknown.
ST2	3	Stratum 2.
ST3	4	Stratum 3.
SMC	5	SONET minimum clock.
ST4	6	Stratum 4.
DUS	7	Do not use for timing synchronization.
RES		Reserved; quality level set by user.
SSM Generation 2 Message Set Message		
PRS	1	Primary reference source (Stratum 1).
STU	2	Sync traceability unknown.
ST2	3	Stratum 2.
TNC	4	Transit node clock.
ST3E	5	Stratum 3E.
ST3	6	Stratum 3.
SMC	7	SONET minimum clock.
ST4	8	Stratum 4.
DUS	9	Do not use for timing synchronization.
RES		Reserved; quality level set by user.

- **Quality of RES**—If your timing source supports the reserved S1 byte, you set the timing quality here. (Most timing sources do not use RES.) Qualities display in descending quality order as ranges. For example, ST3<RES<ST2 means the timing reference is higher than a Stratum 3 and lower than a Stratum 2.

- **Revertive**—If the box is checked, the device reverts to the primary clocking source after the conditions that caused it to switch to the secondary is corrected. This enables the WTR timer.

- **Revertive Time**—If the Revertive check box is checked, this is the amount of time the device waits before reverting to its primary clocking source after it has been restored.

In the BITS Facilities section, you provide the following information:

- **State**—If this node has a BITS clock attached to it or if the node is providing a BITS clock to a third-party source, set the BITS reference to IS (in-service). Otherwise, the State should be set to OOS (out-of-service). The default is IS; so take care to set the state to OOS when using internal timing.

- **Coding**—Set to the coding used by your BITS reference, either B8ZS (binary 8 zero substitution) or AMI (alternate mark inversion). Recall that BITS timing is distributed via a DS1.

- **Framing**—Set the DS1 framing used by your BITS reference, either ESF (extended super frame) or SF (D4) (super frame). SSM is not available with super frame.

- **Sync Messaging**—Check to enable SSM. This needs to be set to In if your system can use the S1 byte and you want to avoid timing loops.

- **AIS Threshold**—Sets the level where a node sends an alarm indication signal (AIS) from the BITS 1 out and BITS 2 out backplane pins. When a node times at or below the AIS threshold quality, AIS is sent (used when SSM is disabled or frame is SF).

In the Reference Lists section, there are three levels of references for the node. There are also three levels of references for the BITS 1 out and the BITS 2 out. Provide the following information for each reference via the drop-down menus:

- **NE Reference**—Enables you to define three timing references (Ref 1, Ref 2, Ref3). The node uses Ref 1 as long as the signal is good. If a failure occurs, the node uses Ref 2. If that fails, the node uses Ref 3, which is typically set to Internal Clock as a last resort.

  The options available change depending on how you configure the Timing mode in the General Timing section.

  — External—The options are BITS1, BITS2, and Internal Clock.

  — Line—The options available include the node's in-service optical cards or the internal clock. Select the card that is pointing in the direction that will be used for primary timing as Reference 1 (clockwise or counterclockwise) from the BITS source. Choose the card pointing in the secondary timing direction (counterclockwise or clockwise; should be the opposite of the primary) from the BITS source.

— Internal—Used for internal (Stratum 3) timing.

In a typical ONS network, one node is set to External. The external node derives its timing from a BITS source. The BITS source, in turn, derives its timing from a primary reference source (PRS) (normally a Stratum 1 clock). The other nodes in the network are set to Line. The line nodes derive timing from the externally timed node through the OC-n cards. All nodes have the tertiary references set to Internal.

Figure 5-17 shows an example of how the time might be configured in an ONS network. Node 1 is set to External timing for both BITS timing references. Give some thought to this design because both clock sources are connected to a single node: node 1. If node 1 were to fail, the network would be without a strong timing source. An option is to move the BITS 2 clock to a different node. The BITS facilities are Stratum 1 timing sources connected directly into node 1. The third reference is set to Internal clock. On an ONS 15454, it is possible to use the BITS output pins to provide timing to outside equipment as depicted with node 3. Slots 5 and 6 of each node contain OC-n cards that are part of the ring. Timing at nodes 2, 3, and 4 is set to Line, and the timing references are set to the trunk cards based on distance from the BITS source. Notice that the primary and secondary reference points of each node point in the same direction around the ring. Reference 1 is set to the OC-n card closest to the BITS source.

Node timing setup and configuration would be complete at this point. The Cisco ONS 15454 supports a number of different configurations for different SONET applications. The next section describes the configurations that the ONS 15454 products support and provides example configurations.

Figure 5-17   *Timing Example*

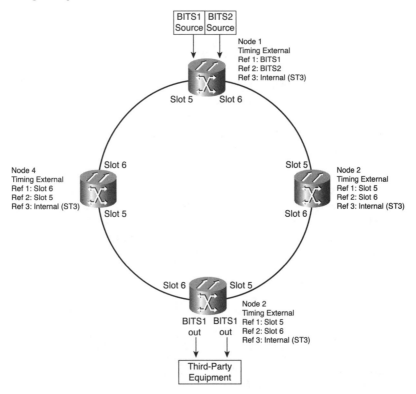

Activating Cards

Whether the network calls for a UPSR or BLSR ring configuration, the first step is always to put the ports utilized In Service:

Step 1   In Node view, double-click the card you want to activate. The Card view appears in the Cisco Transport Controller window.

Step 2   Select the **Provisioning** tab. The Line subtab is automatically selected, and the line on the selected card displays.

Step 3   In the Status field for each port on the card, choose **In Service** from the Status drop-down menu.

Step 4   When finished, click **Apply**.

Step 5   The port on the selected card should change from gray to green. This indicates that the port is In Service/Active.

You must complete these steps for every card participating in the ring.

Ring Configurations

The method implemented to physically interconnect the nodes with fiber in a ring fashion should follow a consistent standard throughout the network. The general rule is to think of the ONS 15454 chassis as having two sides (West and East). Between nodes, the East side (slots 12 through 17) connects to the West side of the next node (slots 1 through 6). This is not a limitation of the ONS 15454; the device has the flexibility of having the West and East cards reside in any slot, but this makes it much easier for troubleshooting problems. Many legacy devices had architecture limitations that forced the West and East orientation. With the ONS 15454, West and East cards can be next to each other on the same side of the chassis. This allows flexibility, but a company-wide procedure should be put in place nonetheless.

If you were building a ring and decided to use slots 6 and 12 for your ring cards, one option available would be to run fiber from the card in slot 12 (East) to the card in slot 6 at the next node in the ring. This would provide consistency when you build your rings, and if you were ever to change your ring designs later from UPSR to BLSR, determining West and East orientation would be simple. Figure 5-18 shows an example of this configuration.

Figure 5-18   *15454 Fiber Attachments for Rings*

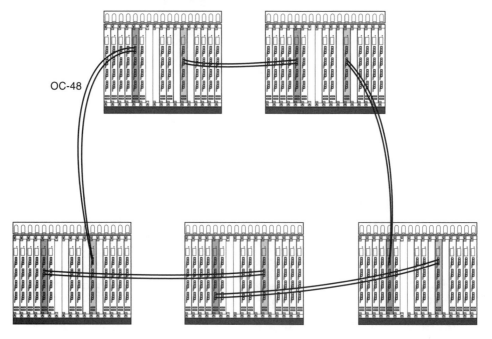

The 15327 is architecturally different from the 15454. A generic standard could be to define the higher-numbered slots as East and the lowered-numbered slots as West or vice versa. (See Figure 5-19.) Remember, this is just a convention and not a requirement.

Figure 5-19   *15327 Fiber Attachments for Rings*

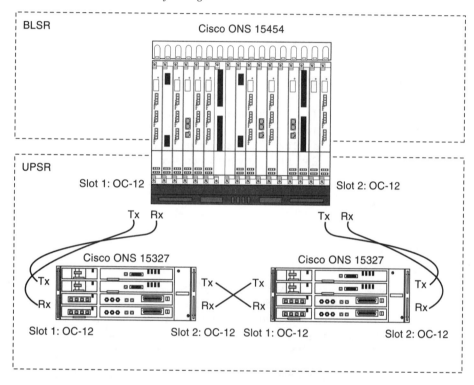

UPSR supports the following features:

- Traffic travels through all nodes on the ring and on both paths simultaneously. The working path is chosen by the receiver. The receiver looks at both signals and chooses the best signal (the one with the lower BER). The path that was not chosen is deemed the protection path.

- Each circuit created consumes that exact number of STSs throughout the entire ring.

The Cisco ONS BLSR configuration supports the following features:

- Up to 16 nodes in a ring are supported in a 4-fiber ring and up to 24 nodes in a 2-fiber ring. The 24-node configuration is a Cisco extension using the K3 byte. This is not recognized by any of the standards bodies, but Cisco devices support remapping the K3 byte for vendor and standard interoperability. If failover time within 50 ms is a major concern, the 24-node implementation may lead to longer switch times.

- Bandwidth is only consumed between the source and destination nodes and can be reused as traffic is added or dropped at various locations.
- Two-fiber BLSR architecture.
- Four-fiber BLSR architecture.

Two-Fiber BLSR Architecture

In a two-fiber ring, one-half of the bandwidth (lower-numbered STSs) is reserved for working circuits, and half of the bandwidth is reserved for protection circuits (higher-numbered STSs). If an OC-48 ring were implemented with 2-fiber BLSR, there would be only 24 STSs available for working traffic (for example, 1.25 Gbps). The other 24 STSs would be reserved for protection traffic that could be used during an outage. Protection bandwidth can be resold as "extra" traffic but would receive no protection during outages. This type of service is normally sold at a discounted rate.

BLSR rings must have the East and West ports explicitly defined when the BLSR ring is activated. If the fiber attachment procedures have been followed, configuration of the BLSR ring should be intuitive.

Four-Fiber BLSR Architecture

In a four-fiber BLSR ring (available since CTC 3.1 on the 15454 and CTC 3.2 on the 15327), one set of fibers (transmit and receive) are used for working traffic, and the other two fibers are reserved for protection traffic. If an OC-48 ring were implemented, there would be 96 STSs available (2-fiber pairs both running at OC-48 speeds). For working traffic, 48 STSs are used (2.5 Gbps), and 48 STSs are used for protection traffic (2.5 Gbps).

Large interexchange and some metro service providers leverage four-fiber BLSR technology (Telcordia GR-1230-CORE) for their interoffice facility (IOF) networks. BLSR networks are ideal for IOF networks that carry large amounts of mission-critical traffic. Four-fiber BLSR networks require double the fiber of a two-fiber BLSR restoration scheme, but service providers gain twice the bandwidth capacity per span with a significant increase in reliability because four-fiber BLSR supports line and span switching. Recall that two-fiber BLSR supports only line switching because of the reuse of bandwidth throughout the ring. Figure 5-20 shows a high-level view of a four-fiber BLSR configuration.

Figure 5-20 *Four-Fiber BLSR Ring*

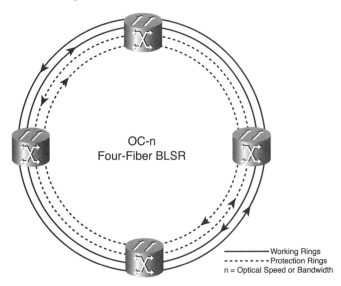

OC-n
Four-Fiber BLSR

— Working Rings
- - - Protection Rings
n = Optical Speed or Bandwidth

BLSR restoration is the preferred protection mechanism for IOF networks due to its capability to reuse bandwidth on each fiber span. This enables efficient bandwidth use in networks that support distributed traffic patterns. To achieve bandwidth reuse, the BLSR restoration reserves half the bandwidth of each span for circuit protection, which is available to transport any traffic requiring span bandwidth. All circuits can share the protection bandwidth.

BLSR rings are best used when traffic flows in a fully meshed environment where traffic originates and terminates at many different nodes. It is important to traffic engineer BLSR rings appropriately. If all the traffic on the network were going to one central site, BLSR would not be advantageous because the traffic cannot be reused to its full potential. Central-site architectures are more ideal for UPSR ring technology.

Creating UPSR Rings Through the SONET DCC Subtab

Ring configuration on the ONS 15454 and ONS 15327 is identical. To configure a UPSR ring, follow these steps:

Step 1   Put the appropriate ports In Service.

Step 2   Activate the SONET DCC. (Cisco's SONET DCC labeling is an implementation of the Section DCC [D1–D3 bytes].)

Step 3   Configure timing.

The SONET DCC enables the system to exchange management information that allows for node discovery and remote management. To provision a SONET DCC, click the **Provisioning** tab, and then select the **SONET DCC** subtab. Notice that no provisioned SONET DCCs currently display.

Click the **Create** button and choose the card that will participate in the UPSR ring. Hold down the **Ctrl** key to select multiple cards.

When you finish, click the **OK** button and SONET DCC will be activated. If the ports are not In Service at this time, CTC 3.3 and later implements a check box at the bottom of the Create SONET DCC screen to put the ports In Service. This can reduce provisioning time. Repeat this process for every node that is on the UPSR ring. The Create SONET DCC dialog box displays.

On the ONS 15454, tunnel DCCs can be created to tunnel third-party networks through your network. This was created to allow the ONS 15454 to be incorporated in existing networks. Most vendors' DCC channels do not speak to other vendors because provisioning is accomplished a little differently in each vendor's device. Up to three tunnel DCCs can be created and use the following LOH bytes:

- Tunnel 1: (D4–D6)
- Tunnel 2: (D7–D9)
- Tunnel 3: (D10–D12)

Verifying a UPSR Ring Configuration

To verify the UPSR ring, go to the Network view. The network nodes display, although the first view always stacks all the icons on top of each other (which makes it appear that there is only one node). Separate these nodes on the topology map by holding down the **Ctrl** key and simultaneously clicking a node icon. Drag it to the desired location while holding down the **Ctrl** key. Use the same technique to separate all discovered nodes. To verify the UPSR node configuration, enter Node view and click the **Provisioning** tab and then the SONET DCC subtab. The SONET DCC configuration should display.

To verify UPSR ring operation, go to Network view by clicking the up arrow or choosing Network view from the View menu. If multiple nodes are connected by green lines, the UPSR ring configuration is active. Figure 5-21 shows the window that displays.

Figure 5-21 *Network View with UPSR Ring Configured*

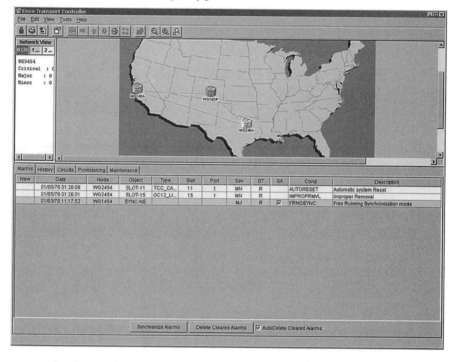

The window should show green lines connecting your nodes that are participating in the UPSR ring. Verify that all configured nodes appear and are connected by the green lines. Placing your cursor over the green line gives you information about the span. Double-clicking the green line gives you information about the circuits running through the ring.

Creating Two-Fiber BLSR Rings

A BLSR configuration is commonly used to connect MAN to the core network that connects separate MANs.

To create BLSR rings, you must first activate the SONET DCCs. Here are the BLSR configuration steps:

Step 1   Install the optical cards that will be used for the BLSR ring.

Step 2   Create the BLSR DCC terminations.

Step 3   Put the BLSR ports In Service.

Step 4 If a BLSR span passes through equipment that cannot transparently transport the K3 byte, remap the BLSR extension byte on the trunk cards on each end of the span.

Step 5 Set up the timing of the ring.

Step 6 Provision the BLSR ring ID, node ID, West slot, and East slot.

Provisioning the BLSR

After all prerequisite steps have been implemented, choose the **Ring** subtab of the **Provisioning** tab. The BLSR dialog box displays, enabling you to create a BLRS ring.

In the BLSR area, click **Create**. The Create BLSR dialog appears.

You need to fill in the following fields:

- **Ring Type**—Select the BLSR ring type, either two-fiber or four-fiber.
- **Ring ID**—Assign a ring ID (a number between 0 and 9999). Nodes in the same BLSR must have the same ring ID.
- **Node ID**—Assign a node ID. The node ID identifies the node to the BLSR. Nodes in the same BLSR must have unique node IDs.
- **Ring Reversion**—Set the amount of time that will pass before the traffic reverts to the original working path. The default is 5 minutes. All nodes in a BLSR ring should have the same Ring Reversion setting, particularly if Never (for example, nonrevertive) is selected.
- **West Port**—Assign the West BLSR port for the node from the pull-down menu.
- **East Port**—Assign the East BLSR port for the node from the pull-down menu.

You should be able to easily determine the West and East port assignments from the agreed-upon network procedures in place.

The BLSR Ring Map Change dialog appears on one node's CTC session when all nodes have "properly" configured the BLSR information. If one node configuration is wrong, the BLSR Ring Map dialog box will not appear.

Click **Yes** to view the new ring map. The BLSR Ring Map dialog box appears.

Verify that the proper ring ID, node IDs, and node IP addresses appear. Click **Close** to accept the changes. The new BLSR ring appears in the Ring subtab.

Verifying a BLSR Ring Configuration

To verify a BLSR ring configuration, return to Node view and click the **Ring** subtab on the Provisioning tab. Figure 5-26 displays this screen without the configured ring. Click the ring configuration that appears on your node, and then click the **Ring Map** button above it to manually display the ring map. Notice the other operations available:

- Create a new ring.
- Delete the selected ring.
- Display the Squelch table. (Squelching is an AIS sent around the ring during BLSR line switching. This verifies that no other nodes in the ring are using the same STS. Squelching is important when there are multiple fiber breaks.)
- Upgrade the two-fiber BLSR ring to a four-fiber BLSR ring.

Test the BLSR using testing procedures normal for your site. Here are a few steps you can use:

Step 1 Run test traffic through the ring.

Step 2 Log in to a node, click the **Maintenance > Ring** tabs, and choose **Manual Ring** from the East Switch list. Click **Apply**.

Step 3 In Network view, click the **Conditions** tab and click **Retrieve**. You should see a Ring Switch West event, and the far-end node that responded to this request will report a Ring Switch East event.

Step 4 Verify that traffic switches normally.

Step 5 Choose **Clear** from the East Switch list, and then click **Apply**.

Step 6 Repeat Steps 1–4 for the West switch.

Step 7 Disconnect the fibers at one node and verify that traffic switches normally.

Four-Fiber BLSR Ring

When configuring a four-fiber BLSR ring, only the working fiber path should have SONET DCC terminations. If the protect path has SONET DCC terminations, the creation will fail. However, these ports should be In Service. To create a four-fiber BLSR ring, you also need to complete these "extra" steps:

Step 1 In the Ring Creation dialog box, select 4-Fiber BLSR ring under the Ring Type option.

Step 2 You then need to fill in the following additional fields on the right side of the screen:

- **Span Reversion**—Set the amount of time that will pass before the traffic reverts to the original working path following a span reversion. The default is 5 minutes. Span reversions can be set to Never. If you set a ring reversion time, the times must be the same for both ends of the span. That is, if node A's West fiber is connected to node B's

East port, the node A West span reversion time must be the same as the node B East span reversion time. To avoid reversion-time mismatches, Cisco recommends that you use the same span reversion time throughout the ring.

- **West Protect**—From the pull-down menu, assign the West BLSR port that will connect to the West protect fiber.

- **East Protect**—From the pull-down menu, assign the East BLSR port that will connect to the East protect fiber.

  You can find the procedures for upgrading a two-fiber BLSR to a four-fiber BLSR at Cisco.com.

Summary

This chapter walked you through all the steps that enable you to initially configure your node to participate in a ring-based network. You learned how to set the node name, how to secure access to the system, and how to set up protection groups for your cards.

You also learned of the importance of accurate timing in your SONET network and how to set up and configure up to three different timing sources for each node.

You also learned how to create two-fiber UPSR and BLSR rings and four-fiber BLSR rings.

In the next chapter, Chapter 6, "Metro Ethernet Services," you learn how transparent LAN services can be used across these rings.

Review Questions

1 In a four-fiber BLSR network, _____ fibers are dedicated to working traffic and _____ fibers are used for protection bandwidth.

   A two, two

   B four, zero

   C zero, four

   D one, three

2 A four-fiber BLSR ring is best for what type of traffic patterns?

   A Meshed

   B Metro area

   C Point-to-point

   D Add/drop linear

   E Centralized

3 The internal clock on the ONS devices is a

   A Stratum 1

   B Stratum 2

   C Stratum 3

   D Stratum 4

4 To receive timing from the central office via the access line, you should use

   A Internal clock

   B External clock

   C Line clock

   D Access clock

5 In a point-to-point network with no service provider, what clock type is preferred?

   A Internal clock

   B External clock

   C Line clock

   D System clock

6 In SONET, a typical network node located away from the central office is synchronized using _____.

   A Line timing

   B Internal timing

   C GPS

   D None of the above

7 Which Subtabs enable you to set the node name?

   A Network

   B Ring

   C General

   D Node

8 Which Subtabs enable you to create a UPSR ring?

A General

B Ring

C SONET DCC

D Timing

9 Which is the recommended way to interconnect fibers between nodes in a ring?

A East to East

B West to West

C East to West

10 What needs to be configured the same for a BLSR ring to be created?

A Ring ID

B Node ID

C West ports

D East ports

11 What needs to be unique for a BLSR ring to be created?

A Ring ID

B Node ID

C West ports

D East ports

This chapter covers the following topics:

- Market drivers for metro Ethernet
- Metro Ethernet architectures
- Design goals and constraints
- Cisco solutions and designs
- Cisco metro Ethernet optical equipment

CHAPTER 6

Metro Ethernet Services

Local-area network (LAN) bandwidth prices have plummeted in the past decade with the advent of Fast Ethernet, Gigabit Ethernet, and 10 Gigabit Ethernet. Legacy MAN/WAN connectivity (such as DS1 and DS3) prices have not followed the same exponentially decreasing pricing trend. Customers want to leverage the cost advantages of Ethernet, while receiving a similar level of service that they currently have with their dedicated lines. Carriers can benefit from Ethernet deployments due to the simplicity and ubiquity of the interface type. Legacy service provisioning normally exceeds 30-days, while some areas might take much longer depending on geographical and bandwidth constraints. Service providers own up to the point of demarcation at the customer or co-location facilities. The service provider must learn how to properly plan, provision, manage, monitor, and trouble-shoot the new equipment used with metropolitan Ethernet. Ethernet technologies are much easier to understand, configure, and manage than the robust operations, administration, maintenance, and provisioning (OAM&P) functionality of SONET services. Specialized equipment required for SONET facilities costs much more than the equipment that Ethernet requires.

Scalability is a major concern in both enterprise and SP environments. Ethernet offers bandwidth over subscription and fast reprovisioning that SONET-based services cannot match. If Company X were to buy a DS1 (1.544 Mbps) service from an SP, but soon there-after need more bandwidth, a new line would have to be run by the local loop provider, and end-to-end services would have to be reprovisioned. If the customer decided to upgrade to a DS3 (44.736 Mbps) service, the provisioning delay cycle would be repeated. The customer would normally wait at least 60 days to provision the new service because of the higher bandwidth needs. Within those 60 days, the customer would receive a new customer premises equipment (CPE) device to terminate the DS3 service. If the customer decided later to upgrade to a higher-speed service, multiple lines would need to be managed, which would cause routing problems because it is hard to equally load balance across different circuits. If a new technology were brought in to meet the need, the learning process would be longer still.

Market Drivers for Metropolitan Ethernet

Metropolitan (metro) Ethernet offers an interface that is familiar, ubiquitous, low cost (a result of the economies of scale that high production yields), and high bandwidth (up to 10

Gbps with the ratification of IEEE802.3ae). An Ethernet interface offers an extension to the LAN, rather than a technology offering that has bandwidth inefficiencies (private lines) or a Layer 2 technology (Frame Relay/ATM) that requires a complete MAC layer rewrite. This concept was first introduced as Transparent LAN Services (TLS), called such because the operation is transparent to the networks. The term *metro Ethernet* is used more often and has replaced the TLS terminology.

With an Ethernet interface, service upgrades that used to take months can be done in a matter of hours because bandwidth can be controlled through rate-limiting quality of service (QoS) policies on the SP-managed CPE router or switch devices. This solution allows greater flexibility: Bandwidth can be provisioned easily and expediently in 8-kbps increments. SPs can also sell more robust policies that guarantee a certain amount of traffic, distinguishing excess traffic with a lower QoS marking. This excess traffic is discarded first, when congestion occurs in the network. Customers can benefit from a metropolitan Ethernet solution in much the same way they do in Frame Relay networks, but the policies could be much more robust with the eight levels of prioritization that 802.1Q trunking/802.1p priority allows.

Metropolitan Ethernet Market Drivers

Although point-to-point (P2P) circuits have been dropping in price rapidly, they still remain costly when compared to the bandwidth prices of LAN Ethernet technology. Although companies have maintained separate voice, data, and videoconferencing systems in the past, there is a strong push to converge these services onto one infrastructure. This desire for convergence results mostly from the steadily growing acceptance of Voice over IP (VoIP) systems such as Cisco AVVIDD (Architecture for Voice, Video, and Integrated Data) solutions. Another factor to consider in addition to the centralization of bandwidth needs is the shifting from a legacy mainframe environment to distributed Internet-based applications. Remote connectivity used to be solved with banks of modems, but organizations are taking advantage of virtual private technologies (virtual private networks, or VPNs) at small-office, home-office (SOHO) locations to decrease the price of services. Metropolitan Ethernet offers the bandwidth, flexibility, and low cost that business users need.

Customers want to interconnect their high-speed LANs at LAN-like speeds, but without having to purchase, configure, and manage prohibitively expensive equipment and services. Ethernet is the preferred Layer 2 technology in LANs and is expanding its presence into the metropolitan area. Ethernet technologies currently offer flexible 10/100/1000-Mbps copper- or fiber-based interfaces that can be rate-limited in a granular manner at the CPE device. Currently, 10 Gigabit Ethernet is available only through fiber-optic links. SPs are now looking to offer this technology with services based on the same Ethernet technologies that enterprise customers have been using for years. This Ethernet focus has the advantage of lowering the cost and complexity of demarcation equipment, while providing the customers with an interface with which they are comfortable.

Metropolitan Ethernet offers various delivery mechanisms, depending on distance constraints. Physical media in which these services can be offered vary. The solutions include the following:

- Single-mode fiber (SMF)
- Multimode fiber (MMF)
- Unshielded twisted-pair (UTP)

Each medium has different distance limitations to consider. UTP implementations are limited to 100 meters (312 feet) and are typically found only in small multiple-tenant units (MTUs) or multiple-dwelling units (MDUs). Heavy machinery or lighting that generates electromagnetic interference (EMF) can greatly affect UTP. Multimode fiber and single-mode fiber Ethernet implementations are limited by the physical characteristics of the fiber. Although the fiber's core diameter largely determines possible distances, as seen in Table 6-1, actual distances are also a function of the laser, receiver, and fiber nonlinearities, which are covered in further detail in Chapter 8, "Implementing DWDM in Metropolitan-Area Networks." Table 6-1 lists Cisco Gigabit Ethernet fiber/laser distance constraints.

Table 6-1   *Cisco Gigabit Ethernet Fiber/Laser Distance Constraints*

Transceiver Fiber Lambda	Diameter	Bandwidth (MHz × km)	Range (m)
1000BASE-SX MM 850 nm*	62.5 um**	160	2–220
1000BASE-SX MM 850 nm	62.5 um	200	2–275
1000BASE-SX MM 850 nm	50.0 um	400	2–500
1000BASE-SX MM 850 nm	50.0 um	500	2–550
1000BASE-LX MM 1310 nm	62.5 um	500	2–550
1000BASE-LX MM 1310 nm	50.0 um	400	2–550
1000BASE-LX MM 1310 nm	50.0 um	500	2–550
1000BASE-LX SM 1310 nm	9.0 um	N/A	2–5000

*nm = nanometer

**um = micrometer

Ethernet technology can lower SP costs by reducing the need for specialized staff, specialized equipment inventories, specialized monitoring, and test equipment for technologies as varied as DS1, DS3, fractional DS1/DS3 services, Frame Relay, ISDN, ATM, and potentially Switched Multimegabit Data Service (SMDS). Ethernet interfaces, experts, test and monitoring equipment, and other tools are much cheaper than that of SONET and ATM equipment that offers similar bandwidth scalability. At the time of this writing, a single-port OC-192 (optical carrier) Packet over SONET (PoS) card for the Cisco 12400 series router costs twice as much as a single-port 10 Gigabit Ethernet card.

Metro Ethernet Architectures

In any service provider circuit, both CPE and point of presence (POP) devices are required. This book covers two POP devices used for metropolitan Ethernet: ONS 15454 and ONS 15327. These devices are add/drop multiplexers providing Ethernet over SONET services. Some other types of metropolitan Ethernet implementations are available through dense wavelength division multiplexing (DWDM) wavelengths and are briefly discussed in Chapter 8. This book focuses on the add/drop multiplexer transport devices and does not discuss the myriad of CPE choices.

Ethernet architectures are flexible in the respect that customers can choose to use Layer 2 or Layer 3 interface implementations on Cisco routers or switches. A Layer 3, IP-enabled solution terminates the Ethernet collision and broadcast domain, whereas a Layer 2 solution switches implementations. SPs are increasingly straying away from the legacy Ethernet Spanning Tree Protocol (STP) implementations due to long convergence times of up to 50 seconds. The architecture of some metropolitan Ethernet cards is flexible enough to allow these STP operations. Multilayer switches that are capable of both switching (Layer 2) and routing (Layer 3) offer the highest degree of flexibility, in that they can offer both switching and routing functions. A requirement of these devices is the rate limiting of traffic. Rate limiting is achieved through committed access rate (CAR) or Modular QoS CLI (MQC) configuration, normally found in Layer 3-capable devices. The ONS 15327 and ONS 15454 devices provide one interface in which other devices can aggregate services from different-speed interfaces and traverse the metro Ethernet circuits. Rate-limiting is supported on the newer ML-Series metropolitan Ethernet cards.

Many SPs offer value-added services, including VPN aggregation, content hosting, and web hosting. These Layer 4 through Layer 7 (OSI reference model) services are applied at the service POP on a device such as the 7600 series multilayer switch platform. The 7600 series Optical Services Router (OSR) is a NEBS3-compliant device that offers multilayer switching, Ethernet over MPLS (EoMPLS), and 802.1Q-in-802.1Q (Q-in-Q) tunneling. Traffic can then be routed to other metropolitan area rings to implement a complete, feature-rich, end-to-end solution. Figure 6-1 shows a network where the Ethernet clients are aggregated into one interface on CPE devices, and the aggregated interface is connected to either an ONS 15454 or ONS 15327. The Optical Networking Systems (ONS) devices have circuit connectivity to the other sites and aggregate at the service POP OSR for off-ring connectivity or to have high-touch services applied. Notice that the diagram has four links between the POP ONS 15454 and the 7600 series OSR. This design aggregates all four of the sites over different optical links at the 7600. The 7600 would route traffic between the circuits for intra-ring traffic.

Figure 6-1 *Metropolitan-Area Networking Service POP Aggregation*

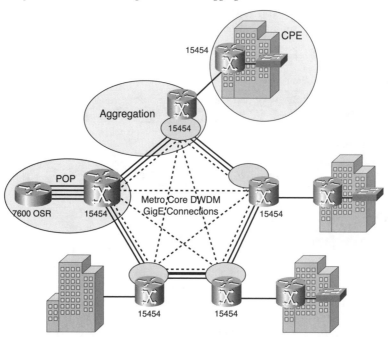

As Ethernet traffic enters the SP's domain, the SP can isolate the traffic through 802.1Q VLAN tagging. An 802.1Q Ethernet frame is usually characterized as an Ethernet giant because the maximum transmission unit (MTU) size of 1522 bytes used by the dot1q frame exceeds the legacy Ethernet MTU limitation of 1518 bytes. With the advent of 802.1Q encapsulations, Cisco started enabling baby-giants on most ports and interfaces. A baby-giant is a frame size that exceeds the 1518 MTU size limitation, but falls under a certain size. Although no mandated standard exists for the size of baby-giants, some newer hardware allows baby-giant sizes up to 1548 bytes (G-series cards). Jumbo frames (9-kb MTU) are supported on most new platforms as well. Jumbo frames allow frame sizes up to 9216 bytes.

Figure 6-2 depicts a standard Ethernet frame, an Ethernet frame with an 802.1Q header, and a breakout of the Tag Control Identifier (TCI) field used in the 802.1Q frame. Notice that the 802.1Q header adds 2 octets/bytes (16 bits) to the Ethernet header. These 2 bytes are composed of a Tag Protocol Identifier (TPID), which is always set to the well-known 802.1Q address of 0x8100 (hexadecimal addressing). The frame also includes a 2-byte TCI. The 16 bits (2 bytes) used in the TCI are composed of 2 User Priority (802.1p) bits, 1 Canonical Format Identifier (CFI), and 12 VLAN Identifier fields. The User Priority field allows 8 levels of priority (0 through 7), mapping directly to the IP precedence bits in the Type of Service (ToS) field of the IP header. The CFI is used for Token Ring compatibility and has no relevance to this material. The VLAN Identifier is 12 bits long and allows for a total of 4096 (2^{12}) VLANs. Although the 802.1Q frame can identify 4096 different VLANs, the E-series cards support only up to 509 VLANs because of the card architecture.

Figure 6-2   *IEEE 802.1Q Ethernet Frame*

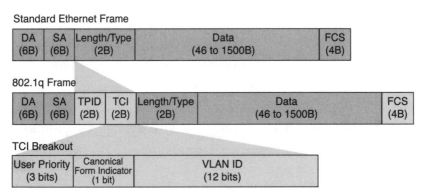

If EoMPLS or Q-in-Q technologies are not being employed, it is essential for SPs to coordinate VLANs between the enterprise customers and the SP because traffic could overlap. The E-series metro Ethernet cards do not support MTU sizes over 1522 bytes and cannot pass EoMPLS or Q-in-Q traffic. Due to this limitation, VLAN coordination is a requirement with the E-series cards. A limitation of 1522 bytes allows only one level of 802.1Q encapsulation; Q-in-Q encapsulation would require an MTU size of 1526 bytes. EoMPLS frame sizes vary depending on the various MPLS technologies employed.

Carrier and enterprise infrastructures are normally independent of each other. The E-series card cannot support technologies that enable SPs to tunnel customer traffic over their Ethernet infrastructures. These technologies are not covered in detail here, but they include EoMPLS, Q-in-Q encapsulation, and Layer 2 Tunneling Protocol version 3 (L2TPv3). You can find more information about these carrier architectures at Cisco.com.

The E-series cards support full line-rate forwarding of unicast traffic up to the circuit size that was provisioned for the interface. The E-series card rate limits broadcast, multicast, and unknown traffic to 9000 frames per second (fps).

The Cisco G-series cards for the ONS 15454 (G1000-4) and the ONS 15327 (G1000-2) allow jumbo frame sizes (9 kb) and do not limit multicast, broadcast, or unknown traffic. The G-series cards support line-rate forwarding of unicast, multicast, broadcast, and unknown traffic. The card provides a P2P circuit and does not run the STP in any way.

Service Level Agreements (SLAs) are an important aspect of metro Ethernet services due to the potential bursting above a customer's committed information rate (CIR). Customers in an MTU, for instance, could share bandwidth that was brought in through an Ethernet circuit by the provider. The SLAs are enforced based on the SP's implemented QoS policy.

Customer Premises Equipment

Customer premises equipment (CPE) is equipment located on the customer's site that is used to establish connectivity to the local exchange carrier (LEC). This equipment normally encapsulates or converts the local LAN traffic (usually Ethernet based) into a technology compatible with the carrier line transport.

The *point of demarcation* determines to which party the responsibility lies for installation, configuration, and management of the CPE. If the point of demarcation is on the telco side of the equipment, the customer has responsibility over the equipment. If the point of demarcation (demarc) is on the LAN side, the carrier has the responsibility to provide the customer with a LAN-compatible port for their network to be plugged into.

Customers can save ongoing monthly fees by controlling this equipment; to do so, however, they must make capital expenditures (capex) to obtain the equipment, and they need to maintain and troubleshoot the equipment themselves. These operations require qualified support personnel that many enterprises do not feel comfortable taking on.

Many SPs sell managed services to maintain the CPE equipment for a recurring cost. All service and configuration changes must be made through the SP's network operation center (NOC) so that the SP can maintain reliability and change control over the devices. This type of arrangement can be prohibitive for a very large organization that needs fast changes but offers benefits in reliability because the SP is in control of the end-to-end connection. After all, it's estimated that 25 percent of outages are caused by user-configuration errors. The decision to purchase managed services might be based on the SP's SLA for customer-maintained equipment.

Ethernet-based transport has many advantages including, but not limited to, the following:

- Ethernet is a scalable technology supporting speeds of 10 Mbps to 10 Gbps without going through any protocol translation.
- Ethernet is an easy technology to manage, operate, and troubleshoot.
- Ethernet chipsets are ubiquitous, and that ubiquity has significantly driven down the cost of Ethernet components in relation to other available network interfaces.

Ethernet-based networks are easily adopted and understood—including how to troubleshoot, design, and operate—within the enterprise network. The migration to Ethernet-based services is currently in its infancy. The E-series card represents the first generation of offerings, whereas the G-series card reflects the research and development in which Cisco and various SPs have invested.

Ethernet-based services can be viewed as an evolutionary step that will require migration support for legacy transports. The ONS 15454 and ONS 15327 reflect this first evolutionary step in that they have integrated metro Ethernet into a device that supports Ethernet transport over SONET rings. The terminology for transporting Ethernet over a SONET frame is referred to as Ethernet over SONET (EoS). This implementation has great migratory potential. The minimum provisionable bandwidth with either of these platforms is at the STS-1 (51.84 Mbps) level.

Platforms such as the ONS 15252/15201 and the ONS 15540/15530 achieve Gigabit Ethernet transport directly over wavelengths in a DWDM ring architecture. The IEEE 802.3ah EFM (Ethernet in the First Mile) committee is looking into carrier-class ways of providing next-generation mechanisms that will enable ETT*x* (Ethernet to the home, office, building, and so forth). Although SONET framing is rich in OAM&P functions that carriers require to provide Five-Nines reliability (99.999-percent uptime), Ethernet currently lacks this functionality.

Prestandard mechanisms, such as Cisco Converged Data Link (CDL) header implemented on the ONS 15530, reflect the work that the IEEE802.3ah EFM standards body is currently doing. This implementation uses the currently unused Ethernet preamble functionality to provide OAM&P functionality. The Ethernet preamble has gone unused because of faster chipsets, but has remained in the standards for the sake of backward compatibility. Evolutionary mechanisms such as this might someday bring Ethernet services into our homes.

Several customers are typically located in a common building in metropolitan environments. CPE in the multiple-tenant units (MTUs) for these locations would preferably be shared for these connections. This piece of equipment would be maintained and owned by the SP and be placed in a secure area. This implementation offers SPs great economies of scale because only one fiber link is needed to the SP infrastructure, and there is only one piece of equipment to manage for multiple customers. The same architecture could be implemented for apartments or multiple-dwelling units (MDUs). Several CPE devices could be aggregated onto one device at either the customer premises or at the POP. This type of aggregation creates a hierarchy that allows the SP to carry more traffic across the core infrastructure.

The 10720 series router was designed with the MTU/MDU market in mind. This router is a 2-slot device that has an Ethernet slot for up to 24 ports of Ethernet and one uplink card that uses Dynamic Packet Transport (DPT) technology. This device takes up only two rack units (RUs) of space, and the DPT interfaces are available with short-, intermediate-, and long-reach optics. All the DPT interfaces run at OC-48 speeds. The 10720 uses a parallel express forwarding (PXF) chip design that employs application-specific integrated circuit (ASIC)-based, wire-speed, high-touch IP services such as IPSec and QoS. Chapter 11, "Dynamic Packet Transport," covers DPT in detail.

SONET Ring Hierarchy

In the same way that LANs follow a three-tier design model of access, distribution, and core, SONET ring hierarchies have access rings, metro rings, and core rings. For instance, an access ring might cover an area in a 10-block radius, aggregating traffic onto a metro ring that covers the entire city for transport between some facilities. Metro rings are connected by core rings that provide national long-haul coverage. Figure 6-3 displays a design where the ONS 15327 is being used for OC-12-speed unidirectional path-switched ring (UPSR) access rings, whereas the ONS 15454s are interconnecting these rings at

OC-48 or OC-192 speeds through UPSRs or bidirectional line-switched rings (BLSRs). If this were a large implementation spanning the country, the ONS 15600 device could be used to aggregate rings onto the core ring. The ONS 15800 series long-haul DWDM platform could also be used for long-haul transport. It is important to mention that the ONS 15327 could be replaced with the ONS 15454 depending on the functionality needed. The ONS 15327 product's target market is small rings where limited port densities are required. For higher port densities and DWDM functionality, the ONS 15454 should be deployed.

Figure 6-3   *SONET Ring Hierarchy*

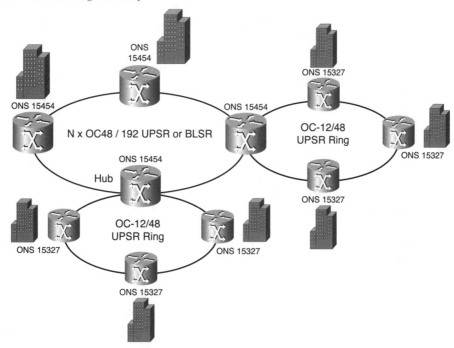

Quality of Service

The amount of LAN bandwidth available to an enterprise will be considerably larger than the amount of bandwidth that the enterprise will purchase from an SP. Because of the burstiness of data traffic, QoS tools—including classification, marking, policing, and congestion management—are necessary to determine which traffic is of highest importance. QoS is an end-to-end solution. After all, congestion can occur at any point in the network, but the WAN is usually the largest bottleneck and source of congestion. Regardless of customer involvement, SPs have begun to employ QoS policies to maintain the customer's binding SLA. SLAs ensure that the customer receives a certain level of service, including throughput, reliability, round-trip times, and delay.

Traffic forwarded by the CPE device could be sent to an aggregation point in the SP's network, depending on the size of the circuit, services offered by the service provider, and so on.

Oversubscription of circuits is a common practice that attempts to take advantage of a technology's capability to statistically multiplex traffic. Time-division multiplexing (TDM) services allocate time slots, and each time slot is serviced in a round-robin fashion. If a particular time slot has no data, padding is sent in the place of user data. This is an inefficient use of the carrier's expensive resources. To avoid the wasteful nature of TDM architectures, technologies such as Frame Relay, which can statistically multiplex traffic, came into vogue. Frame Relay networks can be problematic and complex to transport voice, however, due to the unique requirements of voice transport.

ATM was developed as a convergent technology that would provide inherent QoS to transport voice, video, and data. ATM interfaces are prohibitively expensive and require knowledgeable, dedicated staff. ATM also uses fixed-size 53-byte cells that result in a degree of wastefulness because of something known as *cell tax*. The cell tax is the loss of useable data sent in an ATM cell that results from the extra ATM overhead (fixed 5-byte cell header and ATM adaption layer [AAL] overhead) and cell padding that is needed to pad cells. Cells need to be padded because 1518-byte Ethernet frames do not fit evenly into 53-byte cells. Consider, for instance, that a full-size Ethernet frame of 1518 bytes is encapsulated into the 48 bytes of "potentially" useable space of an ATM cell. This encapsulation results in 31.625 cells that are required to transport the frame. In this scenario, 37.5 percent of the last frame is padded (wasted). The functionality that rewrites Ethernet frames into ATM cells is known as the *segmentation and reassembly* (SAR) layer. SAR chips are expensive and complex to produce.

The simplicity, low cost, and economies of scale inherited through the use of Ethernet technologies enable SPs statistically to multiplex traffic cost efficiently. Ethernet also allows eight levels of prioritization, which maps directly to the ubiquitous TCP/IP's IP precedence.

LAN traffic is aggregated and groomed at a CPE device and sent over a metropolitan-area connection. This traffic is aggregated at an SP aggregation point that accepts both TDM and Ethernet traffic. The Cisco 12000 series routers, 7600 series OSR, 10000 series Edge Service Router (ESR), and the ONS 15454 product lines are well suited to aggregate and groom traffic from multiple sources to be transported across the metro core. These devices provide high port densities, nonblocking backplane architectures, and scalability that is necessary in carrier environments.

After the low-speed traffic has been received at the aggregation and grooming point (historically serviced by SONET infrastructures), the traffic is directed to high-speed circuits that can deliver the traffic to its destination. Traffic that needs to traverse long spans to reach the destination is usually serviced by different POPs in different cities. The connections between these POPs are known as *POP interconnects*.

POP Interconnect

Large SPs typically have multiple locations for the collection and distribution of traffic. These locations represent the POP. If a service provider does not have a point of presence in a particular location where traffic must originate, terminate, or traverse, the provider hands off the traffic to another SP's POP. System compatibility is a must at these interconnection points. SONET has historically provided this compatibility and is leveraged for metro Ethernet services.

Figure 6-4 depicts an environment where TDM and Gigabit Ethernet traffic is aggregated and groomed from a number of CPE devices and then sent through an uplink to devices that are DWDM capable (such as the ONS 15454), where the traffic is transported across the inter-POP interconnects through a DWDM infrastructure.

Figure 6-4   *POP Interconnect*

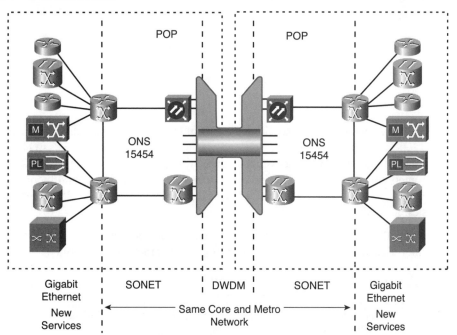

Long-Haul Transport

When transport needs are not confined to a single metropolitan area, equipment that can support longer transmission distances is needed. The issue might not be just one of distance but of fiber capacity and quality as well. The 15454 ELR cards can provide DWDM functionality in the metropolitan area. The DWDM channels from one or more metro rings can then be transported through a long-haul platform. The Cisco long-haul DWDM device is the 15800 series DWDM shelf. Long-haul DWDM networks are normally point to point in nature. Figure 6-5 shows two metropolitan areas that are connected through long-haul equipment in two different geographies (New York City to San Jose).

Particular attention must be paid to the design within the POP. The same level of scalability and redundancy should occur within the SP's POP.

Figure 6-5   *Inter-Metro-Area Transport*

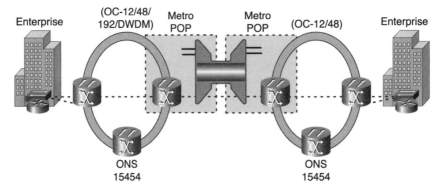

Intra-POP connectivity is not limited to any of the mentioned product lines. The Cisco 12000 series router offers a flexible and scalable solution that allows SPs to build fault-tolerant, highly available networks. Figure 6-6 shows a design where Cisco 12000 series routers are aggregating traffic and transporting the traffic over Gigabit Ethernet interconnects in the POP. The ONS 15327 could be replaced with the ONS 15454 depending on the design requirements.

The 12000 series router can also achieve short-range connectivity in POPs through Very Short Reach (VSR) optics. VSR-1 is the Optical Internetworking Forum (OIF)-approved OC-192 (10 Gbps) interface optimized for interconnection distances of less than 300 meters between routers, switches, and DWDM systems. The VSR interface on the Cisco 12000 series router uses an MTP/MPO fiber ribbon cable that transports 12 parallel fibers in 1 ribbon cable fiber connector. The Cisco implementation of VSR uses SONET or SDH framing. The VSR and 10 Gigabit Ethernet cards are priced similarly.

Figure 6-6   *Intra-POP Connectivity with the Cisco 12000 Series Router*

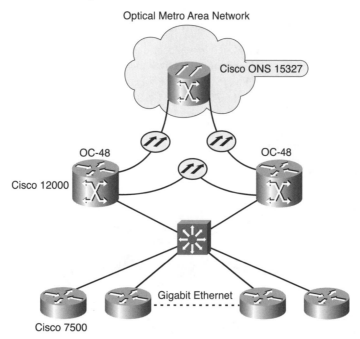

Transporting Metro Ethernet

The mechanisms available for implementing metro Ethernet services are varied depending on the equipment employed. From a price perspective, the best solution is to transport native optical Ethernet interfaces directly to one another. Unfortunately, this approach does not offer any of the interconnection, provisioning, or management capabilities that SPs need in their tool set. When an outage occurs, SPs need some way to diagnose and trouble-shoot the network.

Greenfield (brand new) installations could greatly leverage their fiber investment by using DWDM and transporting their Gigabit or 10 Gigabit traffic directly over a lambda. Several Cisco DWDM products can achieve this. These products include the ONS 15252/15201 and ONS 15540/15530 product lines. (DWDM is covered in Chapter 8.)

If the customer or SP is trying to leverage a legacy SONET infrastructure, the solution must include SONET framing, regardless of whether DWDM is being leveraged in the metro-politan rings. The E-series cards of the ONS 15454/ONS 15327 use two types of Ethernet circuits to transport this traffic. The two circuit types available are point to point and point to multipoint (shared packet ring, or SPR). The G-series cards only use P2P-based circuits because most service providers have chosen to do without the point-to-multipoint functionality that was in the E-series cards.

The E-series card can function either as point to point (Single-card EtherSwitch) or point to multipoint (Multicard EtherSwitch group), but not both at the same time. It is possible to re-create SPR functionality using the P2P nature of the Single-card EtherSwitch mode.

Figure 6-7 depicts a DWDM network where metro Ethernet services are being provided in a hybrid fashion through STS circuits (ONS 15454/15327) and through a lambda (ONS 15252/15201 or ONS 15540/15530). A VPN is also being transported through the high-touch services provided by the 7600 series OSR in the service POP. The difference between dedicated and shared is whether multiple sites in the same ring will be sharing those services (shared) on the same circuit or whether the circuits will be created in a P2P nature akin to leased-line services (dedicated).

Figure 6-7   *Metro Ethernet STS and Lambda-Based Solutions*

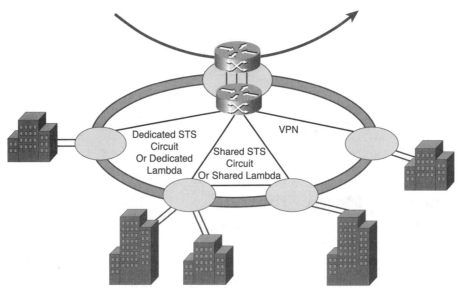

Ethernet over SONET

The core of an Ethernet over SONET (EoS) network is a SONET network with the SP's edge CPE supporting Ethernet. The ONS 15454 and ONS 15327 have direct Ethernet connections and do not require an external aggregation device to uplink the traffic in another format. SONET provides carrier-class management capabilities and sub-50-ms ring-restoration times. The time it takes for metro Ethernet circuits to recover depends on the card generation used. The E-series cards use Spanning Tree Protocol on the trunk (SONET) side that requires up to 50 seconds to converge from changes. The G-series cards do not have this limitation because they are point to point in nature.

In newer versions of the Cisco Transport Controller (CTC) software, you can turn off spanning tree on the trunk side of an E-series card's circuit creation. Prior versions did not offer this option. You can view the spanning-tree state by editing the circuit and viewing the color of the circuit connection in the graphic. The color green indicates that the circuit is in the Forwarding spanning-tree state. A circuit color of purple indicates that the circuit is in the Listening, Learning, Blocking, or Disabled spanning-tree state.

Ethernet over DWDM

Ethernet can be transported over DWDM systems in the form of a lambda. Although this allows great flexibility and ease of use, it could be conceived as wasting bandwidth. Each lambda in a DWDM system can be viewed as having a capacity of 2.5 or 10 Gbps, depending on fiber type. Gigabit Ethernet can currently be transported over a lambda with the ONS 15252/15201 and ONS 15540/ONS 15530 systems. The 15530 system also supports aggregation of traffic into 10-Gbps lambdas to fully use the capabilities of the fiber optics. Depending on application, SPs and users might not need to pay the premium of gaining every ounce of bandwidth capability from their fiber. The 15808 long-haul system can provide 160 wavelengths at OC-192 (10 Gbps) speeds, but it will probably never be found in a metropolitan environment due to the device's high cost. Failure recovery in this system is handled at the optical layer, where the systems can perform switches based on loss of light (LOL) or signal degrade (SD) functionality. The ONS 15252 DWDM system can provide failover in the order of a few milliseconds.

Figure 6-8 depicts a network with 10 ONS 15201 DWDM devices being used as optical add / drop multiplexers, which add / drop one wavelength to each customer. The head end device is the 15252 shelf, which can aggregate all 10 of these sites in a splitter-protected fashion. The 7600 OSR is being used to terminate the Gigabit Ethernet signals sourced from the individual site. DWDM is covered in detail in Chapter 8.

Figure 6-8   *Gigabit Ethernet over DWDM*

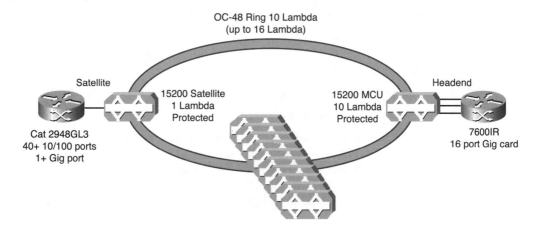

Point-to-Point Dedicated Metro Ethernet Circuits

Dedicated P2P circuits provide the easiest type of metro Ethernet service. Dedicated P2P circuits provide guaranteed bandwidth to the customer in the same fashion as leased lines, but Ethernet works as the transmission medium. Ethernet frames are transported in the synchronous payload envelope (SPE) of the STS circuit. A generic High-Level Data Link Control (HDLC) header is used as a delimiter within the SPE. Figure 6-9 depicts a dedicated Ethernet circuit traversing a SONET ring.

Figure 6-9 *Point-to-Point Circuit*

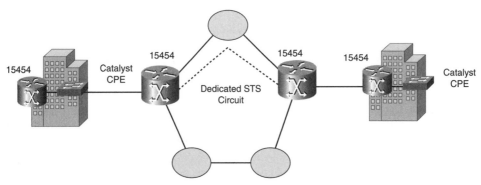

Ethernet services are available in speed variations of 10/100/1000/10,000 Mbps depending on the equipment used. The ONS 15454/15327 solution can provide circuit sizes from a minimum STS-1 (51.84 Mbps) to a maximum size of STS-24c (1.25 Gbps) on the G-series cards. A speed of 1.25 Gpbs is required to transport Gigabit Ethernet due to the 8B/10B encoding used by the technology. Although the technology operates at 1 Gbps, the signaling on the wire is 1.25 Gbps.

Point-to-Multipoint Shared/Shared Packet Ring

Shared point-to-multipoint circuits provide greater flexibility in connecting multiple sites in a metropolitan area. The point-to-multipoint implementation is available only on the E-series cards and accomplishes two goals:

- Multiple drops from one SPR circuit. Circuit sizes vary from STS-1 to STS-6c (311 Mbps).
- Connectivity between E-series cards in the ONS 15454 shelf.

As soon as an E-series card is activated for a Multicard EtherSwitch group operation, an STS-6c of bandwidth is consumed across the backplane STS switching matrix. This STS-6c consumed on the backplane allows E-series cards in the same ONS 15454 shelf to communicate with each other. The card now is left with STS-6c (311 Mbps) of provisioning capability for all ports. All the E-series cards support wire-speed transfer between ports in the same card. Because of these factors, it is very important to provision the E-series card in creative ways that maximize the bandwidth capabilities.

The requirement for cards to communicate with each other mandates the use of a Multicard EtherSwitch group. This requirement can be alleviated by provisioning the same customer on the same E-series card or on E-series cards in different shelves. A good application for the SPR design model is in distributed campus models that need connectivity with each other. The SPR design model enables you to avoid the number of P2P circuits that would be needed to create familiar connectivity. The number of P2P circuits required to build similar connectivity is exemplified by the formula $[n * (n - 1)]/2$, where n = one site.

If connecting 16 sites, a P2P implementation requires 120 circuits. An alternative approach to creating an SPR configuration without using Multicard EtherSwitch group is to put each card in a Single-card EtherSwitch and build P2P circuits between the ONS 15454/15327 devices. This configuration allows flexibility in the size of circuits that are implemented, and the full amount of STS-12c could now be provisioned from the card. One circuit goes into Blocking mode because spanning tree needs to provide for a loop-free topology throughout the network. Allow 50 seconds after the last circuit is provisioned for spanning tree to converge. Recall that the circuit in blocking appears purple when viewing the circuit in Edit mode through CTC. Figure 6-10 displays this configuration.

All CPE devices would have connectivity across the SONET UPSR ring. UPSR rings are recommended when spanning tree is implemented on the trunk side because the BLSR switching behavior can affect spanning-tree calculations and convergence takes longer. Figure 6-10 shows an example of an SPR using a Single-card EtherSwitch mode.

Figure 6-10 *Shared Packet Ring Using a Single-Card EtherSwitch Mode*

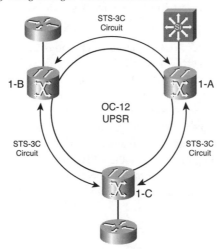

The circuit that goes into blocking follows the rules of spanning tree. According to the STP, the device with the lowest bridge priority becomes the root bridge. The least-efficient path to the root bridge goes into Blocking mode because it is not the best path to the root. With the default configuration, all bridge priorities are set at the same value of 32768. STP next looks for the lowest interface MAC address to make its selection. Because the MAC address is burned into the hardware of the device, this selection might not be the best solution. Fortunately, you can modify the bridge priority in the Etherbridge configuration. You can force a device to be the root bridge by lowering the bridge priority to a value lower than the default of 32676. You could create a hierarchy using a number sequence range in the different devices. Remember that you can shut off STP on the trunk/circuit side in newer versions of CTC.

Customers will most likely have multiple types of traffic (voice, video, and data) with different priority levels traversing the Ethernet cards. A Gigabit Ethernet interface on the ONS 15454/15327 could be limited to a circuit throughput of STS-1. In the event that congestion does occur, a QoS method should be implemented to ensure that high-priority traffic gets delivered first.

Quality of Service

QoS is an end-to-end solution, necessary because congestion can occur anywhere in the network. The QoS implementation on the E-series cards is very simple. Each port has two queues, and the Layer 2 Ethernet traffic marking (802.1p) determines the queue in which the traffic is placed. Software queues are used only when the ingress traffic exceeds the bandwidth available on the egress (circuit). It is easy to exceed the bandwidth available on the circuit when an almost 20-to-1 mismatch occurs between ingress (Gigabit Ethernet port or 1000 Mbps) and egress ports (STS-1 or 51.84 Mbps).

User Datagram Protocol (UDP)-based isochronous traffic such as voice and video do not react to dropped packets, whereas TCP-based traffic is far more resilient to drops. A World Wide Web page using TCP port 80 retransmits any lost packets. With dropped packets, it takes longer to get the page downloaded, but the page looks the same. Compare this with voice, where sections of speech can be missing entirely. Excessive voice packet loss could result in the loss of a call. SPs that guarantee different levels of service for different traffic types should deploy some of the feature-rich QoS options that are available on Cisco routing and switching platforms at the CPE device. Table 6-2 is a synopsis of some of the QoS tools that SPs might implement. QoS is covered in further detail in other Cisco Press titles, such as *IP Quality of Service* (ISBN: 1-57870-116-3), *Cisco Catalyst Quality of*

Service: Quality of Service in Campus Networks (ISBN: 1-58705-120-6), and *Cisco DQOS Exam Certification Guide* (IP Telephony Self-Study) (ISBN: 1-58720-058-9).

Table 6-2   *Implementing QoS*

Campus Access	Campus Distribution	MAN/WAN	Branch
Classification	Layer 3 policing	Multiple queues	Classification and trust boundaries on IP Phone, access layer switch, and router
Trust boundaries on IP Phone/access switches; speed and duplex settings	Multiple queues on all ports; priority queuing for VoIP	Class-based weighted fair queuing/low latency queuing	Multiple queues on IP Phone and all access ports
Multiple queues on IP Phone and access ports	WRED* within data queue for congestion management	Link fragmentation and interleaving	
		Bandwidth provisioning	
		Call admission control	

*WRED = Weighted Random Early Detection

The equipment used for CPE equipment should provide 802.3X flow control, classification, marking, and rate limiting. 802.3X flow control is used with the G-series cards only and is not implemented on the E-series cards. Flow-control operation in the G-series cards is asymmetrical. The cards send flow-control messages to the CPE device to throttle back the traffic sent when the first in, first out (FIFO) queues on the card start backing up. This operation allows the CPE device to start making intelligent packet discard, based on the priority settings of the packets. These intelligent discards ensure that priority traffic is transmitted during times of congestion.

Classification of traffic is the process by which a device inspects traffic. After the inspection has taken place (classification), the device should color (mark) the packets so that they can be identified as having a particular priority value. The closer classification can be done to the source of the traffic, the better.

When a packet is classified, it is marked with the appropriate values for the technology being implemented. Ethernet Layer 2 markings are the 802.1p class of service (CoS) bits in the 802.1Q header. An example of a device that marks packets is the Cisco IP Phone. Voice traffic is delay and drop sensitive. The IP Phones set the 802.1p and IP precedence values of frames and packets to 5 to ensure that lower-priority traffic is dropped before higher-priority voice traffic.

Committed access rate (CAR), or rate limiting, is essential for offerings at sub-STS-1 levels. Rate limiting is implemented on the CPE device to define which percentage of traffic to transmit, drop, or mark to a lower value depending on implementation. This implementation allows SPs to drop an STS-1 circuit to an MxU and define different rates at which each customer can transmit traffic. If 51.84 Mbps is available on the circuit, and the 10 customers in a building have bought 3 Mbps of bandwidth each, an excess of 21.84 Mbps bandwidth is available. SPs can leverage this by selling an extended service to their customers, guaranteeing 3 Mbps of bandwidth to each customer and allowing each customer to burst up to 51.84 Mbps (potentially). The first 3 Mbps from each customer could be marked with a priority value of 1, and traffic that exceeds the 3 Mbps rate would be marked with a value of 0. The traffic marked with a 0 would be dropped before the traffic marked with a value of 1 in times of congestion.

ONS 15454 and ONS 15327 Metro Ethernet Blades

The Ethernet cards on the Cisco ONS 15454 and ONS 15327 products remove the requirement for external Ethernet aggregation equipment to aggregate Ethernet traffic and trunk it with a SONET offering. These interfaces also allow coexistence of TDM traffic with packet-switched data traffic. Each card can operate as an independent switch to switch traffic at line rate between the ports. Depending on the switch mode implemented, the card might or might not be able to switch traffic between cards in the shelf. Recall that Multicard EtherSwitch group on the E-series cards can transmit traffic between cards on the same shelf.

The E-series cards support common features:

- 802.1d STP
- 802.1q VLANs
- 802.1p priority queuing

The following sections describe each feature in more detail.

Spanning Tree 802.1d Support

The ONS 15454 and ONS 15327 products can use the IEEE 802.1D Spanning Tree Protocol (STP) to detect and eliminate loops. Spanning tree was always turned on, on the trunk side in previous releases of CTC. STP can be turned off on the trunk side when provisioning a circuit with later 3.x releases. STP running on the customer side is an optional configuration, which is turned off by default. When STP detects multiple paths between any two network hosts, STP places those ports (client side) into the Blocking state. When STP detects multiple paths between any two circuits, STP places one of those circuits into blocking (trunk side). This single path eliminates possible bridge loops, which are not allowed in Ethernet operation. The ONS 15454 and 15327 can each support up to a maximum of eight spanning-tree instances. Circuits that support the same VLAN have the same spanning-tree instance. You can verify spanning-tree instances through the Ethernet settings of the Maintenance tab. STP can cause convergence time issues in SP networks because spanning tree's default timers take up to 50 seconds to converge. You can alter the bridge parameters that control STP, but the defaults are as follows:

- Maximum Age: 20 seconds
- Forward delay: 15 seconds (used twice)
- Bridge hello time: 2 seconds

You can modify these values to lower values for faster convergence, but take care before doing so. You should acquire detailed STP knowledge before attempting to modify STP timers, and you set the same spanning-tree settings on every device in the network. Rapid Spanning Tree Protocol (RSTP) is one of the latest implementations of STP; it lowers convergence time to less than 5 seconds (as low as 1 second in certain scenarios). RSTP is not available on the ONS 15454 or ONS 15327, but you can find it on other platforms such as the Catalyst 6500 and 7600 series switches.

802.1Q VLAN Support

All traffic entering the E-series cards must belong to a VLAN. If the incoming traffic is not explicitly assigned to a VLAN, the device characterizes it by default as belonging to VLAN 1 as untagged traffic. When using 802.1Q trunking to transport traffic from multiple VLANs, set the native VLAN (VLAN 1 by default) to untagged and all other VLANs to tagged. The E-series cards support a maximum of 509 VLANs per ONS 15454/15327 shelf. When traffic arrives at the customer port, it must be transported across the circuit. The circuit must also be configured to transport those same VLANs; otherwise, the traffic will not traverse the circuit. At the far side of the network, the 802.1Q header is stripped at the destination ONS 15454/15327 port unless the port is an 802.1Q trunk port. The destination customer port must be configured for all VLANs or operation will fail.

VLANs should not be re-used between multiple customers or traffic will leak between the customers. This is a security concern when deploying this solution. Pay careful attention to the implementation of VLAN 1. By default, most switches and routers use VLAN 1 as the untagged VLAN to communicate the tagged VLANs traversing the 802.1Q trunk. If multiple customers use this VLAN, the VLAN traffic bleeds together. If the untagged VLAN is changed at the carrier equipment, it must also be changed in both CPE devices at each end of the connection. The G-series cards do not use VLANs. All connections are point to point with the G-series cards. Everything is transported.

802.1p Priority Queuing Support

The E-series metro Ethernet cards perform queuing operations on traffic based on the 802.1p value. 802.1p is the first 3 bits of the 802.1Q header and provides for 8 different priority values. The E-series cards contain two queues that must be mapped in some way to the eight available priority values. The cards have default mappings, but you can modify these depending on the service design. The default operation maps priority values 0 through 3 to the low-priority queue and 4 through 7 to the high-priority queue. The two queues are serviced in a weighted priority method that allocates 70 percent of the time to servicing the high-priority queue and 30 percent of the time to servicing the low-priority queue.

Order of circuit provisioning is important with the E-series cards. When either E-series card is in Single-card mode, 12 STSs are available. Larger circuit sizes should always be provisioned before smaller circuit sizes; otherwise, larger circuit creation will not be available. If one STS-1 circuit was provisioned from the card, the card has 11 more STS-1s available. At this time, an STS-6c circuit could not be created. An example of circuit size matrices that could be used when the E-series card is set to Single-card mode is as follows:

- $1 \times$ STS-12c circuit
- $2 \times$ STS-6c circuits
- $1 \times$ STS-6c circuit, then (2) $\times$ STS-3c circuits
- $1 \times$ STS-6c circuit, then (6) $\times$ STS-1 circuits
- $4 \times$ STS-3c circuits
- $1 \times$ STS-3c circuit, then (9) $\times$ STS-1 circuits
- $2 \times$ STS-3c circuits, then (6) $\times$ STS-1
- $12 \times$ STS-1

The same provisioning philosophy applies to the E- cards when they are set to Multicard mode, but the aggregate bandwidth is reduced to STS-6. Circuit size matrices for the Multicard mode are as follows:

- 1 STS-6c circuit
- 2 STS-3c circuits
- 1 STS-3c circuit, then $3 \times$ STS-1 circuits
- $6 \times$ STS-1

ONS 15454 E-Series Metro Ethernet Cards

The ONS 15454 E-series cards are available in different varieties, as follows:

- **E100T-12/E100T-12-G card**—This card provides 12 autonegotiating 10/100 Ethernet ports with RJ-45 connectors. The E100T-12-G card is an updated version of the same card that is compatible with the XC10G cross-connect card that is needed for the G-series metro Ethernet cards, OC-192 cards, OC-48 Any Slot Cards, and so forth. The E100TG-12 card is *not* compatible with the previous XC or XC-VT cross-connect cards.

- **E1000-2/E1000-2-G card**—This card provides two slots in which gigabit interface controllers (GBICs) can be placed. This card supports the 1000BASE-SX (short-haul) and 1000BASE-LX (long-haul) GBICs. The 1000BASE-SX is an LED-based, 850-nm interface in which multimode fiber (MMF) is used for connectivity. The 1000BASE-SX can be run up to 550 meters from source to destination. The 1000BASE-LX is a laser-based, 1310-nm optical interface in which single-mode fiber (SMF) is used for connectivity. The 1000BASE-LX can be run up to 10 kilometers from source to destination. The distances these cards can traverse are limited by the core diameter of the fiber used. Table 6-3 contains the different Cisco GBIC specifications.

Table 6-3   *CISCO GBIC Specifications*

GBIC Port	Wavelength (nm)	Fiber Type	Core Size (micron)	Modal Bandwidth (MHz/km)	Maximum Cable Distance
SX	850	MMF[1]	62.5	160	722 ft (220 m)
			62.5	200	902 ft (275 m)
			50.0	400	1640 ft (500 m)
			50.0	500	1804 ft (550 m)
LX/LH	1300	MMF[2] and SMF	62.5	500	1804 ft (550 m)
			50.0	400	1804 ft (550 m)
			50.0	500	1804 ft (550 m)
			9/10	—	6.2 miles (10 km)
ZX[3]	1550	SMF	9/10	—	43.5 miles (70 km)
		SMF[4]	8	—	62.1 miles (100 km)

[1]MMF only.

[2]A mode-conditioning patch cord (Product Number CAB-GELX-625 or equivalent) is required. When using the GBIC-LX/LH with 62.5-micron diameter MMF, you must install a mode-conditioning patch cord between the GBIC and the MMF cable on both the transmit and the receive ends of the link when link distances are greater than 984 ft (300 m).

We do not recommend using the GBIC-LX/LH and MMF with a patch cord for very short link distances (tens of meters). The result could be an elevated bit error rate (BER).

[3]You can have a maximum of 12 1000BASE-ZX GBICs per system to comply with EN55022 Class B and 24 1000BASE-ZX GBICs per system to comply with FCC Class A regulations.

[4]Dispersion-shifted single-mode fiber-optic cable.

ONS 15327 E-Series Metro Ethernet Card

The ONS 15327 has one E-series metro Ethernet card. The E10/100T-4 card provides four RJ-45 autonegotiating, 10/100BASE-T Ethernet interfaces. Each interface can be configured for full (100-Mbps synchronous bandwidth) or half duplex (asynchronous).

E-series in both the ONS 15454 and ONS 15327 support the same features. The only difference is the amount of port density on the card.

The ONS 15327 has four multispeed slots in which you can place the E-series card. Two of these slots are normally used for OC-*n* cards to tie the ONS 15327 into the SONET ring. The remaining multispeed slots can be used for Ethernet cards, which can accept the customer LAN traffic for transport. When more than one E10/100-4 cards is installed in a 15327, each card can be an independent EtherSwitch, or multiple cards can be stitched together to act as a Multicard EtherSwitch group. This card follows the same provisioning rules as the other E-series card on the ONS 15454. The front of this card contains 400 Mbps of available bandwidth for customer connectivity, but there is an aggregate trunk side capacity of STS-12c (622 Mbps) if this card is used in Single-card EtherSwitch mode.

ONS 15454 G-Series Card

The G1000-4 metro Ethernet card provides four GBIC slots. The G-series card represents the second generation of Cisco metropolitan Ethernet cards on the ONS 15454. This card provides Single-card EtherSwitch-type operation through P2P circuits (although it is not explicitly named as such). The G1000-4 card is supported by CTC 3.2 and supports the following GBICs:

- 1000BASE-SX
- 1000BASE-LX
- 1000BASE-ZX

The 1000BASE-SX and 1000BASE-LX GBICs were covered in the E1000-2 card section. The 1000BASE-ZX GBIC provides one 1550-nm laser and can travel up to 100 kilometers across SMF with a core diameter of 8 nm.

Up to STS-48c (2.5 Gb) can be provisioned from the G1000-4 card. This provides enough bandwidth for two wire-speed Gigabit Ethernet interfaces. The G1000-4 supports up to four circuits and has full TL1 provisioning capabilities. The E-series cards do not have TL1 provisioning capabilities. The G-series card supports Ethernet jumbo frames up to a size of 10,000 bytes (providing for giants). This provides a more adaptable solution than the 1522-byte limitation of the E-series card.

Flow Control 802.3X Support   The G1000-4 supports 802.3X flow control and uses frame buffering to help manage congestion. The G1000-4 card does not respond to pause frames received from client devices. When the buffers fill up during times of congestion, the G1000-4 sends an 802.3X-compliant pause frame to the source device requesting that it throttle back the transmit rate for a period of time. If a G1000-4 port has an STS-12c circuit built from it, the G1000-4 card maintains the 622-Mbps rate through flow-control messages even though the card has the capability to run at Gigabit Ethernet speed. The card is considered an asymmetric flow-control device because it transmits pause frames but does not respond to them. 802.3X is only supported with autonegotiation. If the port is set to a static port speed, 802.3X operation is not enabled. Figure 6-11 shows an example of 802.3X flow-control operation.

Figure 6-11   *802.3X Flow-Control Operation*

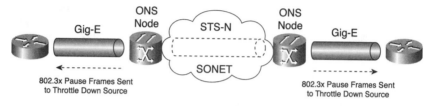

Ethernet Link Integrity and EtherChannel Support   The G1000-4 supports end-to-end Ethernet link integrity: If any part of the end-to-end path fails, the entire path fails. The path failure is accomplished by the ONS 15454 turning off the transmit lasers at each customer port interface, allowing the external devices to see a loss of carrier. This ensures the correct operation of Layer 2 and Layer 3 protocols on the attached Ethernet devices at each end. A figure of the link-integrity mechanism shutting off the transmit lasers because of a loss of signal (LOS) appears in Figure 6-12.

Figure 6-12   *Link-Integrity Operation*

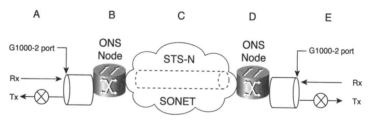

The G1000-4 can also support end devices that are running IEEE802.3ad Link Aggregation and Control Protocol (LACP) and the Cisco proprietary Gigabit EtherChannel (GEC). Although the G1000-4 does not run the link-aggregation systems or GEC directly, its end-to-end Ethernet link-integrity feature allows a circuit to emulate a direct Ethernet link. Therefore, all flavors of Layer 2 and Layer 3 rerouting, and link-aggregation systems and GEC, work correctly.

Two Ethernet devices running GEC connect through the G1000-4 cards to an ONS 15454 network; the ONS 15454 SONET side network is completely transparent to the GEC devices. The ONS 15454 supports any combination of G1000-4 parallel circuit sizes that might be needed.

GEC provides redundancy and protection for attached Ethernet equipment by bundling multiple links and making them active. The bundling of the parallel data links allows more aggregated bandwidth. STP operates as if the bundled links were one link and permits GEC to use these multiple parallel paths. Without the use of GEC, STP permits only a single nonblocked path. GEC can also provide G1000-4 card-level protection or redundancy. It can support a group of ports on different cards (or different nodes) so that if one port or card has a failure, traffic is rerouted over the other port/card. Figure 6-13 shows how two CPE devices could have a 2-Gbps bidirectional connection using GEC/IEEE802.3ad.

Figure 6-13 *Gigabit EtherChannel Support over the G-Series Metro Ethernet Cards*

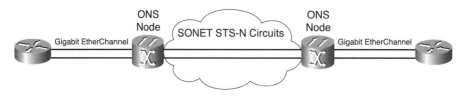

ONS 15327 G-Series Card

The G1000-2 is a G-series metro Ethernet card that fits in any of the multiservice slots of the ONS 15327. The G1000-2 provides two small form-factor pluggable (SFP) optics in which Gigabit Ethernet optics are placed. The SFP transceivers are available in two varieties: a 1000BASE-SX (short wavelength) 850-nm MMF interface, and a 1000BASE-LX (long wavelength) 1310-nm SMF interface. SFP optics take up less real estate on the board because of their small size. The fiber connectors supported on the G1000-2 board are the small LC-style connectors.

The G1000-2 supports a maximum bandwidth provisioning of STS-48c. Both of the interfaces on the card support wire-speed Gigabit Ethernet (STS-24c). Functionality of the card is similar to the ONS 15454 G-series card. The G1000-2 supports the following features:

- 802.3X flow control
- Link-integrity support
- P2P circuits
- TL1 provisioning
- Gigabit EtherChannel/IEEE 802.3ad link aggregation
- Ethernet performance monitoring
- Jumbo frames up to 10,000 bytes

Summary

At this point, you should understand the market drivers, design philosophies, implementation details, and hardware used for metropolitan Ethernet services.

Review Questions

1 ONS equipment can easily support metro Ethernet to the customer site through the addition of which of the following?

   A SONET technology

   B Transfigured LAN services

   C Ethernet service cards

   D DWDM technology

2 Customers have the advantage of advanced traffic shaping and flexibility in logical network design because of the native _____ and _____ support in Ethernet.

   A QoS, VLAN

   B VLAN, MPLS VPN

   C Point-to-point, Gigabit Ethernet

   D QoS, Gigabit Ethernet

3 Metro Ethernet allows the service provider to offer high-speed _____ connectivity in a native _____ technology that the customer understands and already supports.

   A MAN/WAN, LAN

   B LAN, MAN/WAN

   C SONET, LAN

   D SONET, MAN/WAN

4 CPE equipment would most likely be found where?

   A Central office

   B Customer's site

   C Point of presence

   D User's desktop

5 Traffic carried between POPs will most likely run at which speeds? (Choose all that apply.)

   A DS1

   B DS3

   C 10 Mbps

   D OC-3

   E OC-12

   F OC-48

   G OC-192

6 Which three IEEE standards do the E-series ONS Ethernet cards support?

   A 802.1h

   B 802.1c

   C 802.1d

   D 802.1p

   E 802.1r

   F 802.1Q

7 Which two IEEE standards do the G-series metro Ethernet cards support?

   A 802.3x

   B 802.1d

   C 802.1r

   D 802.1p

   E 802.1q

8 What is the amount of bandwidth that can be provisioned if the E-series Ethernet card is configured as a Single-card EtherSwitch?

   A STS-3c

   B STS-6c

   C STS-12c

   D STS-48c

9 What amount of bandwidth can be provisioned if the E-series Ethernet card is configured as a Multicard EtherSwitch group?

   A STS-3c

   B STS-6c

   C STS-12c

   D STS-48c

10 How many VLANs can be configured on the E-series cards?

   A 500

   B 590

   C 509

   D 8192

11 What does 802.1d specify?

   A VLANs

   B Flow control

   C Priority queuing

   D Spanning tree

12 The maximum bandwidth available on a G-series metro Ethernet card is which of the following:

   A STS-6c

   B STS-12c

   C STS-24c

   D STS-48c

13 What is the maximum bandwidth available on an ONS 15454 on a G-series card configured as a Multicard EtherSwitch Group?

   A OC-48c

   B OC-24c

   C OC-12c

   D OC-3c

   E You cannot configure the card with this configuration.

14 The bandwidth provisioning capabilities of the G1000-2 metro Ethernet card is which of the following?

   A   STS-12c

   B   STS-24c

   C   STS-36c

   D   STS-48c

15 Which terminology is associated with the Multicard EtherSwitch group operation? (Choose all that apply.)

   A   Flow control

   B   Shared packet ring

   C   Point to point

   D   Shared bandwidth

This chapter covers configuring Ethernet over SONET circuits.

CHAPTER 7

Configuring Metro Ethernet

This chapter explains how to configure the Cisco ONS 15454 and ONS 15327 optical platforms to support metro Ethernet cards and applications. To complete this chapter, you need to be familiar with the Cisco Transport Controller (CTC) software, which was introduced in Chapter 5, "Configuring ONS 15454 and ONS 15327."

Configuring Metro Ethernet Circuits

The configuration of metro Ethernet circuits assumes that a SONET unidirectional path switched ring (UPSR) has already been built. It is important to provision metro Ethernet circuits only over a UPSR ring if using Multicard EtherSwitch group mode on the E-series cards. Figure 7-1 shows a shared packet ring (SPR) created with point-to-point circuits over an OC-12 (optical carrier) UPSR using Single-card EtherSwitch mode. The card used is the E100T-12 E-series. Configuration steps in this chapter refer to this figure.

Figure 7-1 *Metro Ethernet Shared Packet Ring Configuration*

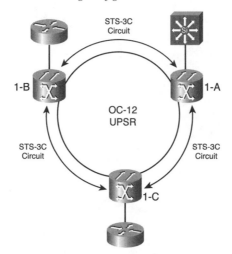

Step 1: Card Mode Configuration

You can provision the E-series cards through CTC or Cisco Transport Manager (CTM). CTM is the Cisco optical network management platform and is not covered further in this book. You can provision the G-series cards through CTC, CTM, or TL1 commands. The default card mode of the E-series cards for the ONS 15454 and ONS 15327 is Multicard EtherSwitch group. CTC enables you to verify or change this. You must set the mode of the card correctly before circuit creation. You must delete and re-create all circuits to change the mode. The G-series cards do not have this configurable option because they operate only in a Layer 1 point-to-point mode that is unique from that of the E-series cards. Most recently, the Cisco ML-series cards, which support routing and Multiprotocol Label Switching (MPLS) capabilities through the Cisco IOS, have been introduced. This book does not cover the ML-series cards, but information on them is available through Cisco.com.

To set the card mode on any E-series metro Ethernet card, follow these steps:

Step 1   From Node view of CTC, double-click the **Ethernet card** that will be the source of the circuit.

Step 2   Click the **Provisioning** tab.

Step 3   Click the **Card** subtab. Figure 7-2 shows what displays.

Figure 7-2   *Card Mode Configuration Screen*

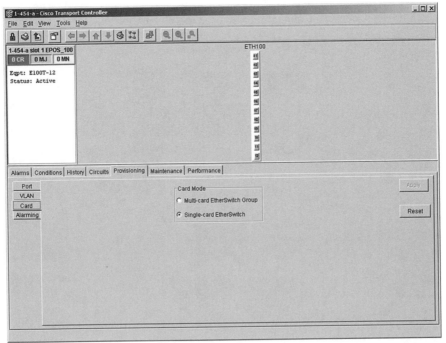

Step 4   Choose **Single-card EtherSwitch mode**. Multicard EtherSwitch group mode will not be used because only one card is used per circuit Documentation that points to setting the card mode to Multicard EtherSwitch group for this operation is not wrong, but it is a different way of implementing the solution. Most implementations follow the guidelines in this book.

Step 5   Click **Apply**. The Apply button should be available only when a change has been instituted. If no changes were made to the screen, the Apply button is grayed out and unavailable. The Apply button commits the changes to the database stored on the TCC (Timing and Communications Card) or TCC+ cards.

Step 6   Navigate to the other ONS 15454 devices on the ring and repeat Steps 1 through 5 on the metro Ethernet cards. All sources and destinations should reflect the same Single-card EtherSwitch operation; otherwise, the card will not be available during circuit creation. Recall that the default operation is Multicard EtherSwitch group.

Step 2: Configuring Port Options

After you have configured the card for the appropriate mode, the next step is to configure the customer side ports. To do so, follow these steps:

Step 1   While in Card view of the E100T-12, click the **Provisioning** tab. Click the **Port** subtab. Figure 7-3 shows this view.

Figure 7-3   *Configuring the Port*

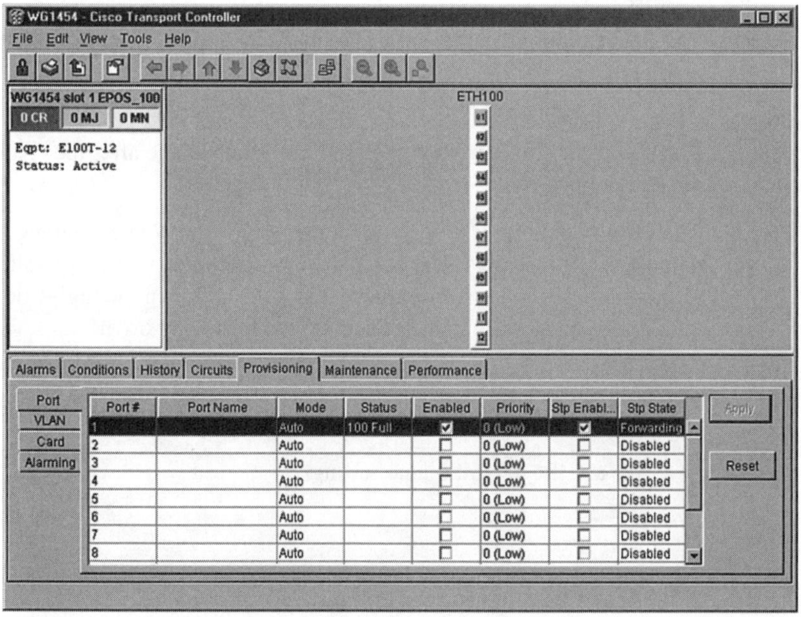

Step 2 Choose the appropriate mode for the Ethernet port in which the customer premises equipment (CPE) device connects. The port is statically configured at 100 Mbps, full duplex, as long as the destination devices support this configuration, and thus alleviates any issues that could come about as a result of autonegotiation. Other valid choices for the E100T-12 card are as follows:

- **Auto**—Configures the port to automatically negotiate speed and duplex settings with the CPE device

- **10 Half**—10 Mbps, half-duplex operation

- **10 Full**—10 Mbps, full-duplex operation

- **100 Half**—100 Mbps, half-duplex operation

- **100 Full**—100 Mbps, full-duplex operation

For the two-port gigabit card, valid choices are 1000 Full and Auto. If 1000 Full is selected, flow control is disabled. Choosing Auto mode enables the port to autonegotiate flow control. The port handshakes with the connected network device to determine whether that device supports flow control. The E-series cards can respond only to 802.3x "pause" frames; it cannot initiate 802.3x pause frames. Flow control on the E-series card operates in the exact opposite manner to flow control on the G-series card. The E-series card is a Layer 2 implementation, whereas the G-series card is a Layer 1 implementation. The E-series card has to support responding to pause frames based on the required standards for autonegotiation.

Step 3 The Status column displays the autonegotiated speed if the Auto option is selected.

Step 4 Click the **Enabled** check box to put the customer-side port in service. The port turns from gray to green to reflect the change after the changes are committed.

Step 5 The Priority column is a drop-down menu in which you can change the default priority settings. The default priority settings put 802.1p values 0 through 3 in the low-priority queue and 4 through 7 in the high-priority queue. The queues are then serviced in the following order:

- **High**—70 percent traffic

- **Low**—30 percent traffic

- We use the default queue settings.

Step 6   The STP Enabled column is used to turn on the Spanning Tree Protocol (STP) on the customer-side interface. Most service providers (SPs) do not run STP with their customers and choose to leave this functionality disabled. This is not an option on the G-series cards.

Step 7   The STP State column displays the active state of the Spanning Tree Protocol instance that is running toward the customer. Although this field is not user configurable, it should display one of the following states:

- **Disabled**—STP is not running on the customer-side port.

- **Listening**—STP is running on the customer-side port and is listening for bridge protocol data units (BPDUs) from other switch devices.

- **Learning**—STP is running on the customer-side port, BPDUs have been received, and the spanning-tree topology is converging.

- **Forwarding**—STP is running on the customer-side port, the topology has been learned, and the port is in the Forwarding state. At this point, there should be a loop-free topology. Any links that could result in bridging loops have been placed in the Blocking state.

- **Blocking**—STP is running on the customer-side port, the topology has been learned, and the port is in the Blocking state because there is a redundant path in the network and the other path has a lower path cost to the root bridge device.

Step 8   Click **Apply** to commit the changes to the database of the TCC or TCC+.

Step 9   Repeat the previous steps for all other cards that will be used in the SPR.

Step 3: VLAN Configuration

In the same way that you could not be assigned to a task that did not exist, a VLAN must be created before it can be assigned. Up to 509 user-provisioned VLANs can exist on one ONS 15454/15327 network. VLANs can be added to the system during circuit creation or as a standalone procedure. To configure VLANs as part of a standalone process, choose the **Manage VLANs** option from the **Tools** menu in CTC. From the All VLANs dialog box, click the **Create** button. (See Figure 7-4.) You can use the VLAN Creation dialog box to configure VLANs and VLAN names. VLANs use the last 12 bits of the 802.1Q header and are assignable in the range of 2 through 4093. As soon as a VLAN is configured, it is propagated to all other nodes on the ring and to any other nodes listed in the CTC Login Node group. Although the Delete button enables you to delete VLANs, it is hard to completely remove a VLAN from the ring without scheduling downtime. As soon as the VLAN is deleted from the node, in milliseconds the node relearns that VLAN information from the neighboring node.

Login Node groups are used to manage disparate ONS 15454 and ONS 15327 devices that are not connected in a ring. These devices might be standalone devices in different geographies. Login Node groups enable you to manage all of these devices from one network view.

Figure 7-4   *VLAN Creation*

Step 1   From Card view of the metro Ethernet card, click the **Provisioning** tab. Click the **VLAN** subtab. VLANs that have been created are available to be assigned to ports. Figure 7-5 shows the screen that displays. Notice that the VLANs that have been created are on the bottom left of the screen capture. Their assignment is listed under the respective port. By default, only VLAN 1 is listed under each port.

Figure 7-5   *VLAN Port Assignment*

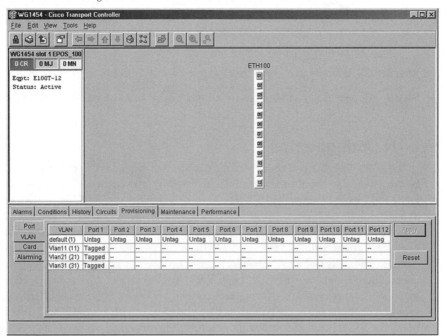

You have three options for any port, VLAN intersection:

- — -- Indicates that the port does not participate in the VLAN listed in the left VLAN column.

- — **Untag** Indicates that the device attached to the port does not support 802.1Q tagging or that the device is using that VLAN as the 802.1Q native VLAN. The 802.1Q native VLAN is used to transfer information regarding the other VLANs trunked by 802.1Q. Only one VLAN can be set to Untag on one port. That one port could trunk the other 508 configurable VLANs on the platform.

- — **Tagged** Indicates that the attached device supports 802.1Q trunking and is trunking the particular VLAN that is set to Tagged. If one VLAN is trunked (set to Tagged) on a particular port, that port must also have Untag defined for the VLAN that is configured as the native VLAN on the attached device. This native VLAN is VLAN 1 by default on most router and switch platforms. Recall that only one customer can use VLAN 1; otherwise, traffic bleeds. This configuration must be communicated to the customer.

Accidentally setting one side of the circuit to Untag and the other side to Tagged results in a mismatch in which traffic cannot be passed. One side is passing an 802.1Q packet, while the other is removing the 802.1Q tag. The customer device on the other side of the circuit is expecting an 802.1Q tagged packet and drops any other traffic.

Step 2 Figure 7-5 shows that VLANs 11, 21, and 31 have been configured to be tagged on port 1. All three VLANs have been set to Tagged because port 1 will be trunking all three VLANs over the 802.1Q traffic that will be coming from the traffic. The actual trunking is being done at the customer site, but the metro Ethernet card must be configured to pass this information because the E-series card is an implementation of Layer 2 functionality in the OSI model. VLAN 1 is used as the native (management) VLAN.

Step 3 Click **Apply** to commit changes to the TCC/TCC+ database.

Now that you have created the VLANs and properly configured the customer-side ports, you must configure the trunk-side circuit that will transport the Ethernet services.

Step 4: Creating Circuits

Circuit creation can take place from one of three of the following views:

- Network view
- Node view
- Card view

The circuit provisioning procedure is similar regardless of the type of endpoints. Minor differences include circuit speed, VLANs (E-series card circuits), and SF/SD threshold values. The E-series metro Ethernet cards are the only cards that have a step for VLANs in the provisioning process. It is important to provision only one circuit at a time. If two users are provisioning circuits at the same time, circuit provisioning will probably fail because of the way the database handles provisioning. Put a process in place to ensure that multiple users never provision at the same time.

Step 1 From Network, Node, or Card view, click the **Circuits tab**. Click the **Create** button. Figure 7-6 shows the Circuit Creation: Circuit Attributes dialog box.

Figure 7-6 *Circuit Creation: Circuit Attributes Dialog Box*

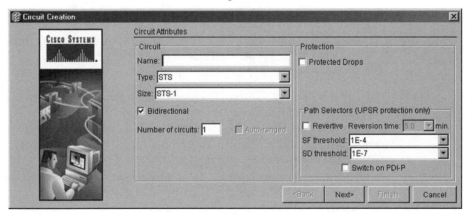

Step 2 Complete the fields in the Circuit Creation: Circuit Attributes dialog box. Table 7-1 lists the fields and their values.

Table 7-1 *Options for the First Screen in Ethernet Circuit Creation*

Field	Value
Name	Name of the circuit to create. The circuit name can be alphanumeric and up to 32 characters. It is a good idea to use the circuit ID and customer parameters when provisioning circuits for future alarms and troubleshooting.

Table 7-1   *Options for the First Screen in Ethernet Circuit Creation (Continued)*

Field	Value
Type	STS, VT, or VT Tunnel is available. Only the STS option is used for metro Ethernet. VT and VT Tunnels are used to map DS1s into the SPE of the STS frame.
Size	The size of Ethernet circuits varies from an STS-1 through STS-24c in the case of wire-speed Gigabit Ethernet with the G-series card. Valid options here are STS-1, STS-3c, STS-6c, or STS-12c. Figure 7-6 displays this provisioning window. Recall that larger circuit sizes must be provisioned first. Because this is the first circuit provisioned, if this circuit is provisioned as STS-3c, the remaining capacity of the STS-12 will be available as STS-3c or STS-1.
Bidirectional	Check this box if you want to create a two-way circuit; uncheck it to create a one-way circuit.
Number of Circuits	Type the number of circuits you want to create. If you enter more than 1, CTC returns to the Circuit Source dialog box after you create each circuit until you finish creating the number of circuits you specified here.
Protected Drops	If this box is checked, CTC only displays cards residing in 1:1, 1:N, or 1+1 protection for circuit source and destination selections. Protection schemes are not used for metro Ethernet cards.
Revertive	Check this box if you want working traffic to revert back to its original path when the conditions that diverted it to the protected path are repaired. If it is not checked, working traffic remains on the protected path.
SF Threshold	In this field, you can set the path level bit error rate (BER) for determining a signal fail (SF) condition. IE-4 indicates that $1 \times 10^{\wedge-4\,(.0001)}$ is the BER. The BER is measured as the number of bit errors divided by the number of valid bits transmitted. This IE-4 rate indicates that there is 1 error for every 1000 bits sent.
SD Threshold	In this field, you can set the path level BER for determining a signal degrade (SD) condition. IE-7 indicates that there is 1 bit error for every 100,000 bits transmitted.
Switch on PDI-P	Choose this option only when you want traffic to switch when the path layer receives a Payload Defect Indicator (PDI-P).

Step 3   Click **Next** to proceed to the next step in the provisioning process. The
Next button is grayed out until something is changed on the Circuit
Creation: Circuit Attributes dialog box. The Circuit Creation: Circuit
Source dialog box 1 appears, as displayed in Figure 7-7.

Figure 7-7   *Circuit Creation: Circuit Source Dialog Box*

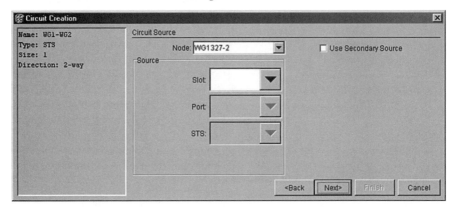

Step 4   Complete the fields in the Circuit Creation: Circuit Source dialog box
using Table 7-2 as your guide.

Table 7-2   *Source Screen Entry Options*

Field	Value
Node	Choose the name of the node that is the source of the circuit. All nodes within your network should appear in the Node drop-down menu.
Slot	This option varies based on the type of circuit you are creating. In our design, the cards are placed in Single-card mode, so the option available is EPoS and the Slot Number. If the card had been placed in Multicard EtherSwitch group mode, Ethergroup would be an option.
Port	Not available for metro Ethernet circuits.
STS	Not available for metro Ethernet circuits.
Use Secondary Source	This field is not used with Ethernet circuits. It will be grayed out after the card type has been selected. This check box is used only if you need to create a UPSR bridge/selector circuit entry point in a multivendor UPSR ring.

Step 5   Click the **Next** button. The Circuit Creation: Circuit Destination dialog box opens, as seen in Figure 7-8. Complete the fields using Table 7-3 as a guide.

Figure 7-8   *Circuit Creation: Circuit Destination Dialog Box*

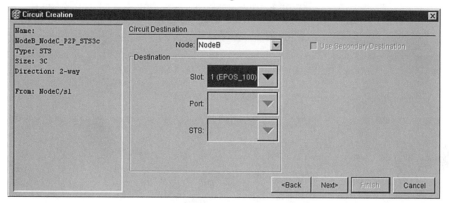

Table 7-3   *Circuit Source and Destination Field Descriptions*

Field	Value
Node	Choose the destination node of the circuit from the drop-down menu. This menu contains all the nodes in the ring.
Slot	This option is the EPoS card (Single-card mode) or Ethergroup (Multicard mode) in which the circuit will terminate. If Ethergroup was selected as the source, the destination is Ethergroup as well. Because Single-card mode is being used for this design, the destination EPoS card slot will be chosen. If CTC does not have the destination available, there could be visibility issues due to DCC termination failures with the other nodes.
Port	Not used for metro Ethernet circuits.
STS	Not used for metro Ethernet circuits.
Use Secondary Source	This field will not be used with Ethernet circuits. It is grayed out after the card type has been selected. This is used only if you need to create a UPSR bridge/selector circuit destination point in a multivendor UPSR ring.

Step 6   Click the **Next** button. The Circuit Creation: Circuit VLAN Selection dialog box will display. All previously created VLANs display on the Available VLANs side of the screen. If the VLAN desired has not been previously created, you can click the **New VLAN** button to launch the VLAN Creation dialog box. In this scenario, you don't have to do so

because VLANs 11, 21, and 31 have been previously created and are available. Figure 7-9 displays the Circuit VLAN Selection dialog box. Notice that the VLANs are available, but not selected.

Step 7 Choose the VLANs that should be transported over the circuit, and then click the >> button to transfer the VLANs onto the circuit. When the VLANs are on the Circuit VLANs side, you can use the << button to remove them. If no VLANs are selected at this point, they might be added later, but VLAN 1 will be selected by default. This might be a problem if another customer is already using VLAN 1.

Figure 7-9 *Circuit Creation: Circuit VLAN Selection Dialog Box*

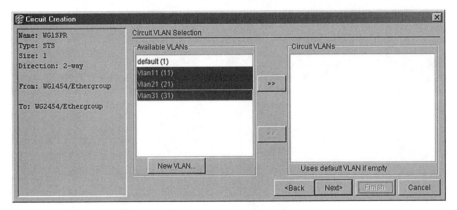

Step 8 Click the **Next** button. The Circuit Creation: Circuit Routing Preferences dialog box will appear. Figure 7-10 shows this screen, and Table 7-4 describes its different fields.

Figure 7-10 *Circuit Creation: Circuit Routing Preferences Dialog Box*

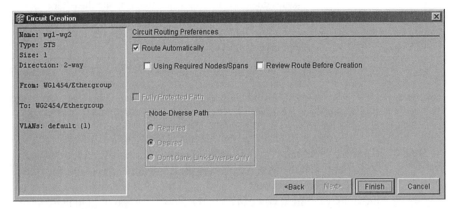

Table 7-4   *Circuit Routing Preferences Field Descriptions*

Field	Value
Route Automatically	When this field is chosen, CTC uses the Open Shortest Path First (OSPF) process 0 running on the trunk side of the ONS 15454/15327 device to route the circuit across the path with the best metric. After this option is selected, the Using Required Nodes/Spans and Review Route Before Creation check boxes become available. If manually cross-connecting the circuit, the Route Automatically check box would be de-selected. This check box would be deselected for this series of steps.
Using Required Nodes/Spans	If selected, you can specify nodes and spans to include or exclude in the CTC-generated circuit route.
Review Route Before Creation	If selected, you can review and edit the circuit route before the circuit is created.
Fully Protected Path	If selected, CTC ensures that the circuit is fully protected. If the circuit must pass across unprotected links, CTC creates a primary and alternate circuit route (virtual UPSR) based on the following three node diversity specifications: Required, Desired, and Don't Care. SPR E-series metro Ethernet circuits should be built as unprotected circuits because STP provides the protection over the redundant path.
Required	Ensures that the primary and alternate paths of the complete circuit path are nodally diverse.
Desired	Specifies that node diversity should be attempted; if node diversity is not possible, however, this field creates link-diverse paths for the circuit path.
Don't Care	Specifies that only link-diverse primary and alternate paths of the complete circuit path are needed. The paths might be node diverse, but CTC does not check for node diversity.

Step 9   Click the **Next** button to proceed to the Circuit Creation: Route Review and Edit dialog box. The Next button is not available until the Route Automatically check box has been deselected. Figure 7-11 displays this dialog box.

Figure 7-11 *Circuit Creation: Route Review and Edit Dialog Box*

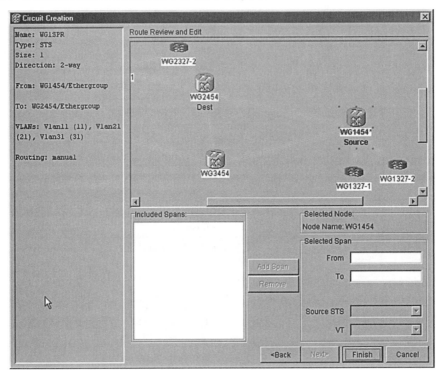

Step 10 Confirm that the information about the point-to-point circuit is correct. You can find this information on the left side of the screen. Verify:

— Circuit name

— Circuit type

— Circuit size

— Direction

— Source

— Destination

— VLAN selection

— Routing preferences

Step 11 Click the source of the circuit and verify that the node icon is surrounded by eight dots, as displayed in Figure 7-11. The dots are used to verify that the node is selected. Clicking the background of the dialog box clears the current selection. Click the green arrow between this node and the destination of the circuit (WG2454 in the case of the graphic). The arrow represents the span. Now that the span has been selected, it displays in the Selected Span section at the bottom of the screen. The Source STS is automatically selected, but you can change this in the Source STS section of the screen. The Add Span button would now be selected to move the Selected Span into the Included Spans section of the screen.

You would complete this process another two times (for the other two nodes in the network) to build the SPR design that appears in Figure 7-1.

You could make any changes to the circuit by clicking the Back button to make the appropriate changes at the necessary screen. If no changes are necessary, click the Finish button to commit the circuit creation to the TCC/TCC+ database.

The circuit now displays in the Circuits tab.

Step 5: Circuit Verification

All circuits that have been made across the SONET ring will appear on the Circuits tab while in Network view. If there are a lot of circuits, you might use the Search button to find the circuit. Other ways to isolate the number of circuits viewed is to display the Circuit from Node view or Card view. Node view only displays the circuits that were sourced or destined for a card in that node. Card view only displays the circuits that were sourced or destined to that particular card.

After you have found the (three) circuits that create the metro Ethernet SPR, select the first circuit and click the Edit button. A display of the circuit appears as it does in Figure 7-12. If the circuit appears green, STP has placed this link in the Forwarding state. If the link appears purple, STP has placed this link in the Blocking state. Click the Close button and open the next circuit. After you have completed this process for all three circuits, notice that two links are in a Forwarding state and one is in a Blocking state. STP always blocks the redundant path. You can alter the path placed in Blocking by modifying the bridge priority of the nodes. If you modify the bridge priorities of the nodes, allow 50 seconds for STP to converge before verifying that the appropriate circuit is now placed in the Blocking state.

Figure 7-12 *Edit Circuit Dialog Box*

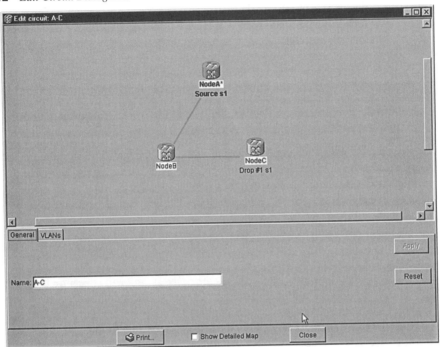

You can also use the Edit Circuit dialog box to modify which VLANs are traversing the circuit. With the Edit Circuit dialog box open, you can select the VLANs tab to add or remove VLANs from the circuit.

After provisioning the circuit, you use Ethernet test equipment to verify the circuit operation, BERs, convergence times, and so forth. After CPE has been attached to both ends and the interfaces come up on either side, you can use ping packets to verify circuit operation. The following is an example of a ping message successfully running across a circuit:

```
Ping 10.1.1.2Sending 5, 100-byte ICMP Echos to 10.1.1.2, timeout is 2 seconds:
!!!!!
Success rate is 100 percent (5/5), round-trip min/avg/max = 1/1/4 ms
```

Troubleshooting Metro Ethernet Connectivity

If a circuit is not passing traffic, troubleshoot the issue through a series of organized steps. The following steps are an example of some troubleshooting procedures and are not meant to be an exhaustive test scenario:

Step 1   Check for any SONET alarms associated with the card, ring, or STS being used by the metro Ethernet circuit. If any alarms are found, clear them by proper troubleshooting. You can look up the alarms in the Help menu of CTC as long as the documentation was installed during the CTC setup process. If the documentation was not installed, you can find it at Cisco.com.

Step 2   Verify that the circuit is in the Active state by clicking the **Circuits** tab and looking at the State column. If the circuit is in the Incomplete state, there is either a problem in the ring or you need to reprovision the circuit.

Step 3   Verify that the appropriate VLANs are included on the circuit by choosing the circuit, clicking **Edit**, and then choosing the **VLAN** tab.

Step 4   Verify that the appropriate VLANs are set up on the client side of the metro Ethernet card ports.

Step 5   At this point, you could use test equipment loopbacks set up on the cards to assist in issue isolation. If you find no problems end-to-end using the test equipment, the issue might involve the Ethernet connections going to the client site.

Step 6   Set up tests between the client site using Ethernet test sets and loopbacks in various areas of the network to isolate problem areas. If the tests come back successfully, the issues are most probably with the 802.1Q configuration of the CPE devices.

Summary

This chapter covered the provisioning of an SPR infrastructure using the E-series metro Ethernet cards. This implementation used the Single-card mode and made use of the available STS-12c. If Multicard mode is used, only STS-6c is available for provisioning. The Multicard allows communications between E-series cards and should be implemented if this functionality is necessary. Proper planning and design should alleviate this need. G-series metro Ethernet card provisioning is point to point based and more basic than the E-series provisioning learned in this chapter. The G-series card does not have VLANs or STP operation, making it easier to provision and troubleshoot. The G-series cards are a Layer 1 implementation, whereas the E-series cards are a Layer 2 implementation.

Review Questions

1 Which two tools enable you to create a metro Ethernet circuit between Cisco ONS 15454/15327 E-series cards?

   A Cisco Transport Manager (CTM)

   B TL-1

   C Cisco IOS

   D Cisco Transport Controller

2 Which three tools enable you to create a metro Ethernet circuit between Cisco ONS 15454/15327 G-series cards?

   A Cisco Transport Manager (CTM)

   B TL-1

   C Cisco IOS

   D Console Cable

   E Cisco Transport Controller

3 In which three places must E-series card VLANs be configured?

   A Circuit

   B Fiber

   C Customer-side port

   D Backplane port

   E Customer premises equipment (CPE)

4 What operation does the native/management 802.1Q default to?

   A --

   B Untagged

   C Tagged

   D Pop Tag

5 Traffic entering a VLAN-configured metro Ethernet port from a customer would be categorized as which of the following?

   A Untagged traffic

   B Tagged traffic

   C 802.1Q

   D 802.1p

6 What tab must you choose to assign Single-card EtherSwitch?

  A Port

  B VLAN

  C Card

  D Circuits

7 If a customer does not have 802.1Q-compatible equipment, how should you configure your circuits' VLANs?

  A All ports should be tagged as VLAN 1.

  B All ports should be tagged as VLAN 2.

  C No ports should be tagged except for the port the customer is coming in on.

  D The port that connects to the customer should be left as the default Untag selection.

  E The port that connects to the customer should have a VLAN configured, and that VLAN should be set to Untag.

8 True or False: The E1000-2 card manages congestion by initiating pause messages.

  A True

  B False

9 True or False: Ethergroup is used to create a Single-card EtherSwitch circuit.

  A True

  B False

10 10.802.1Q-trunked VLANs should be set to which VLAN type?

  A --

  B Untag

  C Tagged

  D Pop Tag

This chapter covers the following topics:

- Challenges in the metro-area network
- Multiplexing optical signals
- Dense wavelength division multiplexing fundamentals
- DWDM components
- Optical impairments in DWDM systems
- Cisco DWDM equipment
- Building a DWDM infrastructure

CHAPTER 8

Implementing DWDM in Metropolitan-Area Networks

Today's explosion in data traffic, fueled in large part by the growth of the Internet, has caused rapid depletion of available fiber bandwidth in the metropolitan area. As a result, service providers have reacted by increasing capacity in their long-haul and core networks. The bandwidth bottleneck is now that of the metropolitan-area networks (MANs), where capacity has not increased as greatly as the long-haul networks. Keeping pace with this demand for greater bandwidth and diversity of services in the MAN would traditionally involve laying new fiber in densely populated urban areas. This solution is both cost and time prohibitive. To cope with larger bandwidth needs, service providers must maximize the use of their optical fiber.

The introduction of dense wave division multiplexing (DWDM) technology allows for the improvement in efficiency of optical transmissions. In addition, the flexibility of these systems allows new, flexible services that can be rapidly deployed. Service providers are no longer inhibited by a single infrastructure. This chapter discusses the issues facing service providers today and the solutions provided by DWDM equipment.

Challenges in the Metro-Area Network

Challenges affecting service providers in the MANs include service diversity, bandwidth scalability, and fiber exhaustion.

Bandwidth scalability deals with the capability of the fiber to carry traffic at exponentially greater speeds. The ONS 15454 provides greater flexibility in the use of fiber bandwidth by allowing the multiplexing of differentiated services onto a common fiber. Historically, each service was transported over its own fiber. The ONS 15454 supports these different services by aggregating and transporting them over OC-48 or OC-192 rings created with DWDM-capable extended long reach (ELR) cards. The result is fewer fibers carrying more traffic. The left side of Figure 8-1 displays a SONET ring hierarchy that would typically be found in a metropolitan area. The right side of Figure 8-1 displays the same hierarchy with DWDM components. The DWDM network has far fewer active components. Active components are those that require power. SONET regenerators providing 3R functionality (re-amplification, reshaping, and retiming) require AC or DC power to provide this operation. Optical add/drop multiplexers (ADMs) and filters do not require power. The magic is in the mirrors.

Figure 8-1   *Metro Network Solution Through DWDM*

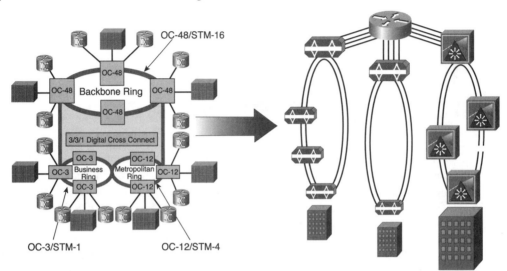

Fiber exhaustion is the principle that the services provisioned have depleted the bandwidth in the fiber. The bandwidth could be increased in some networks (for example, OC-48 to OC-192 upgrade), but this requires the replacement of a lot of equipment. Another solution is to add more fiber to the network, but this could be very expensive and take a long time. DWDM enables the multiplexing of multiple analog signals on the existing fiber. Not all fiber deployed can be used for DWDM transmission because DWDM equipment currently uses the C band window (1520 nanometers [nm] to 1560 nm) of the optical transmission spectrum. Legacy SONET equipment was deployed with fiber that was optimized for transmission in the 1310 nm window. All fiber should be characterized (tested for its light-propagation properties) before purchasing and deploying DWDM equipment.

If adding more fiber is not an option because the cost of laying new cable is prohibitive or the route is constrained as in Figure 8-2, putting more optical signals on a single fiber becomes the only viable solution. Figure 8-2 displays a fiber-optic cable traveling over a small bridge. Certain geographical environments only provide one constrained route in which the fiber can be laid. The constrained route may be providing structural diversity (for example, fiber run in different conduits) but is not providing the resiliency of route or path diversity (for example, fiber routes taking different paths to provide a level of redundancy). Notice, however, that Figure 8-2 is providing loop diversity, which is standard for SONET networks.

Figure 8-2 *Fiber-Constrained Route*

Fiber-Constrained Route

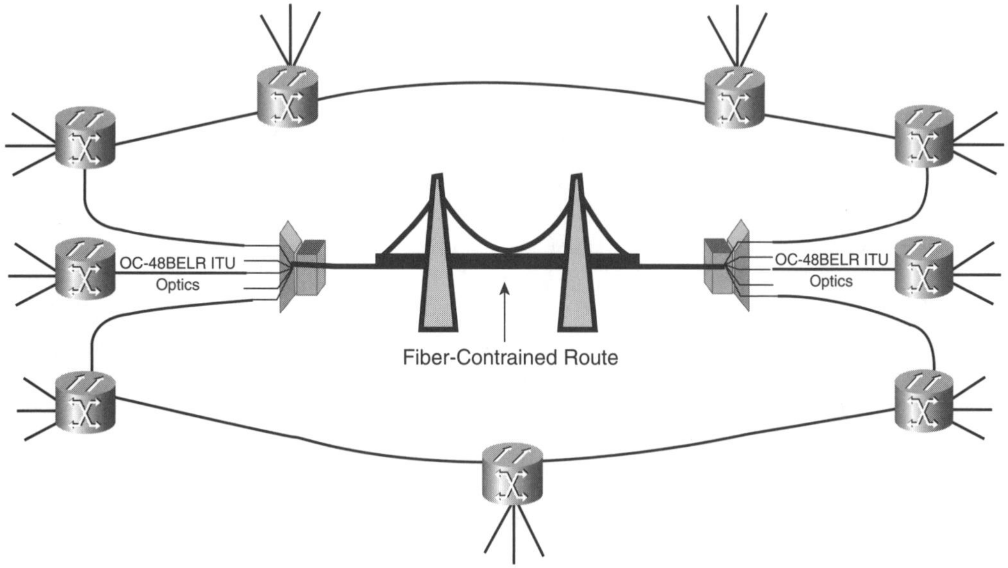

Multiplexing Optical Signals

A few technologies allow the transmission of multiple light frequencies on the same fiber:

- Wavelength division multiplexing (WDM)
- Dense wavelength division multiplexing (DWDM)
- Coarse wavelength division multiplexing (CWDM)

Optical signals used before the introduction of WDM allowed for one wavelength only via a transmitter and receiver. Wavelength is synonymous with lambda. Lambda is represented by the Greek lambda symbol (λ). WDM is the technology that allows optical signals of different wavelengths to be transmitted on the same fiber without interfering with one another.

NOTE   DWDM systems are sometimes represented by their frequencies. Frequencies and wavelength are inverse in nature. If one value is known, the other can be computed via the following formula:

$$F = C \:/\: \lambda$$

F = Frequency

C = Speed of light (3×10^8 m/s)

If we know the wavelength of a component is 1547.47 nm, we can compute the frequency, as follows:

$$F = 300,000 / 1547.72$$

$$F = 193.83 \text{ THz (terahertz)}$$

193.83 THz represents the oscillation of a signal 193.83 trillion times in 1 second.

A fundamental property of light states that individual light waves of different wavelengths will not interfere with one another within a medium. Lasers are capable of creating pulses of light with a very precise wavelength. Each wavelength of light can represent a different channel of information. By combining light pulses of different wavelengths, many channels can be transmitted across a single fiber simultaneously. Basic WDM allows widely spaced wavelengths to coexist on the same fiber, such as 1310 nm and 1550 nm, and is generally limited to just a few wavelengths. These systems are not typically implemented today but represent the history of DWDM systems.

The different lambdas are not typically referred to in frequencies—they are wavelengths as measured in nanometers. Frequencies are measured in degrees of hertz. There are conversion factors that report these wavelengths in THz, but it is not the common reference. Spacing between the channels is referred to in terms of 50, 100, or 200 GHz, but that refers to the spectrum difference between each wavelength as it transmits the medium.

Given that a single-mode fiber's potential bandwidth is nearly 50 Tbps, which is nearly four orders of magnitudes higher than electronic data rates of a few gigabits per seconds (Gbps), every effort should be made to tap into this huge opto-electrical bandwidth mismatch. Realizing that the maximum rate at which an end user, which can be a workstation or a gateway that interfaces with lower-speed subnetworks, can access the network is limited by electronic speed (to a few Gbps), the key in designing optical communication networks to exploit the fiber's huge bandwidth is to introduce concurrency among multiple user transmissions into the network architectures and protocols. In an optical communication network, the concurrency may be provided according to either wavelength or frequency (WDM), time slots (time-division multiplexing [TDM]), or wave shape (spread spectrum, code-division multiplexing [CDM]).

Optical TDM and CDM are somewhat futuristic technologies today. Under optical TDM, each end user should be able to synchronize to within one time slot. The optical TDM bit rate is the aggregation rate over all TDM channels in the system, whereas the optical CDM chip rate may be much higher than each user's data rate. Therefore, TDM and CDM are relatively less attractive than WDM because WDM—unlike TDM and CDM—has no such requirement.

Specifically, WDM is the current favorite multiplexing technology for optical communication networks because all the end-user equipment needs to operate only at the bit rate of the WDM channel, which can be chosen arbitrarily (for instance, peak electronic processing speed). Hence, all the major carriers today devote significant efforts to developing and applying WDM technology in the businesses.

WDM is an important and practical technology to exploit the wide communication bandwidths in a single optical fiber pair. WDM is basically frequency-division multiplexing. A fiber carries multiple optical communication channels, each at different wavelengths, which is essentially the inverse of the carrier frequencies. In this way, WDM transforms a fiber that carries a single signal into multiple virtual fibers, each carrying its own signal.

DWDM means the WDM system is able to multiplex the wavelengths much closer together. This is achieved by using lasers that have been tuned to a particular wavelength and transmit with a narrow spectral width. Current metro systems use the C band optical spectrum because it has the lowest intrinsic attenuation. (The center wavelength in the C band window is the 1550-nm wavelength.) These wavelengths are then multiplexed together optically.

The narrow spectral-width lasers raise the total number of frequencies that can coexist on a single fiber from the 2 to 4 normally found in WDM systems to 16 and higher in DWDM. It is the continued enhancements of the photonic components that allow for the increases in the number of frequencies a single fiber can support. Cisco long-haul 15800 optical platforms can transmit up to a total of 120 wavelengths on 1 fiber.

CWDM is very similar to DWDM with one minor difference: the spectral width of the laser sources used. The spectral width of a CWDM solution is less than .5 nm, whereas the spectral width of most DWDM systems is less than .1 nm.

The Cisco CWDM GBIC solution consists of a set of 8 different Gigabit Ethernet-based Cisco CWDM gigabit interface controllers (GBICs)—1 for each of the 8 different colors or wavelengths—and a set of 10 different Cisco CWDM optical add/drop multiplexers (OADMs).

The CWDM GBICs fit into GBIC ports supporting the IEEE 802.3z standard and are supported across various Cisco switching platforms.

The concepts behind the different types of WDM are very similar, but it is DWDM that offers the most growth potential. This chapter focuses on the DWDM technologies that Cisco provides.

DWDM technology was made possible with the realization of several optical components. Components that were previously an experimenter's curiosity are now compact, of high quality, commercially available, and increasingly inexpensive. It is also expected that several optical functions will soon be integrated to offer complex functionality at a cost per function that is comparable to electronic implementation. The following provides a snapshot of what has enabled the DWDM technology to become reality:

- Optical fiber has been produced that exhibits low loss and better optical transmission performance over the spectrum windows.

- Optical amplifiers with flat gain over a range of wavelengths and coupled in the line with the transmitting fiber boost the optical signal, thus eliminating the need for regenerators.

- Integrated solid-state optical filters are compact and can be integrated with other optical components on the same substrate.

- Integrated solid-state laser sources and photodectectors offer compact designs.

- Wavelength-selectable (tunable) filters can be used as optical ADMs.

- OADM components have made DWDM possible in MAN ring-type and long-haul networks.

- Optical cross-connects (OXC) components, implemented with a variety of technologies (for example, lithium niobate), have made optical switching possible.

In addition, standards have been developed so that many vendors can offer interoperable systems. As DWDM technology evolves, existing standards are updated or new ones are introduced to address emerging issues.

DWDM finds applications in ultra-high bandwidth long haul as well as in ultra-high-speed metropolitan or inner-city networks, and at the edge of other networks (SONET, SDH, IP, and ATM).

As DWDM deployment becomes more ubiquitous, DWDM technology cost decreases, primarily owing to increased optical component volume. Consequently, DWDM is also expected to become a low-cost technology in many access-type networks, such as fiber to the home and fiber to the desktop.

DWDM allows a service provider to offer leased lambda services (dark fiber) to customers, offering individual wavelengths or lambdas as individual dark fibers. *Dark fiber* is a term used in different applications. A fiber optical cable that does not have a transmitter attached to it is not lit up by the transmitting laser; therefore, it is dark. Service providers also call the leasing of a fiber-optic cable dark-fiber services because they are only responsible for providing the fiber-optic cable, not the equipment that will "light" the fiber. This terminology has been extended to DWDM systems because if the service provider leases a customer particular lambda(s), the customer is responsible for transmitting a client (CPE) signal over that wavelength.

Current metropolitan systems use up to 32 wavelengths on the same fiber. As bandwidth needs rise and equipment costs fall, the number of wavelengths (lambdas) used by these systems will increase. (Up to 256 wavelengths have been tested over 1 fiber in lab environments.) All optical, passive components used in DWDM networks are blind to the bit rate of the signal. By producing the wavelengths at specific frequencies, the DWDM system can transport the traffic to its destination regardless of its underlying bit rate. From the perspective of these devices, these are nothing more than pulses of light. Many of these devices (OADM and Erbium-doped fiber amplifiers [EDFAs]) provide scalability and growth because they do not need to be changed when the optical signal fed into them is changed (for example, OC-48 to OC-192). Cisco ONS 15216 devices allow this flexible growth path.

The Technology Behind DWDM

To understand how DWDM works, you must first understand how an optical laser transmits data. Data is transmitted as a series of light pulses in the optical domain. The most popular modulation techniques implemented today are return to zero (RZ) and non-return to zero (NRZ). With RZ optical encoding, a high signal level equates to a 1, and a low signal level over a given period of time (bit period) equates to 0. NRZ does not have to return to 0 during successive bit periods. An optical pulse has an elliptical curve. The pulse from a short-range (SR) or intermediate-range (IR) and long-range (LR) lasers (which consist of several very close wavelengths) looks like that of Figure 8-3 when viewed with an optical spectrum analyzer (OSA). The spectral width of the laser is obtained via the principle of full width half maximum (FWHM). The FWHM principle states that the wavelengths sent by the laser or LED that do not exceed half the amplitude (optical power) of the maximum amplitude will not be used to calculate the spectral width.

Figure 8-3 is a spectrally broad (wideband) laser with a center that can drift. This signal meets the objectives of optical systems with one signal on a fiber. To combine frequencies, a laser with greater precision is needed. The ONS 15454 ELR cards have lasers that have been tuned to specific wavelengths. Figure 8-4 shows a tuned laser with a narrow spectral width.

Figure 8-3   *Wideband Laser Signal*

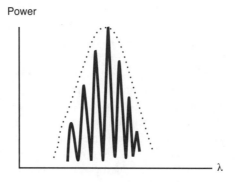

Figure 8-4 *Tuned Laser with a Narrow Spectral Width*

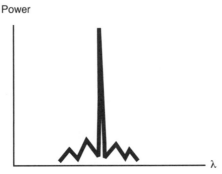

The minimum wavelength separation between two different channels (signals) that are multiplexed on a link is known as *channel spacing*. Channel spacing has to be chosen in such a way that channels do not overlap, causing power coupling between one channel and its neighbor.

Channel spacing is a function of the precision of a laser. The more exact the tuning, the lower the channel spacing. (For example, a 200-GHz-spaced laser generically has one-half the preciseness of a 100-GHz-spaced laser.) The precision of the laser has a linear relationship with the cost of the laser. The spacing that can be used is affected by the existing fiber characteristics. Fiber characteristics fall in the linear and nonlinear domains. Examples of linear fiber impairments would be that of attenuation and dispersion, whereas examples of fiber nonlinearity would be that of four-wave mixing and cross-phase modulation. This is precisely why it is important to always have fiber characterized before procuring equipment. Another factor that affects channel spacing is the optical amplifier's capability to amplify the channel range.

The closer the wavelengths are placed, the more important it is to ensure that the centers are identifiable from other signals on the same fiber. Common spacing increments are 200 GHz (up to 20 wavelengths), 100 GHz (up to 40 wavelengths), and 50 GHz (up to 80 wavelengths). A Cisco DWDM system using 200-GHz spacing will multiplex 18 wavelengths on 1 fiber, whereas the 100-GHz-spaced products will multiplex 32 wavelengths on 1 fiber. 50-GHz spacing is normally only implemented in long-haul systems due to the expense of the lasers. Work is being done to reduce the spacing to 25 GHz as well. At 200 GHz, the spacing of the frequencies is about 1.6 nm; so 100 GHz is .8 nm, and 50 GHz equates to around .4 nm. For manufacturers of DWDM equipment as well as for consumers who want to maintain vendor independence, the ITU has published a wavelength grid as part of International Telecommunication Union Telecommunication Standardization Sector (ITU-T) G.692 to provide standards for companies to work from. This standard was created to allow interoperability between vendors' equipment if the standard is strictly adhered to. Unfortunately, different vendors have used different wavelengths, and standards compliance has not resulted in interoperable equipment. Figure 8-5 displays the ONS 15454 and ONS 15216 support of the G.692 standard.

Figure 8-5 *ITU-T G.692 Channel-Spacing Recommendation*

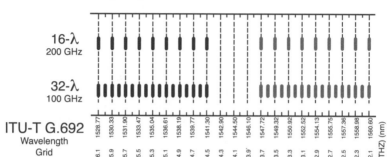

To modularize the DWDM systems, the wavelengths are generally described as being of a color, such as red or blue. The red range and the blue range can then describe the wavelengths that a particular DWDM filter or transponder may manage within the DWDM spectrum. In the Cisco implementation, the lower half of the channels is referred to as the red channels, and the upper half of the channels is referred to as the blue channels. The Cisco implantation modularizes the channel plan into the red base unit and blue upgrade unit, allowing for lower optical loss in the system and greater flexibility for growth.

DWDM Components

DWDM technology requires specialized optical devices that are based on properties of light and on the optical, electrical, and mechanical properties of semiconductor material. Such devices must provide the equivalent functionality of electrical/electrical (opaque) communications systems. These devices include optical transmitters, optical receivers, optical filters, optical modulators, optical amplifiers, OADM, and OXC.

DWDM systems are composed of several parts, as described in the following list:

- **Transmitter**—A transmitter is used to transmit (send) an optical signal. The signal transmitted may be of any optical wavelength. Standard optical signals used are as follows: 850 nm (LED-based transmitter used for short-range multimode fiber systems), 1310 nm (laser-based transmitter used for intermediate-range systems over single-mode fiber), and 1550 nm (laser-based transmitter used for long-haul systems over single-mode fiber). Transmitters of this type are typically referred to as *wideband transmitters*. Narrowband, tuned lasers are orderable by the exact wavelength they generate. Cisco currently has 32 different ELR cards that can be ordered for the 100-GHz-spaced system.

- **Transponder**—A transponder is a device that converts a wideband laser source to a tuned laser with a narrow spectral width (.1 nm). Transponders are much more expensive than standard transmitters because they provide additional functionality.

Each transponder in the ONS 15454 system provides 3R regenerator functionality. The other option is for the customer premises equipment (CPE) card to produce the exact wavelength needed by the DWDM system. This would drive up the cost of customer equipment and extend the service provider's management domain.

- **Filters**—The optical wavelengths generated by the narrowband transmitters on the trunk side of the transponders (DWDM side) are then multiplexed onto a single fiber by a device known as a *filter*. The signal carried over that fiber at this time is referred to as the *composite channel*. The composite channel contains all the constituent wavelengths that created this signal. The individual wavelengths contained in the composite channel can be viewed with an oscilloscope or an OSA. These are the most important pieces of equipment in troubleshooting a DWDM system. At the other end of the link, the filter must demultiplex the signal (for example, break the composite channel out into its constituent wavelengths). The multiplexing and demultiplexing functionalities are the exact opposite, and they are built in to the same filter device.

- **Receivers**—Optical receivers are photodiodes. A photodiode detects an optical signal and generates an electrical signal based on the optical signal. This is the opposite of a semiconductor laser. Receivers are wideband in nature. (For example, any receiver will detect any wavelength transmitter.) It is important to cable the filters correctly because a link light does not necessarily mean the correct wavelength has been connected. This is done primarily for the cost benefits. As long as the system is set up correctly, the wide or narrow range of the receiver does not matter.

- **Optical amplifiers**—The optical power of a transmitted signal attenuates (degrades) linearly with the distance traveled. Depending on the power budget between the transmitter and receiver, the signals may need to be amplified. A power budget is calculated by subtracting the receiver sensitivity level by the transmitter power level. If a laser transmits a signal at 2 decibels per milliwatt (dBm), and the receiver sensitivity is –32 dBm, for instance, the power budget is 34 dB—for example, 2 dBm – –32 = 34. Is 34 dB of loss enough? The actual loss in the system must then be computed based on intrinsic loss of the fiber (*x*dB per km), fiber splices, and connectors. Normally, 3 dB is subtracted from this figure to compensate for the aging of the fiber (in which it will incur more loss) and splices incurred due to unplanned breaks in the fiber. Power budgets are discussed again later in this chapter.

At certain points in the MAN, the signal can degrade to the point where a receiver cannot detect the signal. The way this was dealt with in SONET networks was to provide regeneration sites. Regenerators are 3R costly devices and work on one independent wavelength. All 3R devices do an optical-to-electrical conversion, retime and reshape the signal, and then do an electrical-to-optical conversion to transmit the signal again (re-amplification). It is costly to provide regenerator sites that regenerate 32 wavelengths.

Optical amplifiers remedy this need by providing re-amplification (1R) of all the signals, amplifying all the individual channels at the same time. The signals should input to the 15216 EDFA within 2 dBm (+/– 1 dBm) to get around the gain-tilt phenomenon. Gain-tilt is the amplification of higher-powered channels without the amplification of lower-powered channels.

Optical amplifiers can be located in several different locations within an optical span, as follows:

— **Post amplifier**—The amplification of a composite signal as soon as it is multiplexed onto one fiber. Multiplexers introduce loss into an optical fiber. This would be necessary depending on the length of distance and the optical power budget.

— **Mid-span amplifier**—Boosts the signal strength on long optical spans so that the signals can be detected at the opposite end.

— **Pre-amplifier**—Amplifies the signal just before it reaches the demultiplexers so that the signal strength after demultiplexing loss does not inhibit the receiver's capability to discern the signal.

Optical amplifiers are unidirectional. Two optical amplifiers are necessary for each fiber pair. Optical amplifiers only provide re-amplification; there is no retiming or reshaping of the signal. Any noise that has accumulated in the signal is amplified, too. The amount of noise in the signal is referred to as the optical signal-to-noise ratio (OSNR). This can become a problem for long distances in which noise has accumulated. At some point, regeneration would be necessary. This point is different for every system.

An EDFA is a type of optical amplifier. Its name is used synonymously with optical amplifier because it is the most popular amplifier to date. The EDFA is a length of fiber-optic cable that has been doped with impurities (the rare earth element Erbium). A pump laser operating at 980 nm or 1480 nm is used to take the Erbium ions in the fiber from the fundamental state to the excited state. These excited ions come down to the metastable state, where they release a photon of energy (amplified spontaneous emission, or ASE). The photon of energy interacts with the DWDM signals operating around 1550 nm and amplifies each one respectively. In the case of the 15216 EDFA, the gain realized by each wavelength is 17 dB.

A system should never be thought of in terms of the distance the hardware can span out of the box. The distance a DWDM system can span is the result of many factors:

- Transmitter power
- Receiver sensitivity
- Fiber loss per kilometer (attenuation)
- Splices
- Connectors
- Dispersion

It is bad practice to measure a DWDM system in terms of distance that can be traveled. If a vendor were to say that their system could span 80 km, this would be based on optimal conditions. Could this same system span 80 km with fiber optics that were run 20 years ago optimized for the S band? These are the types of questions a designer needs to examine when preparing to design a DWDM network:

- **Optical add/drop multiplexer (OADM)**—To connect different sites via individual wavelengths, a DWDM system needs to have the capability to add and drop services at distinct locations. The functionality of an OADM is similar to legacy SONET add/drop multiplexers (ADMs), but OADMs are completely passive (no power) and operate entirely in the optical realm. SONET devices provide 3R regeneration and require power. The OADM has the capability to pass through the channels that are not being dropped to the site. The optical losses are different for the add, drop, and pass-through channels. Within the OADM, the placement of the dialectic thin-film filters (TFF) determine how the individual signals travel.

- **Line termination equipment (LTE)**—LTE devices are the devices that transmit or terminate the signal. Remember that LTE is a SONET term that signifies the device is operating at the line layer in SONET's hierarchy.

- **Optical-electrical-optical (O-E-O)**—Transmitters, receivers, transponders, and regenerators are some examples of devices that perform O-E-O functionality. Any device that performs 2R or 3R regeneration is performing some type of electrical function.

- **Optical cross-connect (OXC)**—An OXC is similar in operation to a digital cross-connect (DCS), but in the optical domain. Recall that SONET is an electrical standard, and DCS operates in the electrical realm. An all-optical cross-connect needs to be achieved for DWDM operation; otherwise, you would need many expensive regenerator sites.

Other tools may also be included with the DWDM systems. One of the most important and difficult issues involved with the optical network is network management. Network management is achieved via the monitor out ports on the ONS 15216 red base unit filter and the ONS 15216 OADMs. These ports can be connected to the ONS 15216 Optical Performance Monitor (OPM), which can be accessed remotely. The ONS 15216 OPM displays the individual channel signals in the composite channel (OSA functionality). This is necessary to troubleshoot one lambda without taking down the entire DWDM system. By using multiple ONS 15216 OPM devices, the operator can isolate the source of the issue.

Cisco provides a network management device for its entire line of optical products. This network management platform is Cisco Transport Manager (CTM). You can obtain additional information regarding CTM at Cisco.com.

In addition to these functions, optical supervisory channels (OSCs) are included to communicate local status information for remote management. The ONS 15216 has a 1510-nm OSC OADM and GBIC solution that can be used to monitor the ONS 15216 DWDM components. Other Cisco optical platforms that support an OSC channel include the 15500 and the 15800 product families. Cisco ONS 15252 does not support an out-of-band OSC channel but does support inline management via the Qeyton Proprietary Protocol (QPP).

Figure 8-6 shows a DWDM architecture that uses these components.

Figure 8-6   *DWDM Architecture*

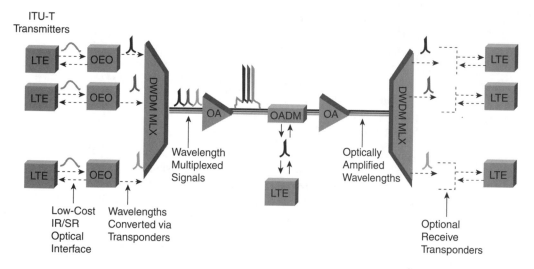

DWDM Optical Limitations

When transmitting DWDM optical signals, you must consider limitations that apply. These limitations are divided into two areas:

- **Linear**—These limitations increase directly with distance. There is a direct relationship between the distance traveled and the amount of one of these linear limitations. These limitations are normally represented per kilometer. (The amount of attenuation in fiber is linear. This is represented as dB/km [decibels per kilometer].) Examples of this type of limitation are attenuation and dispersion. Attenuation is the loss of power (strength) of an optical signal. If the signal strength falls below a certain threshold, the signal cannot be detected. The attenuation level of the fiber is documented by the fiber manufacturer.

  Chromatic dispersion (CD) is the spreading of a signal over distance. It is caused by the intrinsic properties of light propagation. Different wavelengths of light travel at different speeds. Dispersion can be mitigated through the use of dispersion-shifted fiber (DSF). DSF is fiber that has been doped with impurities that have negative dispersion characteristics. They are orderable in the form of dispersion compensating units (DCUs). The 15216 DCUs are currently orderable in different varieties depending on the amount of negative dispersion that should be introduced. This increases the distance the signal can travel by compensating for the dispersion. The amount of dispersion that the fiber introduces should be documented by the fiber manufacturer.

Polarization mode dispersion (PMD) is another linear effect associated with fiber. Light travels orthogonally down a fiber (for example, along both the x and y axis). Single-mode fiber (SMF) is constructed to precise standards, but because of the small size of the core (8 to 10 microns), it is hard to draw fiber in perfect circular symmetry. The actual fiber may be somewhat elliptical in nature, especially in areas where stress has been incurred (for instance, fiber hanging from telephone poles on a windy day or buried fiber on a hot summer day). The light pulses smear as they travel down the fiber because the refractive index (RI) of the fiber varies with the polarity of the signal. PMD compensation requires adaptive optics or some manual mechanism of reversing PMD stress in the fiber.

Another linear limitation is the optical signal-to-noise ratio (OSNR). Loss over distance can be compensated for by amplifying the signal. Because optical amplifiers are amplifying an analog signal, noise as well as the signal is amplified. Over time, the signal can become difficult to distinguish from the noise. Before the signal gets to this level, it should be regenerated (for example, perform 3R on the signal) to ensure that it can be detected at the receiver. Optical amplifiers also add a certain amount of noise to the signals. Figure 8-7 displays the effect of attenuation, slight timing drift, and OSNR.

Figure 8-7   *Noise on an Amplified Optical Line*

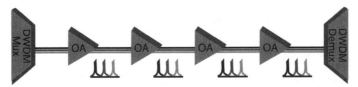

- **Nonlinear**—Nonlinear limitations occur due to other issues (for example, channel spacing and power levels) that do not follow a distinct pattern in relationship to something else.

Fiber-optic communications for SONET originally centered around the 1310-nm range (S band). Fiber optics were created with a zero dispersion point centered at 1310 nm. Through technology advances, it was found that the C band (centered at 1550 nm) had lower intrinsic attenuation than the S band. The C band was perfect to be leveraged for DWDM transmission, but the fiber needed to be altered to change the zero dispersion point to 1550 nm. This is achieved by adding impurities (dopants) during the drawing of the fiber.

DSF was perfect for DWDM operation until a three-wavelength system was deployed and it was found that there were four wavelengths at the receiver. This was a form of cross-talk between the analog signals. It was found that introducing a bit of delay, slightly above zero, mitigated some of these four-wave mixing (FWM) issues. FWM is still a standard way of defining this issue, although systems are transmitting many more wavelengths at this time. Other ways of mitigating FWM are to keep channel counts down or use unequal spacing between certain wavelengths. The new fiber created from this was known as nonzero dispersion-shifted fiber (NZ-DSF).

Rayleigh scattering is the largest contributor to attenuation in fiber optics. Rayleigh scattering causes light to scatter and be attenuated because the fiber production process resulted in density differences in the fiber, causing slightly different RI.

Stimulated Raman scattering (SRS) is an issue in which lower wavelengths pump the higher wavelengths. The higher wavelengths, in essence, cannibalize the lower wavelengths. The best way to mitigate the effects of SRS is to lower the input power. When engineers discovered SRS, they applied it as an amplifier. Raman amplifiers can now be found in various long-haul systems, such as the Cisco 158*xx*.

Stimulated Brillouin scattering (SBS) becomes an issue in bidirectional WDM systems. SBS is the primary reason why a fiber pair is used for optical communications systems. SBS is less of an issue in higher bit-rate systems and can be mitigated by using separate fibers for Tx and Rx and for lowering input power to the system.

Cisco DWDM Equipment

The Cisco optical portfolio includes some transponder-based solutions. The ONS 15252/15201, 155*xx*, and 158*xx* are all examples of systems that use transponders to convert generic client signals into ITU-compatible, tuned DWDM lasers.

The ONS 15252/15201, 155*xx*, and 158*xx* products are not covered in the Building Cisco Metro Optical Networks course. You can find more information on these products at Cisco.com.

Service providers can incorporate DWDM solutions in existing SONET networks with the Cisco ONS 15454 OC-48 ELR cards. These cards work with the DCS of the ONS 15454 and transmit the information with DWDM-compatible optics. All these signals are standalone at this point. To create a composite signal, the individual DWDM lasers must be multiplexed together in some way. This is where the 15216 family products come in. The Cisco ONS 15216 DWDM filter and the ONS 15216 OADM can optically add and drop wavelengths to the ONS 15454 ELR cards. The rest of this chapter describes the product features that the ONS 15454 and ONS 15216 DWDM solution can provide.

The Cisco ONS 15216 DWDM product is a family of products, including the following:

- 9-lambda DWDM red filter (200-GHz spacing)
- 9-lambda DWDM blue filter (200-GHz spacing)
- 16-lambda DWDM red filter (100-GHz spacing)
- 16-lambda DWDM blue filter (100-GHz spacing)
- 1-lambda OADM (100-GHz and 200-GHz spacing)
- 2-lambda OADM (100-GHz and 200-GHz spacing)
- 4-lambda OADM (100-GHz and 200-GHz spacing)
- Optical Performance Monitor (OPM)
- C band EDFA (available in two versions: manageable and unmanageable)
- DCU (available in –350ps/nm, –750 ps/nm, and –1150 ps/nm varieties)
- Optical services channel (ONS 15216 OADM for OSC at 1510 nm; tuned GBICs available for source and destination)

OC-48 ELR ITU Optics

There are 32 distinct ONS 15454 OC-48 ELR, 200-GHz-spaced DWDM cards. The transmit (Tx) interface of each of these cards is then connected to the respective wavelength receive (Rx) interface on the ONS 15216 OADM (200 GHz) or filter (200 GHz). The Rx interface on the 15216 OADM is labeled IN. The Rx interface of the ELR card is connected to the respective wavelength Tx (OUT) port on the ONS 15216 filter.

There are 18 different ONS 15454 OC-48 ELR, 200-GHz-spaced DWDM cards. These cards operate in the same manner as the 100-GHz-spaced cards but need to be connected to the 200-GHz versions of the ONS 15216 OADM or filter.

The OC-48 ELR ITU cards have a dispersion tolerance of 5400 ps/nm (picoseconds per nanometer). A picosecond is one trillionth of a second (1×10^{-12}). If the dispersion exceeds this measurement, the signal cannot be recognized by the photodiode receiver. Recall that an ONS 15216 DCU can be used to compensate for dispersion.

Each OC-48 DWDM card provides one Telcordia-compliant, GR-253 SONET OC-48 interface. The interface operates at 2.488 Gbps (commonly referred to as 2.5 Gbps) over a SMF span and carries both concatenated and nonconcatenated payloads.

The Cisco ONS 15454 OC-48 ELR ITU optics can be configured for operation in the following configurations:

- Unidirectional path-switched ring (UPSR) or bidirectional line-switched ring (BLSR) networks
- Linear point-to-point and hubbed networks
- Mesh configuration

The Cisco ONS 15454 can support up to four OC-48 ELR cards per shelf, depending on user applications. Multiple shelves can be used to terminate more signals. Because the DWDM is normally deployed in a ring configuration, the OC-48 ELR cards are installed in pairs (West and East). If a single card were used, this DWDM network would have no protected route.

All the OC-48 ELR cards are separately orderable components, normally purchased in pairs. This would mean that there would be 64 pairs (32 pairs for the source, and 32 pairs for the destinations) of separate OC-48 ELR cards for a ring-protected network of 32 wavelengths, totaling 128 cards. If onsite spares were procured for this solution, the service provider would have to buy an additional 32 cards for a total of 96 cards. This is not practical, nor is inventorying 32 different versions of a similar card from the vendor's perspective. At this time, tuned lasers are the only option. Much testing is being done to create a tunable laser. This would allow customers to purchase one card that could back up any tuned laser. When tunable lasers become economical, all DWDM systems will be created with tunable lasers, and this may take us into the next era of optical networking: wavelength routing. Table 8-1 describes the wavelength and bands of the ONS 15454 ELR cards.

Table 8-1   *OC-48 ELR Card Sets*

ONS 15454 Card	Band	100 GHz	200 GHz
OC-48/STM-16 ELR 100 GHz, 1528.77 nm	Blue	No	No
OC-48/STM-16 ELR 100 GHz, 1530.33 nm	Blue	Yes	Yes
OC-48/STM-16 ELR 100 GHz, 1531.12 nm	Blue	Yes	No
OC-48/STM-16 ELR 100 GHz, 1531.90 nm	Blue	Yes	Yes
OC-48/STM-16 ELR 100 GHz, 1532.68 nm	Blue	Yes	No
OC-48/STM-16 ELR 100 GHz, 1533.47 nm	Blue	No	Yes
OC-48/STM-16 ELR 100 GHz, 1534.25 nm	Blue	Yes	No
OC-48/STM-16 ELR 100 GHz, 1535.04 nm	Blue	Yes	Yes
OC-48/STM-16 ELR 100 GHz, 1535.52 nm	Blue	Yes	No
OC-48/STM-16 ELR 100 GHz, 1536.61 nm	Blue	Yes	Yes
OC-48/STM-16 ELR 100 GHz, 1538.19 nm	Blue	Yes	Yes
OC-48/STM-16 ELR 100 GHz, 1538.98 nm	Blue	Yes	No
OC-48/STM-16 ELR 100 GHz, 1539.77 nm	Blue	Yes	Yes
OC-48/STM-16 ELR 100 GHz, 1540.56 nm	Blue	Yes	No
OC-48/STM-16 ELR 100 GHz, 1541.35 nm	Blue	No	Yes
OC-48/STM-16 ELR 100 GHz, 1542.14 nm	Blue	Yes	No
OC-48/STM-16 ELR 100 GHz, 1542.94 nm	Blue	Yes	Yes
OC-48/STM-16 ELR 100 GHz, 1543.73 nm	Blue	Yes	No

continues

Table 8-1 *OC-48 ELR Card Sets (Continued)*

ONS 15454 Card	Band	100 GHz	200 GHz
OC-48/STM-16 ELR 100 GHz, 1544.53 nm	Blue	Yes	No
OC-48/STM-16 ELR 100 GHz, 1546.1 2 nm	Red	Yes	No
OC-48/STM-16 ELR 100 GHz, 1546.92 nm	Red	Yes	No
OC-48/STM-16 ELR 100 GHz, 1547.72 nm	Red	Yes	Yes
OC-48/STM-16 ELR 100 GHz, 1548.51 nm	Red	Yes	No
OC-48/STM-16 ELR 100 GHz, 1549.32 nm	Red	No	Yes
OC-48/STM-16 ELR 100 GHz, 1550.12 nm	Red	Yes	No
OC-48/STM-16 ELR 100 GHz, 1550.92 nm	Red	Yes	Yes
OC-48/STM-16 ELR 100 GHz, 1551.72 nm	Red	Yes	No
OC-48/STM-16 ELR 100 GHz, 1552.52 nm	Red	Yes	Yes
OC-48/STM-16 ELR 100 GHz, 1554.13 nm	Red	Yes	Yes
OC-48/STM-16 ELR 100 GHz, 1554.94 nm	Red	Yes	No
OC-48/STM-16 ELR 100 GHz, 1555.75 nm	Red	Yes	Yes
OC-48/STM-16 ELR 100 GHz, 1556.55 nm	Red	Yes	No
OC-48/STM-16 ELR 100 GHz, 1557.36 nm	Red	No	Yes
OC-48/STM-16 ELR 100 GHz, 1558.17 nm	Red	Yes	No
OC-48/STM-16 ELR 100 GHz, 1558.98 nm	Red	Yes	Yes
OC-48/STM-16 ELR 100 GHz, 1559.79 nm	Red	Yes	No
OC-48/STM-16 ELR 100 GHz, 1560.61 nm	Red	Yes	Yes

You can install the OC-48 ELR DWDM cards in any high-speed slot of the 454 (slots 5/6 and slots 12/13) and provision the cards as part of a BLSR or UPSR ring.

Each OC-48 ELR DWDM card uses ELR optics operating at an individual wavelength (100-GHz or 200-GHz spacing). The OC-48 DWDM cards are intended to be used in applications with long unregenerated spans. Transmission distances can be extended with the Cisco 15216 EDFA.

The link-loss budget for the OC-48 ELR cards is 25 dB. Recall that loss is incurred by the type and condition of the fiber, number of splices and connectors, and so forth. A link-loss budget is the same as a power budget. Recall that the Tx to Rx levels result in power-/link-loss budget. It is best practice to use the minimum documented Tx level and the maximum Rx level (not to be confused with the saturation point of the receiver).

Cisco ONS 15216 Filters (200-GHz Spacing)

The ONS 15216 filter-based solution is scalable from 9 to 18 wavelengths using the 200-GHz-spaced solution. The solution is scalable from 16 to 32 wavelengths using the 100-GHz-spaced solution. Filters are normally used at hub sites where all wavelengths should terminate. The ONS 15216 OADMs, on the other hand, are utilized at sites where one to four wavelengths will be added and dropped. Figure 8-8 shows an example of noise on an amplified optical line.

Figure 8-8   *Noise on an Amplified Optical Line*

Base Red Filter

Upgrade Blue Filter

NOTE	The 200-GHz filters have nine channels in the blue band and nine channels in the red band. Not all of these channels line up with the 100-GHz filter ports. Two channels in the blue and two channels in the red are used exclusively in the 200-GHz filters. This explains why there are 36 different orderable cards.

The entire ONS 15216 family is compatible with the spacing of the ONS 15454 ELR cards, as long as the respective spacing is used together (for example, 100-GHz or 200-GHz products). You can internetwork the Cisco ONS 155xx and 15252/15201 product families' 100-GHz-spaced metro products with the ONS 15454/15216 solution.

With the 32-lambda 15216 solution, the ONS 15216 can multiplex up to 80 Gbps across a single fiber (2.5 Gbps × 32 wavelengths). This solution is comprised of the two different ONS 15216 16 lambda boxes. The signals are multiplexed together.

The following material focuses on the 18-wavelength 15216 filter solution, which uses 200-GHz spacing. The material applies to the 32-wavelength 100-GHz spacing as well, but the numbers would be changed: 9 (200 GHz) = 16 (100GHz) and 18 (200 GHz) = 32 (100 GHz). The Cisco ONS 15216 filters are orderable as two separate products. The reason behind this is twofold. Splitting the solution into 2 components allows service providers to purchase the 9-wavelength solution and as their network grows, migrate to the complete 18-wavelength solution. The second reason for the separate boxes is the amount of loss that would be incurred by multiplexing or demultiplexing all of these wavelengths in one device. This allows service providers who need less than nine wavelengths to incur less loss in their DWDM network.

On both systems the multiplexing functionality is handled by the left side of the ONS 15216 filters, whereas the demultiplexing functionality is handled by the right side of the 15216 filter. (Left and right are used from the perspective of an outsider looking at the front of the 15216 filter.) The left side of the filter is responsible for multiplexing the OC-48 ELR tuned lasers to a composite DWDM signal. The right side of the filter is responsible for demultiplexing the tuned laser signals. The COMMON ports on the 15216 filters represent the composite DWDM signals. OUT corresponds to a Tx interface, whereas IN corresponds to an Rx interface. The red base unit integrates two expansion ports, which makes it possible to provide an in-service upgrade from 9 to 18 wavelengths.

The Cisco ONS 15216 base unit incorporates two monitor ports, allowing users access to the composite (DWDM) signal for monitoring or analysis using the ONS 15216 OPM. Figure 8-9 displays the 18-wavelength filter signal processing.

Figure 8-9   *Blue and Red Filter Interconnection and Traffic Flow*

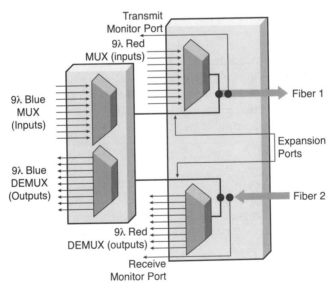

Planning an Optical System Using Cisco Equipment

The Cisco ONS 15216 unidirectional optical filter coupled with the Cisco ONS 15454 and OC-48 ELR ITU optics can be deployed in a wide variety of applications. In the same way that two ONS 15454 optical cards are necessary to complete a ring-based architecture, two 15216 filters are normally used for the West and East sides of the ring. If the 18-wavelength solution were used, there would be two sets of 15216 filters.

A power meter is one of the most important tools to use at the ONS 15454 ELR cards. The power meter can be attached to the Rx fiber to obtain a reading of the signal level. If this signal level falls below the receiver sensitivity threshold, there is a problem in the network. Because a power meter requires removing the cable from the receiver interface, it is more practical to use an OSA or OPM and monitor ports on the 15216 filters, which can be operated remotely. Power meters are a crucial tool for the optical engineer in setting up and troubleshooting a faulty network.

Span-Loss Calculation

After a power budget has been calculated between the optical transmitter and receiver, the actual span loss should be calculated to determine whether optical amplification is needed in the network. If optical amplification is needed, more precise calculations are necessary to determine where the amplification should take place. If the span is long, the design will call for multiple optical amplifiers to be daisy-chained. At this point, OSNR calculations would be necessary to determine the necessity of a regeneration site.

Table 8-2 describes the parameters used in the following equations that would be used to calculate the power budget of the system and then the span loss based on the actual components used in the system. This table includes specifications of the Cisco ONS 15216 filters, the ONS 15454 OC-48 ELR ITU optics, and various fiber-loss specifications.

NOTE   Transmitter and receiver levels are measured in dBm (decibel meters), whereas loss is measured in dB (decibels). dBm is an absolute value that can be explicitly measured, whereas dB is a relative value. 3 dB of loss does not identify the signal level, just what percentage of the signal has been lost. dB is logarithmic in nature. A 3-dB loss represents a loss of approximately 50 percent of the signal, whereas a 10-dB loss represents a loss of 90 percent of the signal (10 percent of original signal left available at the receiver).

Table 8-2 describes the various terminology used in the following calculations:

Span loss = DWDM loss × 2 (Source and destination site) + Power Penalty + Margin + Pf
Span loss = (4.5 dB × 2) + 1.0 dB + 3.0 dB + 12.0 dB
Span loss = 25.0 dB

If the span loss is less than or equal to the power budget allotted, the design will work. Actual numbers should be derived optical card and fiber characterization tests to obtain actual loss in the system. Manufacturers list ranges of optical Tx power. The only way to obtain the true value is to run characterization tests on the equipment. Transmitter levels are documented as maximum and minimum because optics are variable.

If the span loss is considerably less than the power budget, a second set of calculations should be made using the maximum transmitter levels to ensure that the receiver is not oversaturated with power. This occurs at –9 dBm on the OC-48 ELR cards per Table 8-2. When it is determined that the receive level is too high at the receiver, the receiver should be attenuated. Attenuators come in a variety of sizes. The Cisco ONS 15216 OADM includes two (West and East) variable optical attenuators (VOAs) that you can use to attenuate the add/drop signals.

Table 8-2   *Span Budget*

Parameter	Specification	Comment
Transmitter power. Cisco ONS 15454 OC-48 ELR ITU optics cards	–2.0 dBm	Worst-case laser output power (minimum).
Optical path power penalty	1.0 dB	Includes cross-talk, polarization impairments, center frequency drift, reflections, jitter, intersymbol interference, and laser chirp.
DWDM loss. Two Cisco ONS 15216 optical filters (4.5 dB each)	9.0 dB	18-wavelength solution including connectors.
Margin	3.0 dB	Per-company engineering practices. Margin to offset fiber splices, aging, etc.
Receiver sensitivity	-27 dBm	Worst case for 1E-12 bit error rate (BER).
Fiber + Connector attenuation	12.0 dB	Fiber loss is typically around .25 dB/km in the 1550-nm range. Fiber connectors can be estimated at .5 dB per connector. Matched connectors with losses as low as .25 dB per connector are available.
Cisco ONS 15454 OC-48 ELR receiver saturation	–9 dBm	Saturation point of the receiver.

Building a DWDM Infrastructure

To connect the blue upgrade unit to the red base unit, complete these steps:

Step 1   Mount the blue upgrade unit near the red base unit.

Step 2   Using an SMF with SC connectors, connect the red base unit to the blue upgrade unit, route a fiber from the To Upgrade (Tx) port on the base unit to the Common In (Rx) port on the upgrade unit.

Step 3   Using an SMF with SC connectors, connect the blue upgrade unit back to the red base unit, and route a fiber from the Common Out (Tx) port on the upgrade unit to the From Upgrade (Rx) port on the base unit.

NOTE   The syntax Tx and Rx does not appear on the 15216 filters. Tx and Rx in the preceding steps are being used to convey the operation of the port. A Common Out port is similar to a Tx interface in that a measurement device can be connected to it and a reading taken. This is the composite output. A Common In port is similar to an Rx port in that it is meant to receive a signal. This is the composite receiver.

Fiber connectors should *always* be cleaned when connecting to a port. Cleaning should be performed with a CLETOP fiber cleaning device. Using alcohol to clean fiber optics increases the chances of burning impairments into the fiber-optic terminator (especially with high-power lasers). See Figure 8-10 for a detailed view of the red and blue filters. See Figure 8-11 for an example of a connection between red and blue filters.

Figure 8-10   *Detailed View of the Red and Blue Filters*

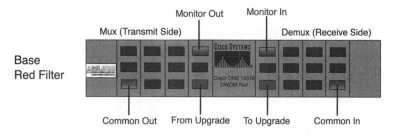

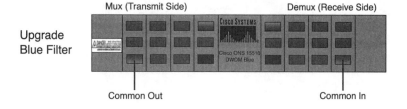

Figure 8-11 *Connecting the Red and Blue Filters*

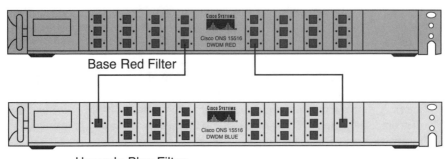

Base Red Filter

Upgrade Blue Filter

DWDM Network Requirements

When deploying the Cisco ONS 15216 on an optical network, consider the type of network: mirrored or nonmirrored. Mirrored basically means the ITU-T optical cards are the same on both ends of the cables. As you can see in Figure 8-12, both systems use the same ITU-T grid wavelength; therefore, one mirrors the other.

Figure 8-12 *DWDM Mirror Network*

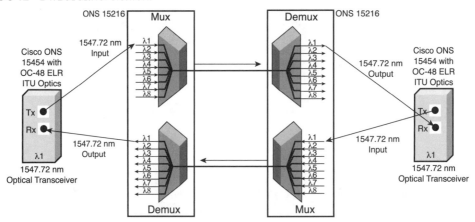

Most optical receivers are wideband in nature and accept any input wavelength between 1310 nm and 1550 nm. Therefore, it is possible to interconnect cards of different wavelengths through DWDM. A nonmirrored system does enhance the complexity of understanding the node relationships for cabling purposes, but it can save on having to spare every wavelength card you use in your DWDM system. Figure 8-13 shows a nonmirrored DWDM system that contains matching Cisco ONS 15454 ITU optics cards connected through an ONS 15216 at each end of the network. Notice, however, that the opposite direction uses a different wavelength.

Figure 8-13 *Nonmirrored DWDM System*

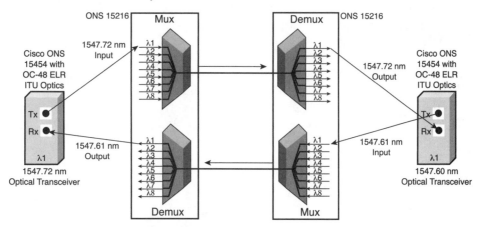

Provisioning the OC-48 ELR ITU Optics Card

When the DWDM hardware cabling is complete, the Cisco ONS 15454 OC-48 ELR ITU optics card in the Cisco Transport Controller (CTC) requires provisioning.

NOTE   There is no arming key used on the OC-48 ELR cards, and the lasers are on by default. When the card is placed in out-of-service status via software, it is still transmitting. Optical components can damage the retina of the human eye. Take great care when handling optical components; high-power lasers increase the risk of incurring eye damage upon exposure. Never look into the end of a fiber-optic cable unless you are wearing the appropriate safety glasses.

To provision the Cisco ONS 15454 OC-48 ELR ITU optics card in the CTC, complete these steps:

Step 1   Access the Node view for the Cisco ONS 15454 platform to configure.

Step 2   In the Node view, double-click the OC-48 ELR ITU optics card. The OC-48_LINE_CARD view appears.

Step 3   Click the Provisioning tab. The screen in Figure 8-14 displays.

Figure 8-14   *Provisioning the OC-48 ELR Card*

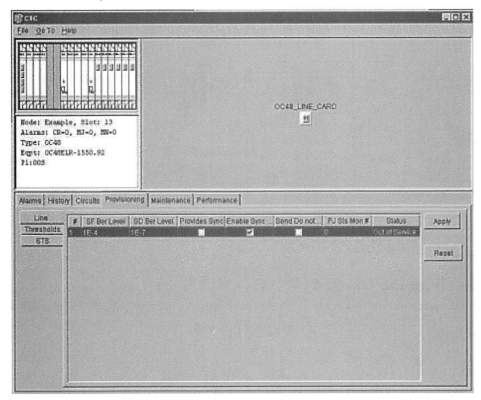

Step 4   Optional: Choose the signal fail bit error rate (SF BER) level.

Step 5   Optional: Choose the signal degrade bit error rate (SD BER) level.

Step 6   Change the Status to In Service/Active. When provisioning a port to In Service/Active, the corresponding number in the OC48_LINE_CARD graphic changes to green, showing that there is a valid signal and no alarms. Cards that are out of service appear as gray interfaces.

Step 7   Click Apply.

If there is no OC-48 optical input signal, the card and circuit go into an alarm.

Repeat this process for all OC-48 cards attached to the DWDM filters. You should now be ready to pass traffic.

Summary

In this chapter, you learned about DWDM and how you can use it when you need more bandwidth but do not have additional fiber available. You learned about the Cisco equipment that can be used to set up and configure a DWDM system.

Review Questions

1 The two largest problems facing service providers today in the metro environment are
   _____ and _____.

   A   Bandwidth scalability, fiber exhaustion

   B   Bandwidth scalability, wavelength multiplexing

   C   Fiber exhaustion, virtual dark fiber support

   D   Fiber exhaustion, wavelength multiplexing

2 Which of the following allows for a composite optical signal on a fiber to be strengthened simultaneously by one element?

   A   Fiber

   B   Microfilter

   C   Splitter

   D   Optical amplifier

3 Generally for a DWDM system, an optical signal can be sent how far before an OA is needed?

   A   1–10 km

   B   20–30 km

   C   40–60 km

   D   100–500 km

4 The wavelength separation between two different channels is known as
   _____.

   A   Dispersion

   B   Attenuation

   C   Channel spacing

   D   Overlap

5 What is the function of the optical amplifier in DWDM?

   A Demultiplex wavelengths

   B Individually regenerate a specific wavelength

   C Amplify all wavelengths simultaneously

   D Multiplex wavelengths

6 Which specification defined the standard channel-spacing wavelengths for a DWDM system?

   A ITU-T G.253 wavelength grid

   B ITU-T G.962 wavelength grid

   C ITU-T G.692 wavelength grid

   D ITU-T G.893 wavelength grid

7 _____ are used to equalize the output signal when it leaves the demultiplexer stage.

   A Post-amplifiers

   B Mid-span amplifiers

   C Pre-amplifiers

   D Wavelength amplifiers

8 Which two components comprise the Cisco DWDM solution?

   A Cisco ONS 15216 optical filter

   B Cisco ONS 15327 with OC-12 ELR ITU optics

   C Cisco ONS 15454 with OC-48 ELR ITU optics

   D Cisco 12000 GSR router with OC-48 ELR ITU optics

9 What is the maximum network bandwidth over a single fiber provided by the Cisco ONS 15454 OC-48 ELR ITU card with 200-GHz spacing?

   A 1.544 Mbps

   B 24 Gbps

   C 45 Gbps

   D 80 Gbps

10 Into what slots on the Cisco ONS 15454 can you install the OC-48 ELR ITU optics cards?

   A   Into slots 1 through 6

   B   Into any high-speed card slot

   C   Into any multispeed card slot

   D   Into any high-speed or multispeed card slot

11 How many wavelengths in the red and blue band does the ONS 15216 with 100-GHz spacing support?

   A   9 in blue optical band, 9 in red optical band

   B   16 in blue optical band, 16 in red optical band

   C   9 in blue optical band, 18 in red optical band

   D   18 in blue optical band, 18 in red optical band

12 When connecting the Cisco ONS 15216 red base unit to the blue upgrade unit, what port on the base unit connects to the Common In port on the upgrade unit?

   A   To Upgrade

   B   Monitor Out

   C   Common In

   D   Common Out

This chapter covers the following topics:

- Evolution of voice and data networks
- Packet over SONET applications
- Packet over SONET operation and specification
- Packet over SONET efficiencies
- Packet over SONET network designs
- Packet over SONET protection schemes

CHAPTER 9

Packet over SONET

SONET is a time-division multiplexing (TDM) architecture that was designed to carry voice traffic. All traffic in SONET is broken down into slots of 64-kbps DS0 increments. A DS0 is the voice line that is typically hard-wired into homes. TDM architectures are not ideal solutions for transporting data. Cable and DSL providers have shown this with their high-data-throughput broadband offerings that do not incur the same costs as comparable TDM services would incur.

When it was discovered that computer data could be transported over telephone circuits, service providers (SPs) leveraged their existing SONET rings. SONET rings were designed and deployed to transport voice but could transport voice by breaking down the data needs into manageable pieces and transporting in 64-kbps increments. When anything less than 100 percent of a TDM circuit was used, the remainder is stuffed with arbitrary data and therefore wasted from both the customer's and SP's perspective. Frame Relay technology offered statistical multiplexing, which offered a solution to the inefficiencies of SONET-based services. Unfortunately, the designers of Frame Relay did not have quality of service (QoS) in mind with the design of the technology. Many carriers also offered zero committed information rate (CIR) services only, which guaranteed the end user absolutely no class of service (CoS). Customers found this unacceptable.

ATM offered a solution to the QoS issues of Frame Relay and offered scalability in the optical carrier (OC-n) domain. ATM relied on a fixed-size cell that is not compatible with the Ethernet technologies that most LANs employ. ATM-designed hardware includes a segmentation and reassembly (SAR) layer to translate Layer 2 frames (Ethernet, Token Ring, FDDI, and so on) into ATM cells. The SAR functionality introduced slight delays in the network, but it is prohibitively expensive and complex to design. Because of the issues associated with ATM, many vendors have not deployed OC-192 ATM interfaces at this time. There is also a concept known as *cell tax* with ATM deployments. ATM introduces extra overhead into each transmission because of its fixed size of 53 bytes. If a Layer 2 (Ethernet) frame does not fall on a cell boundary, the rest of the cell is padded to meet the 53-byte cell requirement. ATM cells might be efficiently multiplexed into a SONET frame, but the architecture has delays and inefficiencies that must be accounted for.

Packet over SONET (PoS) is a highly scalable protocol that overcomes many of the inefficiencies of ATM, while providing legacy support to internetworks with existing SONET architectures. PoS provides a mechanism to carry packets directly within the SONET synchronous payload envelope (SPE) using a small amount of High-Level Data Link Control (HDLC) or PPP framing.

Evolution of Voice and Data Networks

Voice and data networking is constantly evolving as the technology evolves. After the telegraph, telecommunications networks evolved to transport the spoken word. The next evolutionary step, data networks, occurred in the mid-1900s with the advent of computers. Although data networking started out small because only the largest corporations could afford computers, computers have fallen to such a low entry-level price that most people can afford to have a computer now and to be connected to the Internet. Data networks have evolved to the point that the benefits of converging voice and data networks into the same data infrastructure can no longer be ignored. SPs that have legacy SONET infrastructures can still offer customers high-speed alternatives with technologies such as PoS.

The first communications systems were mainframe computers linked to dumb terminals. The Synchronous Data Link Control (SDLC) protocol, developed by IBM, made this system possible by allowing communication between a mainframe and a remote workstation over long distances. This protocol evolved into the High-Level Data Link Control (HDLC) protocol, which provided reliable communications and the ability to manage the flow of traffic between devices. HDLC is an open industry standard protocol, whereas SDLC is an IBM proprietary protocol that must be licensed by IBM. Industry standard protocols such as TCP/IP drive the adoption and low costs of telecommunications equipment.

Cisco offered an enhanced multiprotocol version of the HDLC protocol to enable various protocols over the HDLC (High-Level Data Link Control) Layer 2 framing. This Cisco HDLC protocol is proprietary and exclusive to Cisco. The HDLC standard was loose at the time of Cisco's creation and left too much room for interpretation. When this was standardized in RFC 1619 with PPP in HDLC-like framing, Cisco's HDLC protocol was not compliant. Point-to-Point Protocol (PPP) evolved from HDLC; it offers an industry-standard way to provide multiprotocol networking abilities, as well as many enhancements such as authentication, multilink, and compression. PPP is used for many other technologies, including ISDN. HDLC and PPP are scalable to architectures with fast speeds. Figure 9-1 shows this networking evolution.

Figure 9-1 *Voice and Data Network Evolution*

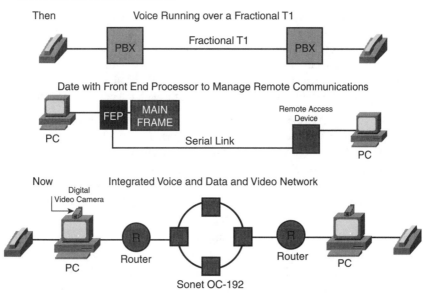

Applications for PoS

PoS is a Layer 2 technology that uses PPP in HDLC encapsulation, using SONET framing. The PoS solution lowers the cost per megabyte when compared to other Wide Area Networking architectures. The PoS interface supports SONET level alarm processing, performance monitoring, synchronization, and protection switching. This support enables PoS systems to seamlessly interoperate with existing SONET infrastructures and provides the capability to migrate to IP+Optical networks without the need for legacy SONET infrastructures. PoS is used in a point-to-point environment, much like the legacy T-carrier architectures, but without the need for TDM.

PoS efficiently encapsulates IP traffic with a low-overhead PPP header. When encapsulated, the traffic is placed inside an HDLC-delimited SONET SPE and transported across SONET. Voice, video, and data can be carried within the IP packets using Layer 3 QoS mechanisms to control priority when bandwidth contention occurs.

PoS can be used in tandem with other technologies carried over SONET architectures. PoS is not compatible with these other technologies, but is not aware of them because they are being transported over different time slots. PoS, TDM voice, ATM, and Dynamic Packet Transport (DPT) can each use their required synchronous transport signals, not interacting with each other. PoS interfaces are available in concatenated and nonconcatenated (channelized) options. Channelized interfaces are more costly than concatenated interfaces.

ATM and PoS

ATM and PoS can be used within the same network. ATM technology provides an effective, flexible provisioning mechanism for low- to high-speed network access. ATM switches can be used to aggregate digital subscriber line (DSL), cable, and customer traffic by using permanent virtual circuits (PVCs) or switched virtual circuits (SVCs), which can then feed into compatible downstream routers. This traffic is then fed into higher-speed links attached to the Cisco 12000 series router for transport through the core through PoS interfaces.

An advantage of ATM is its innate support for QoS. PoS is fully capable of supporting the transport of time-sensitive data, using Layer 3 mechanisms. Technologies such as Resource Reservation Protocol (RSVP), committed access rate (CAR), and Weighted Random Early Detection (WRED) enable providers to offer QoS solutions in a more cost-effective manner than ATM. These technologies are Layer 3 implementations for QoS. This book does not focus on QoS, but other Cisco Press books covering QoS are available (such as *IP Quality of Service*; ISBN: 1-57870-116-3).

Figure 9-2 is an ATM aggregation design in which switches are placed on the ATM network edge and translate ATM traffic into PoS traffic. Notice the amount of PoS interfaces required to create a resilient network design with no single point of failure. PoS is a point-to-point technology, regardless of the distance traveled. A point of presence (POP) environment where equipment might be closely located would be perfect for Very Short Reach (VSR) optics using PoS technology. VSR optics are lower in price than normal PoS interfaces because they are not meant to travel long distances. They can be manufactured with lasers that are weaker in strength and photodiodes that are not as sensitive as photodiodes required for long spans.

Figure 9-2   *ATM Aggregation over PoS Networks*

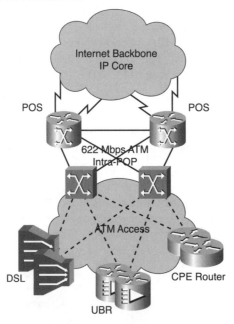

PoS Transport

PoS does not require SONET transport but works in tandem with such as a result of the SONET framing that PoS employs. Two PoS devices can be connected directly with duplex fiber. Because PoS interfaces are Layer 3 enabled, PoS interfaces are an example of an IP+Optical architecture.

Figure 9-3 displays three different ways in which PoS traffic can be transported. The three mechanisms are explained as follows:

- **Connectivity to SONET ADMs**—SONET circuits are provisioned as point-to-point circuits over SONET rings. Routers with PoS interfaces can be attached to SONET add/drop multiplexers (ADMs). As long as the proper number of STS are provisioned, the PoS interfaces will have connectivity. The PoS traffic is multiplexed with the other traffic that the SONET ADMs are carrying.

- **Connectivity to transponders in a DWDM system**—PoS traffic can be translated to a DWDM ITU-grid wavelength using a transponder. Most transponders support SONET framing. Through the DWDM system, 32 PoS circuits can be multiplexed onto one fiber.

- **Dark-fiber connectivity**—PoS interface can be connected directly over dark fiber using PoS interfaces. Dark fiber is fiber that is leased from a service provider; the customer provides the source (Laser or LED)and destination (photodiode receiver). This process is normally referred to as *lighting the fiber*. Long spans can be accommodated through standard SONET regenerators that provide regeneration, reshaping, and retiming (3Rs) of the signal. The Cisco 15104 is an OC-48 SONET regenerator that fits this application.

Figure 9-3   *PoS Transport Options*

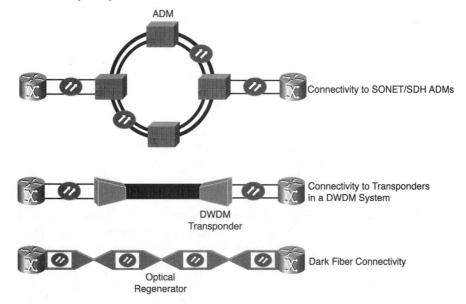

Multiaccess Protocol over SONET

Multiaccess Protocol over SONET (MAPoS) is a high-speed, link-layer protocol that provides multiple access capability over SONET/SDH.

MAPoS is defined in RFCs 2171 and 2176. The MAPoS frame format is based on HDLC-like framing for PPP. MAPoS is a frame switch that allows multiple nodes to be connected in a star topology to form a LAN using MAPoS.

NOTE

You can find all RFCs online at http://www.isi.edu/in-notes/rfc*xxxx*.txt, where *xxxx* is the number of the RFC. If you do not know the number of the RFC, you can find it by doing a topic search at http://www.rfc-editor.org/rfcsearch.html.

MAPoS can be used to allow SONET connectivity directly to the desktop. MAPoS is much more costly than Ethernet connectivity and has not been deployed, albeit a little in European markets. Most ATM-to-the-desktop environments have migrated their infrastructures to Ethernet technologies. MAPoS will probably never gain the market acceptance that Ethernet has. With SPs looking for more ways to leverage the low cost of Ethernet, it would be unlikely that enterprise environments roll out expensive SONET interfaces to their desktops in place of the ubiquitous, cost-effective Ethernet interfaces they currently use.

Packet over SONET Operation and Specifications

The current Internet Engineering Task Force (IETF) PoS specification is RFC 2615 (PPP over SONET), which obsoletes RFC 1619. The PoS RFCs define the requirements that are needed to transport data packets through PoS across a SONET network. These requirements are summarized as follows:

- **High-order containment**—PoS frames must be placed in the required synchronous transport signals used in SONET. An example of this is an OC-12 concatenated PoS interface. This interface requires an STS-12 circuit to contain the required payload of the PoS traffic.

- **Octet alignment**—This refers to the alignment of the data packet octet boundaries to the STS octet boundaries. An octet (byte) defines an arbitrary group of 8 bits. The word *byte* is defined as usually containing 8 bits. IBM used to define a byte as containing 7 bits. Although both byte and octet are used interchangeably, octet is a more accurate representation for 8 bits because its meaning is a series of eight.

- **Payload scrambling**—Scrambling is the process of encoding digital 1s and 0s onto a line in such a way that provides an adequate number for a 1s density requirement. The ANSI standard for T1 transmission requires an average density of 1s of 12.5 percent (a single 1 in 8 bits meets this requirement) with no more than 14 consecutive

0s for unframed signals and no more than 15 consecutive 0s for framed signals. The primary reason for enforcing a 1s density requirement is for timing recovery or network synchronization. However, other factors such as automatic-line-build-out (ALBO), equalization, and power usage are affected by 1s density. RFC 1619 inadvertently permitted malicious users to generate packets with bit patterns that could create SONET density synchronization problems and replication of the frame alignment. RFC 2615 provides a more secure mechanism for payload scrambling.

High-Order Containment

End stations at customer sites are predominantly TCP/IP-enabled devices. At the edge of the customer's network, the IP packet is encapsulated into a Layer 2 format that will be supported on the SP's network. The Layer 2 protocols supported by Cisco are PPP and Cisco HDLC, but the PoS standards specify PPP encapsulation for PoS interfaces. The Layer 2 PPP or Cisco HDLC frame information is encapsulated into a generic HDLC header (not Cisco proprietary HDLC) and placed into the appropriate SPE of the Whereas frame. This can be a confusing concept at first. Although HDLC and PPP are different, mutually exclusive Layer 2 protocols, HDLC is used as a SPE delimiter in the SONET frame. The encapsulation process of an IP packet to a SONET frame is illustrated in Figure 9-4.

Figure 9-4   *Encapsulating IP into a PoS Frame*

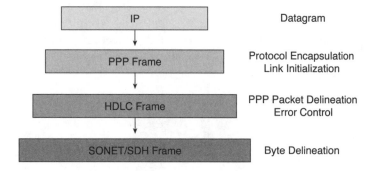

PPP Frame

RFC 1548 defines a PPP frame that contains the following three components:

- Protocol field
- Information field
- Padding field

The Protocol field is used because PPP was designed to be multiprotocol in nature. Multiprotocol encapsulations transport multiple protocols, including IP and IPX. The

Information field is the protocol data unit (PDU) transmitted, and can be from 0 to 64,000 bytes. The Padding field is used to pad the PPP frame if the Information field does not contain enough data. The Padding field might receive padding up to the maximum receive unit (MRU), which will fill the Information field. The default value for the MRU is 1500 octets but can be up to 64,000 octets if negotiated in the PPP implementation. It is the responsibility of the protocol to determine which bits are used as padding. You can find more information about the PPP protocol in RFC 1548 and RFC 1661 at www.ietf.org. Figure 9-5 illustrates the PPP in HDLC-like frame format.

Figure 9-5   *RFC 1662: PPP in HDLC-Like Framing*

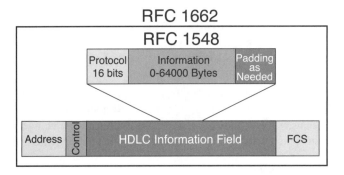

Figure 9-6 illustrates the values used in the PPP in the HDLC-like framing process. Notice that frame delimiters of hexadecimal 0x7E (126 in decimal) are used to denote the beginning and ending of a frame. The transmitting device generates flags as a time fill when there are no data packets.

Figure 9-6   *Packet over SONET Frame Information*

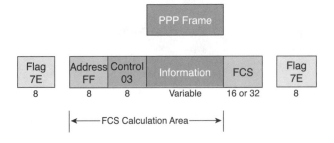

The Address field is always set to 0xFF (255) because every frame is a broadcast frame in PoS. There are only two ends of the point-to-point connection, and the frame always needs to get to the other side. There is no reason to have more than one address because there are no other addressable destinations. The Layer 2 mechanism is terminated at the other end of the link because PoS interfaces are Layer 3 enabled.

A Control field of 0x03 (3) is used to denote an HDLC frame. The Information field is where the PPP frame is inserted and is variable in nature due to MRU variability. A 16- or 32-bit frame check sequence (FCS) is used as a trailer to the frame. The FCS can be 16 or 32 bits long, but 32-bit CRCs are highly recommended due to the enhanced error recovery that is available using 32 bits. Most interfaces that run at speeds greater than OC-12 use FCS-32 as the default. The FCS is a configurable option, and FCS 32 is always recommended. The FCS field needs to match on both ends of the connection; otherwise, the Layer 2 protocol will never come up.

NOTE	Although this book does not specifically deal with SDH, all Cisco PoS interface card framing can be changed from the default SONET framing to that of SDH.

Payload Scrambling

SONET has a default scrambler that was designed for voice transport. The 7-bit SONET scrambler is not well suited for data transport. Unlike voice signals, data transmissions might contain long streams of 0s or 1s. If the change from a 0 to 1 is not frequent enough, the receiving device can lose synchronization to the incoming signal. This can also cause signal-quality degradation resulting from distributed bit errors. The solution to this synchronization and bit error problem is to add an additional payload scrambler to the one normally found within SONET environments. This scrambler is applied only to the payload section, which contains the data. The SONET overhead bytes do not need this additional scrambling because they continue to use the existing 7-bit SONET scrambler. Certain overhead bytes, including the A1/A2 SONET framing bytes and the J0 section trace byte, are never scrambled with any type of scrambling.

The two versions of scrambling that are supported by PoS are defined in the Telcordia GR-253 and ITU-T I.432 documents. The Telcordia GR-253 standard defines a basic $1 + x^{\wedge 6} + x^{\wedge 7}$ algorithm that scrambles the transport overhead of the SONET frame (with the exception of certain overhead bytes). This scrambler cannot be disabled and is adequate when the SONET frames carry phone calls in the payload.

The ITU-T I.432 standard defines an ATM-style scrambling. This scrambler uses a polynomial of $x^{\wedge 43 + 1}$ and is a self-synchronous scrambler, meaning that no state needs to be sent from the sender to the receiver. With this scrambler, only the data SPE of the SONET frame is scrambled.

The scrambling function is performed in hardware and does not affect performance in any way. Scrambling is performed directly in the framer application-specific integrated circuits (ASICs) on newer line cards and in a separate adjacent ASIC on older line cards. As technology evolves, more functionality is integrated into the same ASICs to lower the real estate (space) needs of hardware. Cisco supports port densities as large as 16 OC-3 ports on 1 Cisco 12000 series line card.

The path overhead C2 byte (path signal label) is used to instruct the receiving equipment that payload scrambling is turned on. If the traffic is carrying PPP with scrambling turned on, the value is set to a hexadecimal value of 0x16 (22). If scrambling is turned off, the original hexadecimal value of 0xCF (207) is used. Figure 9-7 shows the scrambling functions used in PoS environments.

Figure 9-7   *SONET RFC 2615 Payload Scrambler*

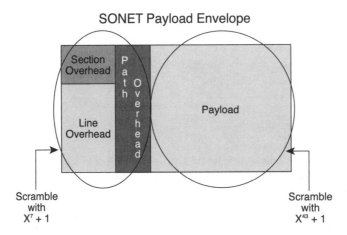

PoS Efficiencies

Overhead efficiency is a critical topic for SPs that charge for the amount of bandwidth capacity customers use. ATM was introduced to the SP market as the technology that would enable converged voice, video, and data traffic to reside on the same infrastructure (because of the intrinsic QoS parameters built in to the technology). The technology is widely used by Internet service providers (ISPs) today, but many of the original intentions behind ATM have not been used due to the complexity of configuring, selling, and maintaining such features. ISPs want to maximize their profits and minimize the costs associated with transporting IP over ATM.

ATM uses fixed-sized ATM cells of 53 bytes. Each cell's composition includes 5 bytes of fixed overhead and 48 bytes of data. Depending on the ATM adaptation layer (AAL) used, the amount of AAL overhead can be as high as 4 bytes in addition to the 5 bytes of fixed overhead. This extra AAL overhead could result in as little as 44 usable bytes in a 53-byte cell. This equates to an approximate efficiency level of 83 percent (or 17-percent overhead). SONET's TOH and POH combined equal 36 bytes per the calculation that follows:

> TOH (Transport Overhead)
> 9 rows × 3 columns = 27 bytes
> POH (Path Overhead)
> 9 rows × 1 column = 9 bytes
> TOH + POH = 36 bytes

The 36 bytes of overhead used in SONET represent approximately 4 percent of the total 810-byte STS frame. The ATM inefficiencies are further compounded when the variability of data sizes is calculated. Because of the nature of web pages, most Internet browsing traffic consists of many small-size packets. Most traffic that is generated originates from workstations connected to a LAN with Ethernet technology. The smallest PDU available in Ethernet networks is 46 bytes but can vary up to the maximum transmission unit (MTU) size of 1500 bytes. If a packet does not fall neatly on a cell boundary, the rest of the cell is padded.

In the case of a frame sent with a frame size of 64 bytes, two ATM cells are needed to transport the data. The first cell would be fully used at 48 bytes of payload (assuming that an AAL with no extra overhead is in use), and the second cell would be nearly empty with only 16 bytes of payload. This scenario results in a low efficiency level (approximately 43 percent). Figure 9-8 illustrates the PoS efficiency over ATM in both a line graph and table, which compares efficiency based on packet size.

Figure 9-8   *PoS Efficiencies Compared to ATM*

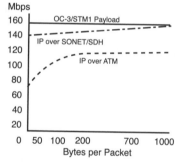

Packet Size (byte)	POS Eff. (SPE %)	ATM Eff. (SPE %)
64	86.8	43
128	94	69
256	97	75
512	98.6	85
1024	99.3	86
1518	99.5	88
2048	99.6	89
4352	99.8	89

PoS Network Designs

Resiliency is an important concern in SP networks. Outages result in lost revenue and might cause customers to cancel their service. SPs enter into Service Level Agreements (SLAs) with their customer. These SLAs guarantee certain levels of service. SLAs differ in many respects depending on the amount of risk the customer is willing to take. The more risk the customer is willing to take, the looser the SLA is and the cheaper the cost to the customer. The downside is that the customer is not guaranteed the same level of service as the customer who was not willing to take as much risk and paid more money for a stringent SLA.

PoS provides support of the optical 1+1 automatic protection switching (APS) mechanism. A customer desiring this level of protection orders two circuits from the SP: one for working traffic and one for protect traffic. SPs offer discounts for circuits that are used for protect traffic. The CPE router in this design could be a single point of failure, depending on how the circuit terminates at the CPE. Both circuits terminating on one line card of one router would result in a single point of failure from both the line card and router perspective.

A method of slightly higher resiliency is to still use one router, but use separate line cards for the working and protect circuits. This scenario provides fault tolerance in the case of a line card failure, but not a router failure. A higher level of fault tolerance might be achieved if each circuit terminates on a separate router.

All of these survivability network designs are connected to one ADM at the service provider. APS 1+1 protection schemes are normally implemented per add/drop multiplexer. Ring failure is handled by the SP's robust SONET ring protection mechanisms. ADMs are carrier-class devices and must maintain a level of Five-Nines reliability. Five-Nines reliability refers to the amount of uptime a customer should expect from that network. Five-Nines reliability represents an uptime of 99.999 percent.

One Router

Figure 9-9 shows a design where there is one router at the customer premises with two optical interfaces used for APS 1+1 protection. Although a one-router CPE design does not provide the highest level of resiliency, this design does offer some advantages, including the following:

- No routing convergence upon failure of the working circuit or optical interfaces.
- 1+1 APS optical protection. Convergence time can be achieved in sub-60-ms time.
- Low-complexity network configuration.

Figure 9-9   *One-Router CPE Design*

Two Routers

In a two-router design, each router has one optical connection to the SP's add/drop multiplexer. Fault tolerance has been increased with this design because the CPE router is no longer a single point of failure and the routers can be located in different areas of the building to facilitate fault tolerance associated with issues that could arise in isolated areas.

Figure 9-10 illustrates the two-router design philosophy. Although each router has one optical interface to the ADM, one link is working (active) while the other link is protecting (standby) the working link.

Figure 9-10   *Two-Router CPE Design*

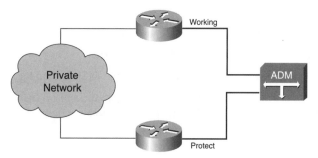

Resources are wasted if only one router is actively forwarding traffic. To fully use both routers, you could use a design including four circuits. The design requires twice the number of interfaces and circuits, but this might still be cost advantageous depending on the amount of bandwidth required and the router hardware employed. Figure 9-11 displays an environment that includes two routers and four circuits in which both routers are in a working state for one circuit. The optical protection scheme is logically divided into APS protection groups that the routers monitor. A large router such as the Cisco 12000 can accommodate hundreds of APS groups.

Figure 9-11   *Two Routers with Four Circuits*

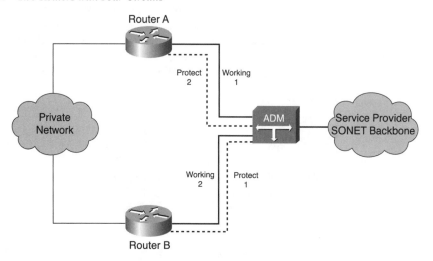

Advantages of having two PoS-connected routers include the following:

* Router redundancy in addition to circuit redundancy
* Load balancing of traffic

Disadvantages of having two routers connected to the single ADM in the SONET/SDH network include the following:

- **Convergence time**—The Layer 3 routing protocol must converge to optical circuit failure.
- **Complex network configuration**—APS groups.
- **Cost**—The costs associated with setting up and maintaining the design.

Failure recovery using two routers cannot achieve the sub-60-ms time that the one-router alternative offers. The Layer 3 routing protocol implemented in the infrastructure needs to reconverge around the failure. This is not an issue with one-router designs because both of the PoS interfaces on one router can have the same IP address with the PoS APS 1+1 configuration commands. This feature is allowable because only one of the interfaces is active at any one time.

PoS Protection Schemes

Packet over SONET protection uses the SONET APS 1+1 protection scheme. APS 1+1 looks at the K1 and K2 bytes of the SONET line overhead to determine whether issues exist with the SONET ring. A failure in the SONET network that affects the customer's working path causes a failover at the client site. This failover time occurs under 50 ms in the SONET Layer 1 network. The router interface uses a keepalive to determine whether the other side of the connection is alive. Keepalives are sent every 10 seconds by default, and the loss of 3 subsequent keepalives results in the interface going to an up/down state. After Layer 2 is lost, the Layer 3 routing protocol must converge around the link failure. Waiting for Layer 2 and Layer 3 to go through this procedure can take a long time (more than 30 seconds). Configuring the keepalives to 1 second lowers the convergence time to 3 seconds. Because PoS interfaces are Layer 3 implementations, the interfaces need to rely on a hierarchical error-recovery method such as that shown in Figure 9-12.

Figure 9-12 *PoS Hierarchical Error Recovery*

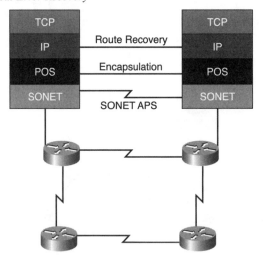

Layer 3 provides rerouting decisions during network failures to provide intelligent resiliency to the network. Layer 3 routing might be needed during a link failure if the Layer 3 IP address is changed. The Layer 3 IP address in the one-router design would be identical and there would never need to be Layer 3 routing protocol reconvergence. PoS interfaces in different routers require different IP addresses and always result in routing protocol reconvergence.

APS 1+1 Protection

SONET APS 1+1 is used for any PoS design that has more than one optical interface. APS provides optical protection during times of optical failure in the SP network. This information is carried over the K1 and K2 bytes of the SONET overhead. The CPE listens to the K1 and K2 bytes generated by the SONET network and generates K1 and K2 bytes when a failure occurs on the customer side. If the working interface of an APS 1+1 group fails, the protect interface can quickly assume its traffic load. The Layer 1 APS 1+1 recovery mechanism operates in 60 ms.

NOTE   SONET rings have a 50-ms switchover time rather than the 60-ms switchover used in the Bellcore APS 1+1 specification. The APS 1+1 specification provides 10 ms for failure detection and 50 ms for switch initiation, which collectively equal 60 ms.

SONET APS works at Layer 1 providing switchover times significantly faster than any protocols operating at Layer 2 or 3.

The SONET protection mechanism used for PoS on Cisco products uses APS 1+1 with either unidirectional or bidirectional switching. (You can read more about 1+1 uni- and bidirectional switching in Chapter 3, "SONET Overview.")

The SONET APS 1+1 architecture designates that there will be two circuits and each will carry the same traffic. One circuit is considered the working circuit; the other is the protect line. This differs from 1:1 or 1:*n* electrical-protection schemes because the backup equipment in electrical-protection schemes only carries traffic upon failure. The working and protect lines of APS 1+1 are both always transporting traffic. The receiving device(s) only process the traffic being received on the working circuit.

Protection mechanisms are more complex when circuits are terminated on different routers. The protection router must somehow be identified of the failure situation. An additional protocol is needed to provide for this signaling. This protocol is a Cisco proprietary mechanism called the Protection Group Protocol (PGP).

If a signal fail (SF) or a signal degrade (SD) condition is detected, the hardware switches from the working circuit to the protect circuit. APS 1+1 has reversionary capabilities allowing the hardware to switch back to the working circuit automatically when the original signal is restored for the configured time interval. The configurable reversion time is used to prevent the system from switching back to the working circuit if it is flapping (repeatedly going up and down). Flapping is sometimes referred to as *switch oscillation* and should be avoided at all costs so that the SP equipment can meet the SLAs. If the revertive option is not used, after a switch has moved to the protect circuit, the hardware does not automatically revert back to the working circuit. A system administrator must manually perform this function. Bidirectional switching is the default operation in Cisco routers. A circuit that automatically switches back to the original facility is called a *reversionary circuit*.

The K1/K2 bytes from the line overhead of the SONET frame indicate the current status of the APS connection and convey any requests for action. In standard APS, the two ends of the connection use this signaling channel to maintain synchronization.

With Cisco PoS, the working and protect channels are synchronized through an independent communications channel that is not part of the standard SONET APS system. This independent channel works whether the interfaces are on the same or different routers. This low-bandwidth connection is the Cisco PGP.

Cisco Protect Group Protocol (PGP)

PGP is the Cisco proprietary APS communication channel that is used between routers to complement APS 1+1 protection signaling. APS 1+1 is normally only done on the same router, but PGP enables this functionality to span multiple routers for added resiliency.

Performing APS 1+1 operation between routers creates some Layer 3 convergence issues. The standard Layer 2 mechanism used to determine whether an interface is down is the keepalive function. To accommodate fast reconvergence times, the keepalive update timer should be changed to 1 second and the hold timer changed to 3 seconds. PGP is the signaling channel used to inform the router with the protect facility about the failure. PGP operation closely resembles that of Cisco Hot Standby Router Protocol (HSRP) performing a heartbeat operation over a low-speed interface that tracks the status of certain ports. You can configure different protection groups to monitor multiple ports. The PGP protocol is a connectionless protocol that uses User Datagram Protocol (UDP) port 172 for message transfer. Figure 9-13 displays two routers that are configured in the same APS group. Notice that PGP updates are propagated bidirectionally between the working and protect routers to exchange information regarding the status of the PoS interface.

Figure 9-13 *Protection Group Protocol Operation*

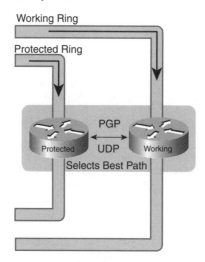

Figure 9-14 displays a network in which an outage occurs between POP B and POP C on the working facility. The routers at POP B and POP C will have knowledge of this outage through a loss of signal (LOS) condition, and PGP will notify the other router that it will now become the working interface. The other routers in the network will learn of this occurrence through the K1/K2 byte signaling occurring throughout the network.

Figure 9-14 *PGP Link Selection*

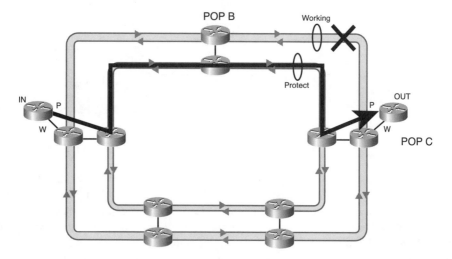

PoS Convergence

Convergence time is the amount of time required for all routers in a network to learn of changes in the network topology. Routers must propagate new route information from one end of the network to the other. Routing protocols are implemented to exchange this information. The routing protocol implemented should provide an ample amount of scalability to meet the future needs of the networks used in the environment. The faster the routing protocol can converge, the less downtime that will occur.

Scalable IP network routing protocols, such as Open Shortest Path First (OSPF), Integrated IS-IS, and Border Gateway Protocol (BGP), are responsible for recovering from error conditions in the network. Although the SONET APS 1+1 protection switching mechanisms guarantee a restoration time of 60 ms, the PoS interfaces are Layer 3 implementations and require some deal of routing protocol convergence. Typical convergence times for scalable routing protocols are several seconds or more depending on the environment and routing protocol design.

Figure 9-15 displays a design in which one router is used for the PoS interfaces. With this design, both of the PoS interfaces in the router can be configured with the same IP address. If a failure occurs, the router can perform switchover in the APS 1+1 switchover time of 60 ms. The Layer 3 routing protocol has not changed in any way on the LAN or WAN side of the router. The Layer 2 keepalive mechanism might not be aware of this switchover because it occurred in less than the lowest keepalive timer of 1 second. Regardless, three keepalives must be missed before an interface is determined as down.

Figure 9-15   *1-Router APS 1+1 Convergence*

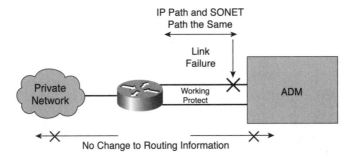

Figure 9-16 displays an environment that requires a higher degree of fault tolerance. This design uses two routers to implement the APS 1+1 group to protect the design from a router failure. The added resiliency creates some Layer 3 convergence issues because the interfaces used cannot have the same IP address if they reside on different physical routers. When the failure occurs, PGP is used to determine that the working interface has gone down, and the protect interface takes over. After this switchover has occurred, the Layer 3 routing protocol must communicate this information on both the LAN and WAN side so that the end to end network learns of the failure and solution. It is best to use HSRP on the LAN side if the PoS routers represent the default gateways out of the network. HSRP update and dead timers should be configured to match those of PGP.

Figure 9-16   *2-Router APS 1+1 Convergence*

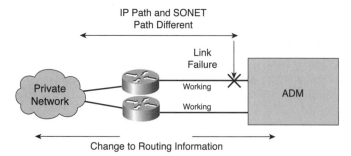

Flapping

Flapping is the operation of a transmission line regularly transitioning from an up/up to an up/down state in a short period of time. Intermittent failures can result in the APS protection mechanism switching between the working and protection traffic repeatedly, causing many fluctuations in the network. If a two-router PoS model is implemented, the Layer 3 routing protocols will flap, too. You can see this issue in Figure 9-17.

APS switches traffic upon failures, but the routing protocol must send out routing updates. If another failure happens (Failure 2), the failure results in another APS switchover and more routing updates. Subsequent failures (Failures 3 and 4) repeat the process. The result of this flapping is that the network could end up spending all the time sending routing updates and reconverging around repeating failures instead of sending data across the network.

Figure 9-17   *Flapping in a 2-Router PoS Design*

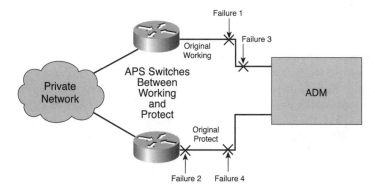

The issue is manageable by tweaking the reversion timer to a time greater than that necessary for the Layer 3 routing protocol to converge. The interfaces would not bring down the network because they must be stable in that amount of time before any switchover will take place.

PoS Reflector Mode

PoS Reflector mode is a process that is used to inform the remote router of a change in the network topology due to a line failure. Figure 9-18 displays an environment with two routers where a failure has occurred on the working line. As soon as the protect router receives information of the down interface through PGP, the protect router initiates a packet to the other side of the connection to speed up convergence. The packet contains the router ID information needed by the routing protocol to create the new Layer 3 adjacency. The remote router can now change the IP adjacency information immediately and reduce the convergence time dramatically.

Figure 9-18   *PoS Reflector Mode*

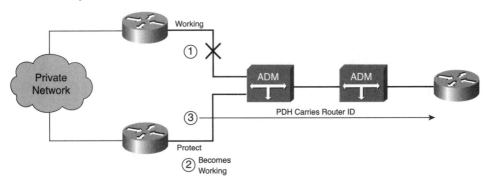

Load Balancing

Load balancing refers to the capability to have traffic traverse two separate paths simultaneously to maximize the resources at the site. Load balancing is possible in a PoS APS 1+1 environment where four circuits are present. APS groups are configured on each router. One router is the working router for Group 1, and the other router is the working router for Group 2. Each of these routers protects each other using the PGP mechanism to alert the other side of failures. Figure 9-19 shows this design. You can use Multigroup HSRP (MHSRP) on the LAN side to actively forward traffic to both of these devices while providing the resiliency necessary. Layer 3 convergence is an end-to-end solution.

Figure 9-19   *PoS Load Balancing*

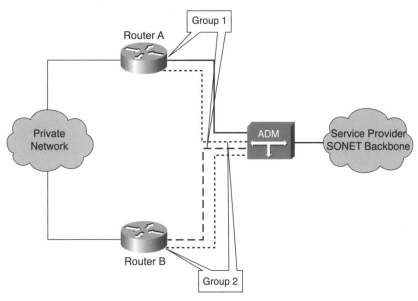

Alarms and PoS

Customers want to be notified of problems and errors that occur on their lines. PoS uses the same alarming of that used for SONET alarm reporting. The information that is carried in the overhead bytes of the Section, Line, and Path overhead layers are used by PoS to determine and report errors. This includes such items as the following:

- **Loss of signal (LOS)**—Signal failure due to a loss of light on the receive interface. A loss of light can also be thought of as receiving an all-0s pattern before descrambling. A downstream AIS should be sent when an LOS is detected.

- **Loss of frame (LOF)**—Issue created by receiving A1 and A2 bytes that do not indicate the 2-byte code of F628 in hexadecimal. An LOF condition is registered after no valid framing information has been received in 3 ms. The receipt of two subsequent valid A1/A2 frames clears this condition. A line alarm indication signal (AIS) must be sent downstream when this condition occurs.

- **Bit interleaved parity (BIP) errors**—BIP-3 errors occur at the path layer. The PoS interface is a path terminating equipment (PTE) device. The B3 byte carries the path parity errors in this byte.

- **Loss of pointer (LOP)**—When a pointer processor cannot obtain a valid pointer condition, an LOP state is declared, and a downstream AIS must be sent. Recall that the H1 through H3 bytes of the LOH are used for the pointer functionality.

Threshold registers record all the normal SONET counters for errors that occurred over the past 15 minutes and past 24 hours. You can view these by using IOS **show** commands. When the threshold register exceeds the threshold register settings, a threshold crossing alarm (TCA) indication occurs, meaning the device needs to notify the management station of the alarm.

Summary

This chapter covered PoS operation, encapsulation, protection, and convergence. You should be able to describe the most popular uses for PoS. You should be familiar with the PoS frame structure and encapsulation and be able to describe the efficiencies of PoS over other technologies. PoS design models provide advantages in the area of resiliency but need the PGP mechanism to decrease convergence times. You can achieve load balancing in PoS networks by creating multiple APS groups and using MHSRP on the LAN side of the connections. PoS is implemented as a point-to-point technology.

Review Questions

1 True or False: Packet over SONET was developed because there was no other way to transport data over a SONET network.

   A   True

   B   False

2 PoS can be directly encapsulated onto the network media. Which of the following is not a method for connecting PoS to network media?

   A   Connectivity to SONET/SDH ADMs

   B   Connectivity to transponders in a DWDM system

   C   Dark-fiber connectivity

   D   ATM connectivity

3 Which of the following are the three requirements for data to be successfully transported over SONET/SDH?

   A   The use of high-order containment is required.

   B   The PoS frames must be placed inside of the SONET containers aligned on frame boundaries.

   C   The x^{43+1} scrambler must be used in addition to SONET native scrambling.

   D   The PoS frames must be placed inside of the SONET containers aligned on the octet boundaries.

4 What is the hex value of an HDLC delimiter flag byte?

A 0x7D

B 0xE7

C 0x7E

D 0xFF

E 0x03

5 What does the protocol field inside of the PPP frame indicates?

A The protocol that is carrying the PPP frame

B The protocol used to decode the FCS field

C The protocol used to detect the number of padding bytes found in this frame

D The protocol used to format the data in the Information field

E None of the above

6 The C2 byte value of a PoS interface that is using the payload-scrambling function is set to which two of the following values?

A 0xCF%

B 0x16

C 22

D 207

E 0xFF

7 It is recommended that the Layer 3 protocol and the SONET protocol configurations should _____.

A Have both Layer 3 and SONET in a bidirectional ring configuration

B Have Layer 3 in a point-to-point configuration and SONET in a bidirectional ring configuration

C Both be in a point-to-multipoint configuration

D Both be in a point-to-point configuration

8 HDLC frames in a PoS environment contain which four fields?

A Address, Data, Destination, and Frame Check Sequence

B Location, Data, Destination, and Protocol

    C   Location, Control, Information, and Protocol

    D   Address, Control, Information, and Frame Check Sequence

9 What is the purpose of PGP?

    A   Transport data packets across SONET/SDH links

    B   Overcome routing problems between Layer 3 and the SONET network layer

    C   Reliable end-to-end communication and error-recovery procedures

    D   Achieve adequate transparency, protection against malicious attacks, and enough zero-to-one transitions to maintain synchronization between adjacent SONET/SDH devices

10 What is the ideal configuration for APS 1+1 to reduce the need for routing updates due to a failure?

    A   One SONET line between the private network router and the service provider ADM

    B   Two SONET lines between one private network router and the service provider ADM

    C   Two routers with one line each to the service provider ADM

    D   PoS Reflector mode

11 PoS Reflector mode is used for what purpose?

    A   By the working router to keep the distant router up to date

    B   By the protect router to notify the distant router when it takes over for the working router

    C   By the Layer 3 protocol to send routing updates

    D   To send AIS downstream

12 In which three ways can you interconnect PoS interfaces?

    A   SONET

    B   Dark fiber

    C   Gigabit Ethernet

    D   DWDM

    E   Bidirectional path switched rings (BPSRs)

13 One of the advantages to PoS is that when there is a network failure, _____ can restore the network connection before the Layer 3 routing protocol even realizes that there is a problem.

    A ATM

    B SDH

    C APS

    D IPS

This chapter covers the following topics:

- Supporting PoS with Cisco equipment
- Cisco 12000 series PoS line cards
- Configuring the SONET controller
- Configuring the PoS interface
- Sample PoS interface configurations
- PoS **show** commands

CHAPTER 10

Configuring Packet over SONET

This chapter reviews the equipment that can support Packet over SONET (PoS) and the commands that are necessary to configure and support PoS interfaces.

PoS enables core routers, such as the Cisco 12000 series routers, to send native IP packets directly over SONET frames at speeds from OC-3 through OC-192.

A number of platforms in the Cisco product line support PoS, and a number of devices can support the PoS port adapters. These adapters are designed to provide various routers and switches with PoS capabilities. The following Cisco platforms accept port adapters:

- Catalyst RSM/VIP2-40 in the Catalyst 5*xxx* family of switches
- Catalyst 6500 family of switches with the FlexWAN module Cisco 7100 series routers
- Cisco 7200 series routers
- Cisco uBR7200 series universal broadband routers
- Second-generation Versatile Interface Processor (VIP2) used in the Cisco 7500 and 7000 series routers
- Fourth-generation Versatile Interface Processor (VIP4) in the Cisco 7500 and 7000 series routers

The Cisco 10000 series router can support PoS interfaces at line rates up to OC-12. As of the time of this writing, the 7600, 6500, 15190, and 7300 series devices support PoS interfaces at line rates up to OC-48 speeds. You can obtain additional product information at Cisco.com.

The configuration commands necessary to configure PoS on any Cisco platform are nearly identical. The 12000 series router was the platform used for all commands that appear in this chapter. The commands should be similar on other platforms. Be sure to check Cisco documentation for the particular platform and IOS version that you are using.

The Cisco 12400 series routers support bidirectional line rates at up to OC-192 speeds (10 Gbps). The older 12000 series routers support bidirectional rates of 2.5 Gbps per line card. You can calculate the aggregate speed needed by the line card by taking the line speed of each port and multiplying that number by the number of ports on the line card. If the result equals more than 2.5 Gbps, the card would result in a blocking backplane architecture on the 12000 series router. Cisco's intent is to create nonblocking architectures for the backbone 12000 series routers. As a result, cards that require more than 2.5 Gbps of bandwidth are supported only on the 12400 series router.

Figure 10-1 shows an example of how the Cisco 12000 series routers could be included in an end-to-end network traversing a SONET ring. PoS interfaces utilize SONET framing for interoperability with existing SONET infrastructures.

Figure 10-1 *Cisco 12000 Series Routers in PoS Network Designs*

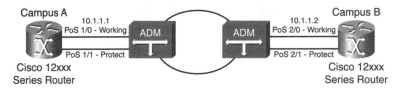

Cisco 12000 Series PoS Line Cards

PoS cards used in the Cisco 12000 series are available in multimode (MM) and single-mode (SM) varieties. The SM optics are available in short-, intermediate-, and long-reach flavors. Most of the cards use a 1310-nanometer (nm) laser, although some of the long-reach cards use 1550-nm lasers. Because lower optical signal loss occurs on fiber at the 1550-nm wavelength, the signal can be transported farther. The following PoS line cards are available for the Cisco 12000 series router:

- 16-port, OC-3c
- 8-port, OC-3c
- 4-port, OC-3c
- 4-port, OC-12c
- 1-port, OC-12c
- 1-port, OC-48c
- 1-port, OC-192c

There are also channelized versions of the OC-3, OC-12, and OC-48 line cards that allow each PoS card to create variable circuit sizes down to the STS-1 level. You can use one OC-48 PoS interface for up to 48 circuits of STS-1 size. Channelized PoS interfaces are considerably more expensive than unchannelized PoS interfaces.

Cisco PoS Line Card Features

The Cisco PoS line cards offer many value-added features:

- **Multiple virtual-output queues**—This feature eliminates head-of-line blocking. Head-of-line blocking occurs when traffic destined for another card stalls because the destination card is busy. Traffic that is behind the stalled traffic must wait until this traffic is cleared. The multiple virtual-output queues provide a system in which traffic can be directed to virtual queues, and each queue has access to the switch fabric. If one queue is blocked from sending traffic, the other queues can still place traffic on the bus for delivery to available cards.

- **Cisco Express Forwarding (CEF)**—CEF tables can accommodate up to one million forwarding entries. The Internet routing tables are currently at approximately 140,000 entries. The CEF table can accommodate approximately eight Internet routing tables, allowing for growth for technologies such as IPv6. CEF contains an adjacency table and is not a demand-driven caching system like fast switching. With fast switching, the first packet of any flow is process-switched by the processor. CEF does not have this limitation because the CEF table information is gleamed from the routing table.

- **Application-specific integrated circuit (ASIC)-based queuing**—ASIC-based queuing provides a Weighted Fair Queuing mechanism to implement type of service (ToS) and differentiated services quality of service (QoS).

- **Distributed Cisco Express Forwarding (DCEF)**—DCEF distributes the routing table to line cards in forwarding information bases (FIBs). Using FIBs, each line card can make forwarding decisions independent of the gigabit route processor (GRP). This allows for the rapid processing and forwarding of traffic and does not create backplane traffic or interrupt the route processor. This also adds to the resiliency of the Cisco 12000 series router because the line card can forward traffic if the GRP goes down.

- **Standards-compliant SONET interface**—Detect and report alarms on the following SONET information:
  - **Section**—Loss of signal (LOS), loss of frame (LOF), threshold crossing alarms (TCAs), and bit interleave parity errors (B1)
  - **Line**—Alarm indication signal (AIS), (line and path) remote defect indication (RDI), (line and path) remote error indication (REI), TCAs, and B2
  - **Path**—AIS, RDI, REI, B3, new pointer events (NEWPTR), positive stuffing event (PSE), negative stuffing event (NSE)

- Other reported information includes the following:
  - Signal fail (SF) bit error rate (BER)
  - signal degrade (SD) BER
  - C2 path signal label (payload construction)
  - J1 path trace byte

- B1, B2, B3 are performance-monitoring parameters. LOS, LOF, PLOP (path loss of pointer), and L-AIS are alarms. Performance monitoring refers to advance alerts, whereas an alarm denotes a failure.

- K1/K2 byte status is reported for SONET automatic protection switching (APS).

Packet over SONET Physical Fiber Configuration

PoS fiber configuration is a point-to-point configuration with a duplex fiber connector. Figure 10-2 displays a linear SONET ADM configuration with a circuit created for the PoS interfaces on the Cisco 12000 series routers.

Figure 10-2 *PoS Between 12000 Series Routers*

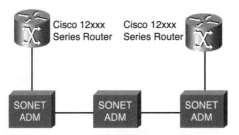

Figure 10-3 displays a SONET ring in which PoS interfaces of different devices are being aggregated by the Cisco 12000 series router at the top of the diagram. Notice the three connections between the top Cisco 12000 and the top ADM displaying the three circuits terminating on this device. The three devices at the bottom of the figure source one circuit each. Each connection is an OC-3 circuit going over a SONET unidirectional path-switched ring (UPSR). Because the SONET UPSR is running at OC-12 speed, there is a capacity of STS-3 left on the ring. The three OC-3 PoS point-to-point connections are using STS-9 of the capacity available.

Figure 10-3   *OC-3 PoS Circuits Traversing a SONET OC-12 Ring*

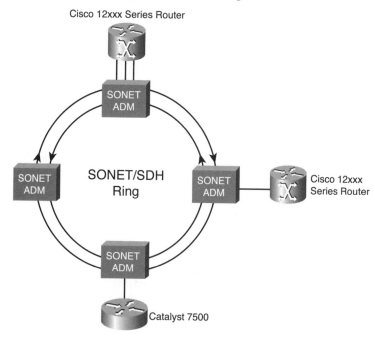

The following section introduces the commands necessary to configure PoS interfaces.

Configuring PoS Interfaces

To set up a PoS interface, you follow three basic steps:

1   Access the Interface Configuration mode of the PoS interface.

2   Configure the appropriate options, including the following:

ip address—Provide addressing for traffic delivery and reception

encapsulation—How PoS packets will be encapsulated, either HDLC or PPP in HDLC-like encapsulation

framing—SONET or SDH

scrambling—To provide 1s density

crc—Cyclic redundancy check for packet integrity

clocking—How timing is to be handled

APS protection—Redundancy protection

- Activate the interface with the **no shutdown** command.

| NOTE | Following standard Cisco command documentation, this book's command descriptions use the following conventions: |

- **Boldface** applied to commands and keywords that are entered literally as shown.

- *Italics* applied to arguments for which you supply values.

- Vertical bars (|) separate alternative, mutually exclusive elements.

- Square brackets [] indicate optional elements.

- Braces {} indicate a required choice.

- Braces within square brackets [{}] indicate a required choice within an optional element.

Accessing the PoS Interface

To access the PoS interface, you need to enter Privileged Exec mode of the IOS. Enter Interface Configuration mode. The **show ip interface brief** command is useful to determine which interfaces are available for configuration. If the PoS interfaces are not available, ensure that the card is being recognized and that the appropriate IOS version that supports the microcode needed for that card is being used. Example 10-1 shows the commands required for this process.

Example 10-1 *Initial Configuration Commands*

```
Router>enable
Router#
Router#configure terminal
Router(config)#
Router(config)#interface POS 0/0
Router(config-if)#
```

The command **interface pos 0/0** represents the slot/interface of the PoS line card on the Cisco 12000 series routers. The Cisco 7500 series router using a VIP card would address an interface as **interface pos 0/0/0** (*slot/port-adapter/interface*). The 7500 VIP is available in different varieties (VIP2, VIP4, and VIP8) with different line-card memory capacities. Each VIP supports up to two port adapters. The Catalyst 6500 and 7600 support FlexWAN modules, which emulate the VIP operation and accept port adapter modules. Similar to standard slots, the first VIP adapter is addressed as VIP 0.

IP Address

The IP address of the PoS interface is required because all Cisco PoS interfaces are Layer 3 implementations. The IP address needs to be on the same subnet as the other PoS interface that it is connecting to (no different from any other point-to-point technology). This chapter does not cover routing protocol configuration, but be aware that it is necessary to include the proper statements in the routing protocol to advertise the PoS interface.

Example 10-2 is an example of configuring the IP address of a PoS interface.

Example 10-2 *Configuring the IP Address*

```
Router(config-if)#ip address 10.11.1.1 255.255.255.0
Router(config-if)#end
Router#
```

Encapsulation

Cisco High-Level Data Link Control (HDLC) is the default encapsulation mechanism. The PoS RFC does not mention HDLC because Cisco HDLC is proprietary to Cisco equipment. To be compliant with other vendors' equipment, you need to use PPP. You can use Cisco HDLC when connecting two Cisco routers together using PoS. You can also implement Frame Relay encapsulation if necessary.

The syntax for the **encapsulation** command is as follows:

```
Router(config-if)#encapsulation (hdlc | ppp | framerelay)
```

You can use the output of the **show interface pos slot/interface** command to verify the encapsulation.

Framing

You have two options available for framing:

- SONET (default)
- SDH

SONET is the default (North American standard). SDH is used throughout most of the rest of the world, including Europe. All interfaces connected to the circuit need to have the same framing type configured. If the SONET framing is being used, no configuration of this option is necessary. If the interface is not coming up, you can verify the framing with the **show controller pos 0/0** command.

The syntax for the **pos framing** command is as follows:

```
Router(config-if)#pos framing (sdh | sonet)
```

Payload Scrambling

The PoS scrambling feature is off by default. To turn on the PoS scrambling feature, use the following command syntax:

```
Router(config-if)#pos scramble-atm
```

The **pos scramble-atm** command automatically inserts the **pos flag c2 22** command into the configuration. This is necessary so that the other side of the connection is told through the path signal label (C2 byte in POH) that the interface is using ATM style x^{43} scrambling. If the PoS flag is changed to 22 and the **pos scramble-atm** command is not issued, the interface will not be scrambling, but the C2 byte will be reporting that scrambling is turned on. Only the **pos scramble-atm** command turns on scrambling. Two connected PoS interfaces must use the same scrambling options; otherwise, the interface will not come up. You can verify the PoS scrambling feature with the **show interface pos** *slot/port* command.

CRC

The cyclic redundancy check (CRC) is an error-checking technique that uses a calculated numeric value to detect errors in transmitted data. The sender of a data frame calculates the frame check sequence (FCS). The sender appends the FCS value to outgoing messages. The receiver recalculates the FCS and compares it to the FCS from the sender. If a difference exists, the receiver assumes that a transmission error occurred and sends a request to the sender to resend the frame. Notice that FCS and CRC terminology are used interchangeably.

You have two options for the **crc** command:

- CRC-16
- CRC-32

Depending on the port speed, the default value will vary. CRC-16 is mainly used for backward compatibility with legacy equipment. The CRC-32 is the recommended CRC value for all configurations for its superior error detection. The option adds two extra octets of overhead, which is small when considering that the minimum rate of an unchannelized PoS interface is 155 Mbps. All endpoints must be configured with the same CRC value.

The syntax for the **crc** command is as follows:

```
Router(config-if)#crc (16 | 32)
```

You can verify the CRC information with the **show interface pos** *slot/port* command.

Clocking

Timing is a critical issue of digital transport. Each bit that is sent over the fiber has to be identified as a unique entity from the bit before it and the bit after it. An accurate clock is needed to precisely identify the time increment where a bit should be detected.

PoS interfaces are usually connected over SONET networks, where timing is synchronized to a Stratum 1 clock. The PoS interfaces can extract timing information from the incoming data stream. When timing from the incoming data stream, the interface is line-timed. The timing can be distributed and synchronized throughout the rest of the network.

If the PoS interface is not connected to a network that has a Stratum 1 timing source, the timing can be internally derived by a Stratum 3 (20 parts per million-ppm) internal oscillator. This timing method refers to internal timing.

The clock source can be set to one of two values:

- Line
- Internal

The default value is line and should be used when the derived clock source is of higher quality than the internal Stratum 3 clock. If the link is a direct point-to-point connection, both ends of the link could internally time. The point-to-point configuration can also be achieved by line-timing one side of the link and internally timing the other end of the link.

Configuring both sides as line is a misconfiguration that leads to timing loops and frequency drifts, which will create intermittent problems on the link.

The syntax for the clocking configuration is as follows:

```
Router(config-if)#clock source (internal | line)
```

PoS ais-shut Command

The PoS **ais-shut** interface configuration command configures the PoS interface to send a line alarm indication signal (AIS-L) to the other end of the link after a **shutdown** command has been issued on the **ais-shut**-configured PoS interface. By default, the AIS-L is not sent to the other end of the link. You can stop transmitting the AIS-L by issuing either the **no shutdown** or the **no pos ais-shut** commands.

The PoS **ais-shut** command is useful in an APS 1+1 environment when one interface is taken down for maintenance. If the PoS **ais-shut** command is not in the router, the APS automatic switchover will not occur and will cause outages unless the switchover is manually initiated with the **aps manual** (*working* | *protect*) command.

```
Router(config-if)#pos ais-shut
```

Alarm Thresholds

Alarm activation thresholds can be set when the registers associated with the alarm exceed the set threshold. Each threshold has a default value that might be used, or you can configure a custom threshold. Table 10-1 lists available PoS alarms. (TCA refers to threshold crossing alarm.) The alarms that have an asterisk (*) next to them in Table 10-1 are on by default.

Table 10-1   *Alarms for Reporting*

Alarm	Description
B1-tca *	Report B1 BER threshold crossing alarms
B2-tca	Report B2 BER threshold crossing alarms
B3-tca *	Report B3 BER threshold crossing alarms
Lais	Report line alarm indication signals
Lrdi	Report line remote defect indicators
Pais	Report path alarm indication signals
Plop *	Report path loss of pointer errors
Prdi	Report path remote defect indicators
Rdool	Report receive data out of lock errors
Sd-ber	Report signal degradation BER errors
Sf-ber *	Report signal fail BER errors
Slof *	Report section loss of frame
Slos *	Report section loss of signal

The command syntax for changing alarm reporting parameters is as follows:

```
Router(config-if)#POS report (b1-tca | b2-tca| b3-tca| lais | lrdi | sd-ber | sf-ber |
slof | slos)
```

Manually configuring the alarm thresholds enables the network administrator to be notified when BERs meet a level that is no longer acceptable for that environment. Every network is different, and there is no one-size-fits-all solution. Some might want to raise the default values if alarms are being reported that do not affect their traffic. Based on the quality of the fiber, some environments could consistently have a BER level that is not deemed as acceptable to the default values, but no noticeable degradation occurs in network traffic. Some customers might have a strong contract with stringent SLAs. These customers could modify their parameters so that alarms are generated for the smallest of issues. The values can be set to any value between 3 and 9. This number is multiplied by 10e-6 to obtain the number of errors before an alarm is generated. The default value 6(10e-6) is used for all thresholds except for sf-ber, which is set to a default value of 3(10e-3).

To change alarm thresholds, use the following command syntax:

```
Router(config-if)#pos threshold {alarm type} <3-9>
```

A configuration that changes the PoS B2 TCA threshold to 3 (.01) would set the alarm value to a higher value (more errors) than the default value of 6 (.00001). Alarms would not be triggered as fast as they were before. Careful attention must be paid to the signal SD and SF values. The SF value should always be a greater number than the SD value so that an SD condition will exist before an SF condition.

Sample PoS Interface Configurations

The following are some sample configurations of PoS interfaces with and without protection.

PoS Configuration Without Protection

One form of PoS interface connectivity is the point-to-point connection through a linear SONET connection. The configuration example is a point-to-point PoS connection through the linear SONET configuration without APS protection. Figure 10-4 shows a diagram of the configuration, and Example 10-3 lists the configuration commands.

Figure 10-4 *PoS Connection*

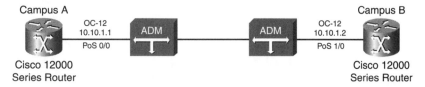

Example 10-3 *Point-to-Point Router Configuration*

```
CampusA(config)# interface pos 0/0
CampusA(config-if)# ip address 10.10.1.1 255.255.255.0
CampusA(config-if)# encapsulation ppp
CampusA(config-if)# pos framing sonet
CampusA(config-if)# pos scramble-atm
CampusA(config-if)# clock source line
CampusA(config-if)# no shutdown

CampusB(config)# interface pos 1/0
CampusB(config-if)# ip address 10.10.1.2 255.255.255.0
CampusB(config-if)# encapsulation ppp
CampusB(config-if)# pos framing sonet
CampusB(config-if)# pos scramble-atm
CampusB(config-if)# clock source internal
CampusB(config-if)# no shutdown
```

To complete this configuration, you need to configure a circuit configured between the two ADMs, one that matches the bandwidth capacity of the unchannelized PoS interface.

PoS Configuration with APS Protection

To provide added network resiliency, you use two fiber connections between the ADM and the CPE. This configuration requires added configuration of the APS 1+1 feature. Figure 10-5 displays PoS connectivity between two campuses (A and B), which are both redundantly connected to the nearest ADM. While the SONET network is providing ring protection for the traffic traversing the ring, the final link between the customer and network will sometimes be protected. In a router configured for APS, the configuration for the protect interface includes the IP address of the router. The address selected should be a loopback IP address configured on the router, because a loopback address is a logical interface that can never go down. Figure 10-5 shows this network design. Example 9-4 shows the configuration for the PoS-connected routers at Campus A and B.

Figure 10-5 *Router Configuration with APS Protection*

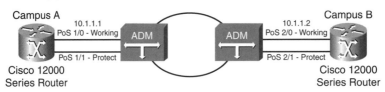

Example 10-4 *PoS Configuration with APS 1+1 Protection*

```
CAMPUS A ROUTER:
CampusA(config)# interface loopback0
CampusA(config-if)# ip address 10.10.1.1 255.255.255.0

CampusA(config-if)# interface pos 1/0
CampusA(config-if)# ip address 10.1.1.1 255.255.255.0
CampusA(config-if)# aps group 10
CampusA(config-if)# aps working 1

CampusA(config-if)# interface pos 1/1
CampusA(config-if)# ip address 10.1.1.1 255.255.255.0
CampusA(config-if)# aps group 10
CampusA(config-if)# aps revert 1
CampusA(config-if)# aps protect 1 10.10.1.1
CampusA(config-if)# end

CAMPUS B ROUTER:
CampusB(config)# interface loopback0
CampusB(config-if)# ip address 10.20.1.1 255.255.255.0

CampusB(config)# interface pos 2/0
CampusB(config)# ip address 10.1.1.2 255.255.255.0
CampusB(config-if)# aps group 10
CampusB(config-if)# aps working 1

CampusB(config)# interface pos 2/1
CampusB(config)# ip address 10.1.1.2 255.255.255.0
CampusB(config-if)# aps group 10
CampusB(config-if)# aps revert 1
CampusB(config-if)# aps protect 10.20.1.1
CampusB(config-if)# end
```

It is important that the APS configuration is done before the IP address is entered into the protect interface. You can use the same IP addresses on both interfaces of the router because they reside on the same device. The configuration of this does not work unless the **aps protect** command is entered before the same IP address is attempted on the second interface.

The **aps group** command enables you to have more than one working and protect interface supported on a router. This option must be configured even if there is only one APS group.

The **aps protect** command enables you to define the protect interface. The IP address of the router that has the working interface must be defined after this. It is important to ensure that the loopback addresses are being advertised by the routing protocol in use if the working and protect interfaces reside on different routers. The two-router model adds a level of redundancy to the configuration, but different IP addresses need to be used on the two devices.

The **aps working** command enables you to define the working interface. The number that appears after **working** defines the circuit number that is associated with the working interface.

The **aps revert 1** command enables and sets the wait-to-restore (WTR) reversionary timer. The number parameter that follows the **revert** keyword sets the number of minutes that the WTR timer will be set to upon working circuit failure.

Various PoS **show** and **debug** commands are available to assist in the management and troubleshooting of PoS environments. Table 10-2 provides a list of the commands.

Table 10-2 *Additional PoS APS Commands*

Command	Command Description
aps authenticate "string"	To enable authentication and specify the string that must be present to accept any packet on the out-of-band (OOB) communications channel, use the **aps authenticate** interface command.
aps lockout	To prevent a working interface from switching to a protect interface, use the **aps lockout** interface configuration command.
aps manual	To manually switch a circuit to a protect interface, use the **aps manual** interface configuration command.
aps timers	To change the time between hello packets and the time before the protect interface process declares a working interface's router to be down, use the **aps timers** interface configuration command.
aps unidirectional	To configure a protect interface for unidirectional mode, use the **aps unidirectional** interface configuration command.
aps force	To manually switch the specified circuit to a protect interface, unless a request of equal or higher priority is in effect, use the **aps force** interface configuration command.

PoS show Commands

Many **show** commands enable you to display information about the PoS system interfaces. Table 10-3 describes the most often used PoS commands. The text following this table displays examples of each command and highlights useful information.

Table 10-3   *PoS Display Commands*

Command	Description
show controllers pos	Displays information about SONET alarm and error rates divided into Section, Line, and Path sections
show protocols pos	Displays status information for the active network protocols
show interfaces pos	Displays detailed information about the PoS interface
show aps	Displays information about the configuration of APS and whether a switchover has taken place

show controllers pos Command

Use the **show controllers pos** command to display information regarding the clock source, SONET alarm and error rates, and register values.

The syntax for the **show controllers pos** command is as follows:

```
Router#show controllers pos slot/interface
```

Example 10-5 shows the output from the **show controllers pos** command.

Example 10-5 **show controllers pos** *Command*

```
CampusA# show controllers pos
POS1/0
SECTION
  LOF = 0              LOS = 1553                    BIP(B1) = 38404325
LINE
  AIS = 1553       RDI = 18        FEBE = 2380930987   BIP(B2) = 1766393764
PATH
  AIS = 1558       RDI = 43087     FEBE = 24938382    BIP(B3) = 102932832
  LOP = 235932     NEWPTR = 11392802  PSE = 3928347      NSE = 3545

  Active Defects: B2-TCA B3-TCA
  Active Alarms: Node
  Alarm reporting enabled for: B1-TCA

APS
  COAPS = 12832782    PSBF = 3887
  State: PSBF_state = False
  Rx(K1/K2): 00/CC  Tx(K1/K2): 00/00
  S1S0 = 03, C2 = 96
CLOCK RECOVERY
  RDOOL = 39233039
```

Example 10-5 **show controllers pos** *Command (Continued)*

```
     State: RDOOL_state = True
   PATH TRACE BUFFER: UNSTABLE
     Remote hostname :
     Remote interface:
     Remote IP addr  :
     Remote Rx(K1/K2): ../.. Tx(K1/K2): ../..
   BER thresholds:  SF = 10e-3  SD = 10e-8
   TCA thresholds:  B1 = 10e-7  B2 = 10e-3   B3 = 10e-6
 CampusA#
```

The main areas in the command are the Section, Line, and Path breakdown of any errors. These headings display the counts associated with various line conditions such as LOF, LOS, and BIP errors. The APS section displays the status of the K1/K2 bytes, which is useful for troubleshooting switch oscillations. The PATH Trace Buffer displays the J1 path trace SONET byte that was learned from the other side of the connection. This information normally carries the IP address and host name of the other PoS interface. The configured BER and TCA thresholds display at the bottom of this output command.

show protocols pos Command

To display status information for the active PoS network protocols, use the **show protocols pos** command.

The syntax for the **show protocols pos** command is as follows:

```
Router#show protocols pos slot/interface
```

Example 10-6 shows the output from the **show protocols pos** command.

Example 10-6 **show protocols pos** *Command*

```
Router# show protocols pos 1/0
POS1/0 is up, line protocol is down
  Internet address is 10.1.2.3/8
Router#
```

This command quickly displays the status of all PoS interfaces and the protocol information running on the interface. The interface in Example 10-6 has an IP address of 10.1.2.3 with an 8-bit (255.0.0.0) subnet mask. The example also displays that there is an issue with the Layer 2 network because the line protocol is down. Both sides of the connection should verify that they have the proper encapsulation set.

show interfaces pos Command

To display detailed information about the PoS interface, use the **show interfaces pos** command.

The syntax for the **show interfaces pos** command is as follows:

```
Router#show interfaces pos slot/interface
```

Example 10-7 shows the output from the **show interfaces pos** command.

Example 10-7 show interfaces pos *Command*

```
CampusA# show interfaces pos 1/0
POS1/0 is up, line protocol is down
  Hardware is Packet Over SONET
  Internet address is 10.1.1.1/24
  MTU 4470 bytes, BW 622000 Kbit, DLY 100 usec, rely 255/255, load 1/255
  Encapsulation PPP, crc 32, loopback not set, keepalive not set
  Scramble enabled
  LCP REQsent
  Closed: CDPCP
  Last input never, output never, output hang never
  Last clearing of "show interface" counters never
  Queuing strategy: fifo
  Output queue 0/40, 0 drops; input queue 0/75, 0 drops
  5 minute input rate 0 bits/sec, 0 packets/sec
  5 minute output rate 0 bits/sec, 0 packets/sec
     0 packets input, 0 bytes, 0 no buffer
     Received 0 broadcasts, 0 runts, 0 giants, 0 throttles
              0 parity
     0 input errors, 0 CRC, 0 frame, 0 overrun, 0 ignored, 0 abort
     0 packets output, 480 bytes, 0 underruns
     0 output errors, 0 applique, 5 interface resets
     0 output buffer failures, 0 output buffers swapped out
     0 carrier transitions
CampusA#
```

This command proves useful to troubleshooting. It enables you to display the counters related to transmitted and received traffic and to view the interface status. This command displays the IP address, maximum transmission unit (MTU), bandwidth, encapsulation, CRC, loopback status, keepalive parameters, and whether scrambling has been turned on.

show aps Command

Use the **show aps** command to display information about the current status of the automatic protection switching configuration. The syntax for the **show aps** command is as follows:

```
Router#show aps
```

Example 10-8 shows the output from the **show aps** command. In this example, the router is configured with a working interface in group 1, and that interface is the active interface.

Example 10-8 show aps *Command*

```
CampusA# show aps
POS1/0 working group 1 channel 1 Enabled Selected
CampusA#
```

Verifying the PoS Configuration Using ping

After the PoS configuration is complete, you can verify that the network is operational using ICMP pings. To verify a PoS configuration using the **ping** command, ping the other side of the network connection and verify that ping replies are being successfully received, as follows.

```
Sending 5, 100-byte ICMP Echos to 10.1.1.2, timeout is 2 seconds:
!!!!!
Success rate is 100 percent (5/5), round-trip min/avg/max = 1/1/4 ms
```

You could use extended pings to fully test a circuit or test particular options available in an extended ping.

You can set loopbacks on the interface to test the transmitting of traffic throughout the network. Loopbacks create network isolation that helps identify faults in the network.

Physical Interface Loopbacks

The following loopbacks are available:

- **Internal**—With internal (or local) loopback, packets from the router are looped back in the framer. Outgoing data is looped back to the receiver without actually being transmitted. Internal loopback is useful for checking that the PoS interface is working. With internal loopback, note the following:

  — When configuring a loopback, ensure that you configure the interface for internal clocking with the **clock source internal** command. When configured for clock source line, the framer waits for incoming valid frames to synchronize with and uses them to time its transmission. With no receive frames, you have no timing to send frames.

  — If you do a hardware loop—in other words, you just loop the fiber back onto the interface—make sure that you use an attenuator to attenuate the optical power level below the saturation level of the receiver.

- **Line**—With line loopback, the terminal end (far side) of the connection is looped back so that the components of one device, the circuit, and the remote interface might be checked.

You should complete the internal and line loopback tests on one end of the connection. If the tests succeed, you should remove all loopbacks, and the opposite end of the connection should perform a series of tests involving both internal and loopback tests to isolate the problem area. This physical interface loopback that is set in Interface Configuration mode has nothing to do with a logical loopback interface that is used for APS configuration.

The syntax for the **loopback** command is as follows:

```
Router(config-if)#loopback [internal | line]
```

Summary

This chapter covered the Cisco equipment that you can use to implement PoS connectivity. The PoS cards available on the Cisco 12000 series router were covered. The configuration commands that are needed to configure the PoS interfaces were covered as well. You also saw sample configurations that showed you how to set up PoS with and without APS protection. This chapter also covered **show** output commands that enable you to display management and troubleshooting options available in PoS environments.

Review Questions

1   Which three features do the Cisco 12000 series PoS interface cards support?

   A   128-Mb burst buffers

   B   ASIC-based queuing

   C   Multiple virtual output queues

   D   Quality of service (QoS) support

   E   Cisco Express Forwarding (CEF) table that accommodates up to 14 million forwarding entries

2   The Cisco OC-3c PoS line cards are available in three different versions, each offering a different number of ports. How many ports are supported by the OC-3c/STM-1 PoS line cards?

   A   1 port

   B   2 ports

   C   4 ports

   D   8 ports

   E   16 ports

3   What two IETF RFCs specify PPP in HDLC-like framing?

   A   RFC 1615

   B   RFC 1619

   C   RFC 791

   D   RFC 1918

   E   RFC 2615

4 What command is used to configure the APS protect interface?

   A   **aps protect**

   B   **aps protect "IP address of protect interface"**

   C   **aps protect "IP address of working router"**

   D   **aps protect "IP address of working router"**

5 Which Cisco IOS command sets framing to SONET STS-3c?

   A   **pos framing 3**

   B   **pos framing SDH**

   C   **pos framing SONET****

   D   **pos scramble-atm**

6 Which Cisco IOS command displays status information for the active network protocols?

   A   **debug interfaces**

   B   **show protocols pos**

   C   **show interfaces pos**

   D   **show controllers pos**

7 Which Cisco IOS command displays information about SONET alarm and error rates divided into Section, Line, and Path sections?

   A   **debug interfaces**

   B   **show protocols pos**

   C   **show interfaces pos**

   D   **show controllers pos**

This chapter covers the following topics:

- Dynamic Packet Transport in metropolitan-area networks (MANs)
- Dynamic Packet Transport fundamental concepts
- Spatial Reuse Protocol
- Dynamic Packet Transport frame format unicast, multicast, and broadcast transmissions across a Dynamic Packet Transport ring
- Quality of service operation in a shared packet Dynamic Packet Transport ring
- Dynamic Packet Transport fault recovery

CHAPTER 11

Dynamic Packet Transport

IP+Optical networking environments lower equipment prices, make efficient use of bandwidth, and lower operational costs when compared to legacy time-division multiplex (TDM)-based systems. SONET is a voice-driven technology that has inefficiencies when transporting data. Although the SONET overhead is not the most efficient use of the SONET frame, it does offer operations, administration, maintenance, and provisioning (OAM&P) functionality that is not readily available in any other technology. Multicast and broadcast traffic in a SONET environment can have disastrous consequences because each packet must be sent around the ring once for each recipient. Although this process sounds counterintuitive, it is the only way of transmitting multicast traffic across a SONET ring. The usable bandwidth is another limitation of SONET environments. Bidirectional layer-switched rings (BLSRs) provide great improvements when compared to that of unidirectional path switched rings (UPSRs), but might not be practical based on the ring's traffic patterns.

Data is normally transported in packets, and this packet-based traffic continues to grow as more users use larger-scale distributed programs, such as remote storage and web-based applications. Circuit-switched voice technologies are rapidly being migrated to Voice over IP (VoIP) data networks, creating greater needs for data-oriented transport systems. Maximizing bandwidth is important to service providers (SPs) and enterprises. SPs have been transporting packet-based services over their circuit-based systems for many years.

Dynamic Packet Transport (DPT) is a resilient, ring-based technology that is optimized for packet-based traffic. DPT is a proprietary protocol developed by Cisco Systems, but DPT is also published as an informational RFC that vendors can use to create their own DPT interfaces. Riverstone has a line of DPT cards that work with the Cisco proprietary DPT. Because DPT is such a successful technology, the IEEE decided to ratify a standard based on it. The standard is in draft form at the time of this writing, but it should be a standard by the time you read this. The IEEE 802.17 Working Group has coined the industry standard version of this technology, Resilient Packet Ring (RPR).

The DPT frame format is closely aligned to that of Ethernet. DPT uses SONET framing for backward compatibility with legacy transport infrastructures, but is physical (PHY) layer agnostic and can run directly over fiber without the SONET framing. SONET framing allows DPT to be transported over legacy SONET infrastructures.

This chapter covers the DPT efficiencies. The Spatial Reuse Protocol (SRP) used by DPT is detailed, as is DPT's ring architecture. Quality of service (QoS) and fairness in DPT environments are also covered.

Background Information

As with any new technology, one of the biggest adoption hurdles is customer acceptance. New technologies that have been introduced into the network have usually come with the requirement for forklift upgrades, extensive training, and even service disruptions. As networks have become more sophisticated, new technologies need to account for issues such as migration paths, existing system interoperability, and implementation costs. Implementation costs can include equipment, installation, and training to achieve widespread deployment.

As customers have expanded the boundaries of their corporate applications and made the services provided by external organizations more integral to their core business operations, they cannot afford interruptions of those services. The success of new technologies depends on both their interoperability with today's protocols and the ability to provide protection in the advent of equipment or fiber failures. Switching from a working facility to a protection facility should be so quick that customer applications are not affected. Voice networks will contain noticeable degradation of quality if switchover takes place in any time greater than 50 ms. Data environments typically don't come close to the 50-ms requirement, but the requirements are changing with the deployment of Voice over X technologies.

Physically connecting to remote sites is not an easy task. To do so, rights of way must be obtained to run fiber optics. The metropolitan area is especially difficult because of the high cost of real estate in some areas. Because the cost of installing cable is high, it is very important to maximize the cabling infrastructure. Figure 11-1 shows a point-to-point architecture on the left side and a ring-based architecture on the right side. To provide the same level of resiliency obtained in a ring architecture, six point-to-point (P2P) optical links would need to be run. In the ring-based architecture on the right side of Figure 11-1, only four links are necessary. The P2P architecture would rely on Layer 3 routing convergence failover mechanisms, whereas the ring-based architecture could failover in sub-50-ms time. The P2P architecture would also require four more optical interfaces than the ring-based architecture. High-speed optical interfaces are not cheap. The savings of fiber and interfaces combined makes ring-based architecture much more affordable.

Figure 11-1   *Point-to-Point Versus Ring Topologies*

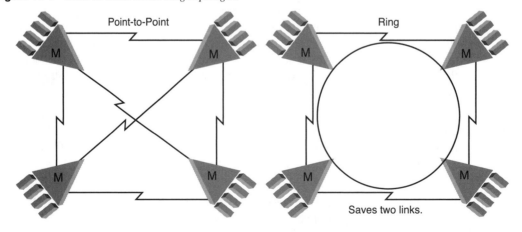

IP+Optical Concepts

Figure 11-2 illustrates different technologies' encapsulation methods that occur in a top-down fashion before traffic is transmitted on the optical interface. Each layer needs separate management and experienced engineers. The goal of efficient transport is to eliminate extra layers of overhead. Benefits include reduced complexity, less overhead, and reduction of cost for equipment, configuration, provisioning, and support.

Figure 11-2 *Overhead Layers of Different Technologies*

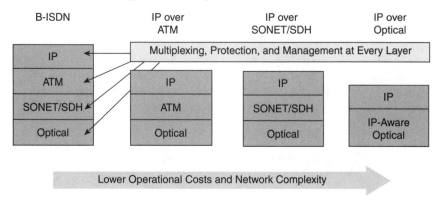

Optical networks are evolving to an architecture that puts IP packets directly onto optical interfaces. IP is designed such that it does not assume sophisticated functions from the underlying network. Lower layers might use checksums and error correction, but TCP/IP has its own error-recovery mechanisms.

The design goal of evolving to IP+Optical networks is to consolidate links into bigger pipes and increase the bandwidth as much as possible on a simplified topology.

Resource management problems in shared infrastructures such as DPT have to be handled at Layer 3 (IP) or above. Managing bandwidth usage by customers or business units is important in shared infrastructures such as DPT. Many QoS tools are available for managing the use of bandwidth in the ring.

DPT: An Efficient MAN Solution

DPT combines the intelligence of IP with the bandwidth efficiencies and protection of optical rings. DPT is a Layer 2 protocol with optical protection schemes that increase efficiency over SONET networks. DPT is based on SONET BLSR technology without the need to reserve half of the ring bandwidth. A DPT ring consists of dual counter-rotating fiber rings referred to as the *inner* and *outer rings*. There is no working or protect facility as there is in SONET rings. Both rings support data simultaneously. Whenever a data packet is transmitted on one of the rings, control packets are sent in the other direction to share this bandwidth-usage information around the ring. Don't be confused. Both rings support data and control packets. Figure 11-3 displays the DPT ring transporting data and control traffic. Notice that control and data packets from the same source always flow in opposite directions, but each ring can transport both data and control traffic.

Figure 11-3 *DPT Ring Transporting Data and Control Traffic*

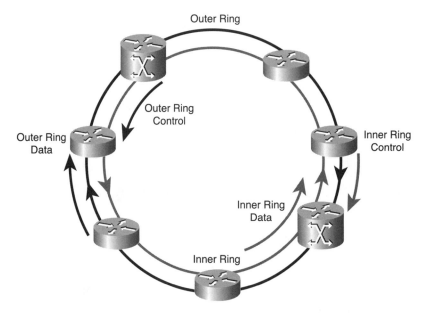

DPT is optimized for transporting TCP/IP traffic. Other Layer 3 technologies are not supported, but they might be tunneled through the DPT ring. The Cisco 10720 supports tunneling Layer 2 connections over a Universal Transport Interface (UTI) tunnel that works in the parallel express forwarding (PXF) ASIC. UTI was a predecessor to the Internet Engineering Task Force (IETF) standard version of Layer 2 Tunneling Protocol version 3 (L2TPv3). With UTI or L2TPv3, Layer 2 and 3 protocols can be transported efficiently using ASIC technology.

DPT borrows its frame format from the IEEE 802.3, while using many of the ring-based concepts of SONET. DPT supports fast failover times that can normally exceed those found in SONET BLSR rings. DPT is advertised to support the same sub-50-ms restoration time of SONET BLSR.

DPT has the following characteristics:

- Bandwidth allocation is not fixed; the system tries to use all available bandwidth as much as possible using statistical TDM.

- Extends IP functionality over the metro area by replacing the legacy TDM nodes with DPT nodes and mapping the physical topology into IP. This allows for dynamic discovery of the proper metro nodes for IP traffic delivery.

- Significantly reduces the configuration management requirements because, from the IP perspective, a DPT ring is as simple to handle as an Ethernet segment.

- Based on SRP, which provides bandwidth-usage information. SRP was developed by Cisco, submitted to the IETF, and published as informational RFC 2892, "The Cisco SRP MAC Layer Protocol." SRP is discussed in detail later in this chapter.

DPT is available on a broad range of Cisco routers, including the following:

- **Cisco 12000 series routers**—The Cisco 12000 series is a routing platform featuring the capacity, performance, and operational efficiencies SPs require for building (core) IP backbone and high-speed provider edge networks.

- **Cisco 10720 router**—The Cisco 10720 router is an Internet-class metro access edge platform that provides optical transport with DPT. It integrates full IP routing and services delivering intelligent Ethernet subscriber interfaces to create simple, scalable, and reliable networks.

- **Cisco 7200 VXR and 7500 series routers**—Cisco 7x00 series high-end data, voice, and video routers combine Cisco Systems software technology with reliability, availability, serviceability, and performance features to meet the requirements of today's mission-critical internetworks. Cisco 7x00 routers are designed to have the flexibility needed to meet the constantly changing requirements at the core, backbone, and distribution points of the internetwork.

- **ONS 15194 IP Transport Concentrator**—The Cisco ONS 15194 enables the creation of logical rings over physical star-based fiber topology for maximum performance and reliability. It offers high port density (64 OC-48/STM-16 ports in a 9-RU chassis) and a broad range of management and monitoring functions.

NOTE   The Cisco 12000 series router DPT devices are covered in this text. If you want more information on the Cisco products that support DPT, you can find additional information at Cisco.com.

Internetworking with SONET

DPT by design is a physical layer agnostic protocol. Although it can run over any technology, the current framer uses SONET framing so that DPT interfaces can be internetworked with the large number of deployed SONET networks. DPT does not require a SONET infrastructure. DPT devices might be connected directly to each other over dark fiber leased from an SP. Dense wavelength division multiplexing (DWDM) can be leveraged with DPT, as long as the transponders support the same SONET line rate as the DPT interfaces. Similar to Packet over SONET (PoS), the SONET framing allows DPT to be used anywhere SONET framing is supported.

DPT implementations support the following physical media:

- Dark fiber
- DWDM transponder-based networks
- SONET networks

Network designers have the freedom to mix and match the implementation according to their various needs. One DPT ring can be transported over a mix-and-match of UPSR and BLSR rings, then a DWDM network, and finally interfaces connected directly with dark fiber.

Many point of presence (POP) interconnects were created with Fiber Distributed Data Interface (FDDI) networks due to their ring architecture resiliency over competing Ethernet technologies. FDDI is limited to 100 Mbps of bandwidth and does not have the bandwidth possibilities of PoS interfaces. For each link used in PoS, two physical interfaces and one new fiber link is required. Optical interfaces and fiber are expensive resources that should be conserved, while still meeting the needs of the network.

Ring-based networks conserve interfaces and fiber resources, while providing fault tolerance with switchover times of 50 ms. Whereas PoS solved the bandwidth issues of FDDI, the ring-based architecture of DPT solves both the bandwidth and architecture limitations of these other technologies. Figure 11-4 displays POP interconnects completed with FDDI and PoS, with DPT, which can solve the POP interconnect issues.

Figure 11-4   *POP Interconnect with DPT*

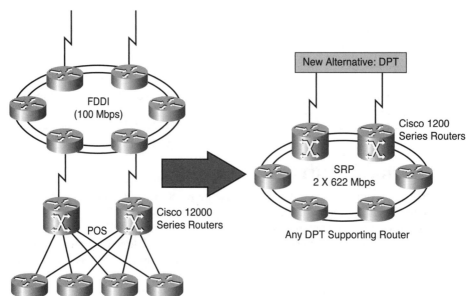

A DPT ring provides resiliency when there is a fiber cut, but it cannot provide protection against multiple simultaneous fiber cuts. Multiple fiber cuts would result in the creation of multiple segmented rings. Most networks designed today balance cost, functionality, and survivability in the advent of problems. Mesh networks support the highest survivability because of their multiple redundant paths but rely on Layer 3 routing protocols and cannot reconverge in less than 50 ms.

DPT in Campus and MAN Environments

DPT is an effective technology for connecting buildings in close proximity. Campus backbone environments and MAN environments are perfect for DPT technologies. Figure 11-5 shows a DPT ring in a campus university backbone. DPT interfaces are currently available at up to OC-192 (10 Gbps) speeds but will follow optical technology trends to higher-bandwidth speeds.

Figure 11-5   *DPT in a Campus LAN*

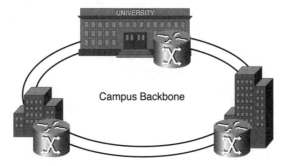

An existing FDDI campus backbone solution could be replaced with DPT. Implementing DPT as the campus backbone solution would provide high speeds in the network core, while still supporting legacy equipment in the campus buildings. DPT rings use the same MAC layer addressing as Ethernet technologies. DPT is complementary to Ethernet technology.

DPT ring interconnection points need to route between IP subnets because each ring is viewed as one logical broadcast domain. Multiple DPT rings can be connected to form a large DPT network. Each ring supports reconvergence of sub-50 ms, but end-to-end connections across rings are limited to the convergence time of the Layer 3 routing protocol used. Interconnected DPT rings can follow a hierarchy such as the one displayed in Figure 11-6. Recall from Chapter 3, "SONET Overview," that access rings link small geographical areas and uplink to a metropolitan ring, which will span the entire metropolitan area. Some traffic might be destined for a site in a remote metropolitan area. Such traffic would have to traverse a backbone network in which long-haul DWDM or SONET rings could be used.

Figure 11-6 *DPT as Part of a Hierarchical Ring Structure*

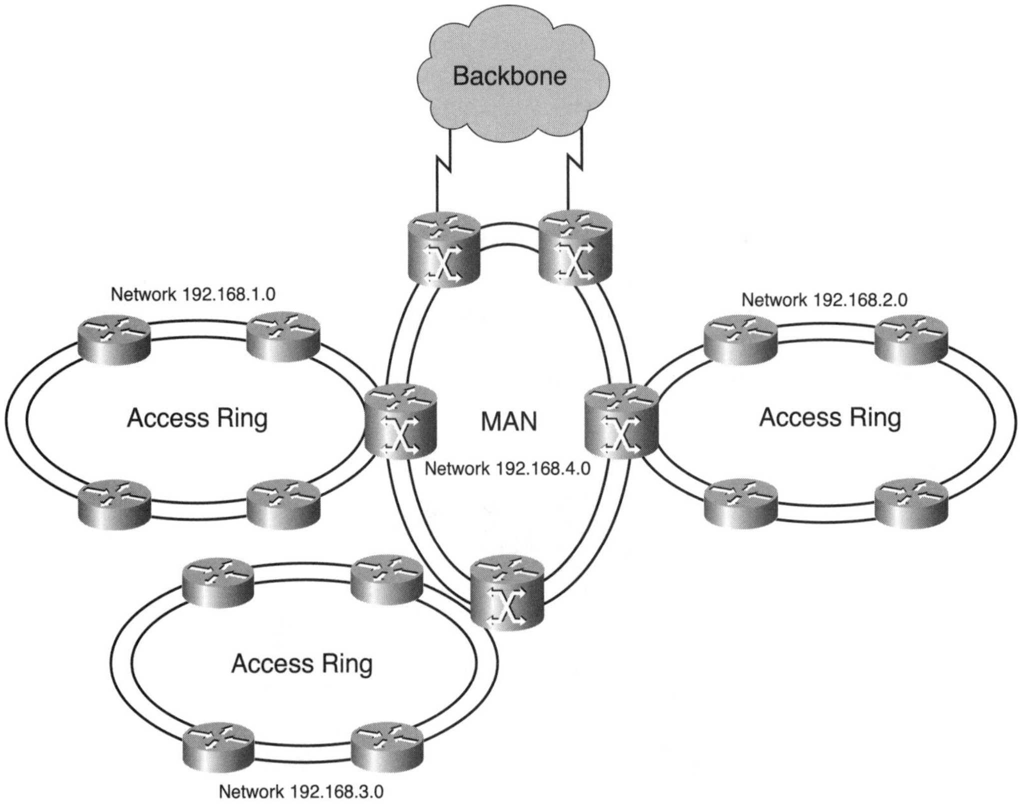

SRP

To fully understand DPT, you must understand the protocol behind the scenes. This protocol is the SRP, which is a Cisco proprietary protocol that has been submitted to the IETF as an informational RFC. RFC 2892 covers SRP in exhaustive detail. For additional information, go to www.ietf.org to obtain a copy of RFC 2892.

SRP was designed to provide a MAC layer for optical networks transporting data traffic. SRP assumes that the logical topology in the MAC layer is based on dual counter-rotating rings. SRP does not use a token to control access to the DPT like Token Ring or FDDI does. DPT uses a fairness algorithm to enforce fair access to the bandwidth of the rings.

Spatial reuse was first covered in Chapter 3 with the introduction to BLSR technology. Spatial reuse is the bandwidth-optimizing features that a technology introduces. SRP can increase the bandwidth available to the nodes on a shared ring. SRP accomplishes this by allowing and controlling oversubscriptions by using a destination packet-stripping mechanism. LAN-based ring technologies (Token Ring and FDDI) used a source packet-stripping mechanism that resulted in the loss of usable bandwidth due to data that traversed the entire ring. DPT allows multiple nodes to transmit packets simultaneously. Unicast traffic is the only type of traffic that is source stripped. Multicast and broadcast traffic are supported by SRP but cannot use destination stripping because the destination of the packet is not predetermined. Source stripping is utilized for multicast and broadcast traffic. Although it might seem that this is not optimal, it is a leap from SONET architectures that would have to replicate each multicast/broadcast packet to each destination. The SONET implementation of multicast and broadcast traffic would result in poor bandwidth efficiencies.

SRP is a Layer 2 protocol that controls bandwidth usage and fairness of resources in the ring. All Cisco DPT interfaces are Layer 3 implementations. Figure 11-7 displays the encapsulation hierarchy of IP over a DPT ring. DPT offers the closest alternative to the IP+Optical concept to date.

Figure 11-7 *IP Encapsulation in DPT Environments*

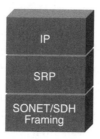

SRP was designed to support high speeds. The Cisco 12000 series routers support DPT cards with speeds up to OC-192 (10 Gbps). The SRP protocol is scalable to higher speeds and feeds. When the optics for higher speeds are available, SRP can easily accommodate.

SRP allows up to 128 nodes in a single ring. Limitations of some of the older hardware DPT framers limited the number of nodes to 32 or 64 nodes. The current hardware implementations support up to 128 nodes.

SRP Protection

DPT offers the same protection switchover times as SONET, but without the reservation of bandwidth. SONET's protection mechanism is automatic protection switching (APS). DPT does not use SONET's APS but instead introduces the concept of intelligent protection switching (IPS). IPS leverages a keepalive packet that is sent out approximately every 106 usec (microseconds). The keepalive timeout interval is set to 16 usage packets that should be received in 1.696 milliseconds (106 usec × 16).

IPS handles the automatic failover process of switching traffic from a failed path on the ring to an operable path. The process is deemed intelligent because it uses the Layer 2 SRP protocol to perform the wrapping function and can differentiate severities. Multiple types of failures can cause a ring wrap to occur, including the following:

- Fiber failure
- Node failure
- Signal degradation

SRP control packets handle tasks such as topology discovery, protection switching, and bandwidth control. Control packets propagate in the opposite direction from the corresponding data packets transmitted. If data packets were being transported by the outer ring for a particular flow, the control packets would be transported by the inner ring. Control packets not related to data packets might be put onto any ring that is appropriate for that function.

SRP Packets

SRP control packets are point-to-point in nature. A control packet is always sourced by a node's upstream or downstream neighbor. The control packet uses a special destination MAC address of all 0s for its control packets (00-00-00-00-00-00). When a destination address of all 0s is received, the node knows it is a control packet and can forgo the address lookup function to see whether the packet should be processed locally. All control packets are transmitted with a Time To Live (TTL) value of 1. Each DPT router decrements this TTL field by 1 and removes the packet from the ring when the TTL field reaches a value of 0. The TTL offers a loop-avoidance mechanism.

The SRP protocol has gone through two iterations (version 1 and version 2). Version 1 has limited deployment and was made obsolete by SRP version 2. RFC 2892 covers the differences between the two versions at the beginning of the RFC. Some of the differences are as follows:

- **Time To Live**—The TTL for version 1 is 11 bits and 8 bits for version 2.
- **Destination strip**—Version 1 used a bit to determine whether the packet should be source or destination stripped. This bit was removed in version 2.
- **Usage information**—In version 1, SRP fairness algorithm (SRP-fa) information was part of the header. In version 2, usage information is carried in a separate control packet.

Version 2 SRP Header

The SRP version 2 overhead consists of at least 20 bytes. There are 16 bytes of header followed by a 4-byte CRC. The control packet header is larger. Figure 11-8 displays the SRP overhead, and Table 11-1 describes the fields you see in the figure.

Figure 11-8 *SRP Data Packet Format*

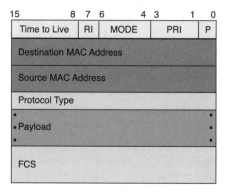

Table 11-1 *Table 10-1SRP Data Packet Definitions*

Field	Definition
TTL	Time To Live. This 8-bit field is used as a hop-count field, which is decremented by 1 at each router. When the TTL reaches 0, the packet is stripped from the ring and discarded. The TTL field supports a decimal value up to 255. The TTL field supports up to 128 nodes with a ring failure.
RI	Ring Identifier. This 1-bit field might have a value of 1 (= inner ring) or 0 (= outer ring).
Mode	This 3-bit field is used to identify the packet content (data or control message). Notice from the following mode values and descriptions that any mode value of 1xx is a control packet. The exception is the mode value of 111, which denotes a data packet. 000 Reserved 001 Reserved 010 Reserved 011 ATM cell 100 Control message (pass to host) 101 Control message (locally buffered for host) 110 Usage message 111 Packet data

continues

Table 11-1 *Table 10-1SRP Data Packet Definitions (Continued)*

Field	Definition
PRI	Priority. This 3-bit field is used to carry SRP priority value. Values range from 0 through 7, with 7 as the highest priority. The value of this field is copied from the IP precedence bits in the Type of Service (ToS) byte in the IP header. The Priority field drives SRP packet transfer and scheduling decisions.
P	Parity bit. This 1-bit field is used for odd parity over the rest of the first 15 bits of the MAC header.
Destination MAC Address	This 48-bit field is used for the MAC address of the destination node on the ring. A control packet has a value of all 0s.
Source MAC Address	This 48-bit field is used for the MAC address of the node sending the packet.
Protocol Type	This 16-bit field is where 0x2007 identifies an SRP control packet. 0x0800 identifies an IPv4 packet, and 0x0806 is for Address Resolution Protocol (ARP). The IPv4 and ARP Ethertype values used in Ethernet are identical to those used in SRP.
Payload	Variable in length.
FCS	Frame Check Sequence. This 32-bit field is used for a cyclic redundancy check (CRC). The CRC is calculated over the whole packet excluding the SRP header (first 16 bits).

The total overhead for each data packet is 20 octets (16 octets for header, and 4 octets for the FCS). The Protocol Type field is used to determine whether a larger header is in use. The Protocol Type field of 0x2007 identifies a control packet. The node should now look further into the header for important control information.

Control Packets

Control packets are always sent point to point. The destination MAC address of control packets is always set to all 0s. The TTL is set to 1 to ensure that the control packet is processed only by the neighbor and not propagated along the ring.

Each node continuously generates control packets to its neighbors to run the distributed IPS fairness algorithm. The destination MAC address of control packets is not necessary because their nature is P2P communication. MAC address table lookups are not needed by this function. In packet processing, the Mode field is used to dispatch packets, so the MAC address of 0 just disables the address table lookup. When a packet is passed to higher-layer processing, the Protocol Type field is used to identify control packets.

Control packets are critical to the operation of the DPT ring. All control packets should be sent with the highest priority. The packet-processing algorithm guarantees that control packets are never dropped.

SRP version 2 specifies three control packet types, as follows:

- **Topology discovery (0x01)**—Control packets that are used to discover nodes on the ring and to automatically build and maintain a current topology map of the network. This topology map is stored on each router, and the network administrator can use it for troubleshooting purposes.

- **Intelligent protection switching (0x02)**—Packets used to implement protection switching in the case of optical failures or signal degradation.

- **Reserved (0x03 through 0xFF)**—Values reserved for future use

Figure 11-9 displays the frame format of a control packet. The control packet format includes additional header information predicated by the protocol type of 0x2007. The additional information is a control version number, control type, control checksum, and a control TTL. Other than these new control fields discussed in Table 11-2, the control packet is similar to an SRP data packet.

Figure 11-9 *SRP Control Frame Format*

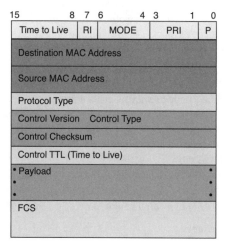

Table 11-2 defines the control packet fields shown in Figure 11-9.

Table 11-2 *Control Packet Fields and Definitions*

Field	Definition
Destination MAC Address (DA)	48 bits. Globally unique IEEE 48-bit MAC address. Always set to 00-00-00-00-00-00 for control frames.
Protocol Type	16 bits. All control packets are identified by a protocol type of 0x2007.
Control Version	8 bits. This field is the version number associated with the Control Type field. Initially, all control types are version 0. The Control Version field might be used to support multiple future versions of SRP in the same DPT ring to make for smooth migrations.
Control Type	8 bits. • 0x00 is reserved. • 0x01 is a topology discovery packet. • 0x02 is an IPS message. • 0x03 through 0xFF are reserved.
Control Checksum	16 bits. This field is the 16-bit 1s' complement of the 1s' complement sum of all 16-bit words starting with the control version. If there is an odd number of octets to be included in the checksum, the last octet is padded on the right with 0s to form a 16-bit word for checksum purposes. The pad is not transmitted as part of the packet. While computing the checksum, the Checksum field is set to 0s. Note that this is the same checksum algorithm used for TCP checksum calculation.
Control TTL	16 bits. This field is a control layer hop count that is decremented by 1 each time a node forwards a control packet. If a node receives a control packet with a Control TTL 1, it accepts the packet but does not forward it. Note that the Control TTL is separate from the SRP Layer 2 TTL, which is always set to 1 for control packets. The originator of the control packet sets the initial value of the control TTL to the SRP Layer 2 TTL normally used for data packets. The Control TTL field size is bigger than necessary. Other functions already limit the number of nodes to 128 nodes. The 16 bits were possibly selected for a better byte alignment in processing on 16-bit or 32-bit CPU architectures.
Payload	Variable.
FCS	32-bit CRC.

The overhead for a control packet is 26 octets (22 octet header and an additional 4 octets for the FCS).

The following sections introduce each of the three control packet types.

The Topology Discovery Packet

Each node performs topology discovery by sending out topology discovery packets on one or both rings. The node originating a topology packet marks the packet with the egressing ring ID, appends the node's MAC binding to the packet, and sets the Length field in the packet before sending out the packet. This packet is a point-to-point packet that hops around the ring from node to node. Each node appends its MAC address binding, updates the Length field, and sends it to the next hop on the ring. If there is a wrap on the ring, the wrapped node indicates a wrap when appending its MAC binding and wraps the packet. When the topology packets travel on the wrapped section with the ring identifier being different from that of the topology packet, the MAC address bindings are not added to the packet.

Eventually, the node that generated the topology discovery packet gets back the packet. The node makes sure that the packet has the same ingress and egress ring ID before accepting the packet. A topology map is changed only after receiving two topology packets that indicate the same new topology (to prevent topology changes on transient conditions).

Intelligent Protection Switch (IPS) Packet

IPS provides SRP with a powerful self-healing feature that automatically recovers from fiber facility, node, or signal degradation overlimit failures. IPS is analogous to the self-healing properties of SONET rings but without the need to allocate protection bandwidth. The basic format of the packet is similar to the other control packets but includes an IPS byte.

The IPS byte consists of 8 bits. The definition of the bits is as follows:

- **Bits 0–3**—IPS request type
    - **1111**—Lockout of protection (LO)
    - **1101**—Forced switch (FS)
    - **1011**—Signal fail (SF)
    - **1000**—Signal degrade (SD)
    - **0110**—Manual switch (MS)
    - **0101**—Wait to restore (WTR)
    - **0000**—No request (0)
- **Bit 4**—Path indicator
    - **0**—Short (S)
    - **1**—Long (L)
- **Bits 5–7**—Status code
    - **010**—Protection switch completed (W)
    - **000**—Idle (0)

The first 4 bits of the IPS packet borrow nomenclature and functionality from SONET APS. The IPS process is discussed in greater detail later in this chapter.

Usage Control Packet

The usage control packet is used by the distributed SRP fairness algorithm (SRP-fa) to govern access to the ring. Traffic that is being sourced by a DPT node is considered transmit traffic because the node is the originator of the traffic. Traffic that is sourced by another node and merely passing through the DPT node is considered transit traffic. The transit node cannot be the source or destination of the traffic. The usage packet is the only control packet that does not contain all the normal packet fields. See RFC 2892 for detailed information regarding the usage frame format.

The SRP-fa requires three counters, which control the traffic that is forwarded and sourced on the SRP ring. The counters are my_usage (tracks the amount of traffic sourced on the ring), forward_rate (tracks amount of traffic forwarded on to the ring from sources other than the host), and allowed_usage (the current maximum transmit usage for that node).

The usage control packet passes this information from node to node so that the distributed SRP-fa can govern access to the ring in a fair manner. The usage packet is covered in detail later in this chapter.

Unicast Packets

When unicast packets are placed on legacy LAN data rings (Token Ring or FDDI), source-stripping and token-handling procedures are used to control ring access. Packets circulate around the entire ring before being stripped by the source.

SRP performs destination stripping of unicast packets. The efficient use of destination-based stripping allows much better usage of the ring resources. Figure 11-10 shows a DPT ring with three active flows of unicast traffic. Node 1 is sending data to nodes 2 and 3, while node 3 is transmitting data to node 4.

1   Node 1 transmits a unicast packet for node 2 and another unicast packet for node 3.

2   Node 2 receives the packet from node 1 and strips it from the ring because no other node needs this packet. Node 2 also receives a packet destined for node 3. Because the destination address is for a different node, node 2 is a transit device. The packet is propagated to its destination of node 3.

3   Node 3 receives the packet from node 1 and removes the packet from the ring. Node 3 transmits data to node 4. The traffic only travels from node 3 to 4.

4   Node 4 receives the packet from node 3 and strips the packet from the ring.

Figure 11-10 *Unicast Traffic on an SRP Ring*

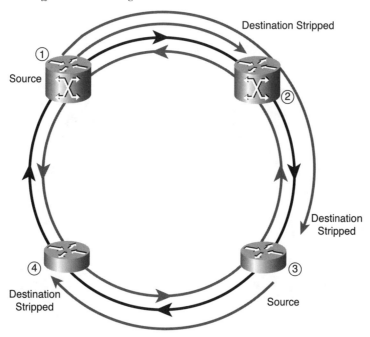

Multicast Packets

Multicast traffic differs from unicast traffic in that the destination is not an individual address but a group of nodes whose membership can change. A multicast packet is sourced onto the ring using a destination MAC address built around the multicast standards. The source node is responsible for the multicast-to-MAC address mapping that takes the low-order 23 bits of the IP Class D multicast address and maps it directly to the low-order 23 bits of the MAC address. Nodes that are enabled for multicast create a special entry in the content-addressable memory (CAM) of 0100.5Exx.xxxx with a mask of 0000.007F.FFFF. If the receiving node has the appropriate entry in the CAM, the receiving node checks whether the multicast address fits the group membership definitions. If it does, the packet is copied out for processing by the local host. Figure 11-11 displays a DPT ring with multicast traffic.

Figure 11-11 *Multicast Packets on an SRP Ring*

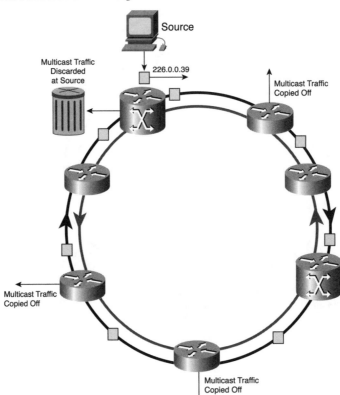

Node A is transmitting a multicast packet onto the ring with nodes B, E, and F subscribing to the multicast group. When node B receives the packet, it copies the multicast traffic for host processing and forwards the original multicast packet along the ring to the next node.

The multicast packet is always forwarded to the next hop on the ring because the receiving node cannot know whether there are other members of this multicast group farther down on the ring. When node C receives the packet, it forwards the packet back onto the ring because it has no subscribers for that particular multicast group. When the packet arrives back at the source of the transmission, the packet is stripped from the ring.

Packet Processing

Each node participating in a DPT ring has two SRP MACs per SRP interface. These two MACs are the Ring Access Controller (RAC) chips. RAC and MAC can be used interchangeably for Figure 11-12. Do not confuse this with addressable Media Access Control

(MAC) interfaces. Although a DPT node has two optical interfaces to connect to the ring, it only contains one MAC address and one IP address. Figure 11-12 displays the Layer 2 and Layer 3 architecture of DPT nodes in a ring. The Layer 2 MACs receive the traffic and pass the traffic to the Layer 3 processing. The neighboring SRP MAC is typically referred to as the mate. Each SRP MAC is connected to a fiber pair, which is either designated as Outer Receive (RX) / Inner Transmit (TX) or Outer Tx (O-Tx) / Inner Rx (I-Rx). This is similar to the A and B ports used in FDDI rings. In Figure 11-12, you can see that the Outer Rx / Inner Tx is Side A, and the Outer Tx / Inner Rx is Side B. Side A should always be connected to Side B or there will be a ring wrap. A ring wrap can be found using the IOS topology map or the wrap LED indicator on some DPT interfaces. By switching Side A and Side B, you should be able to alleviate the wrap. The IOS provides a higher level of detail than LEDs through the topology map.

Figure 11-12 *DPT Packet Flow Architecture*

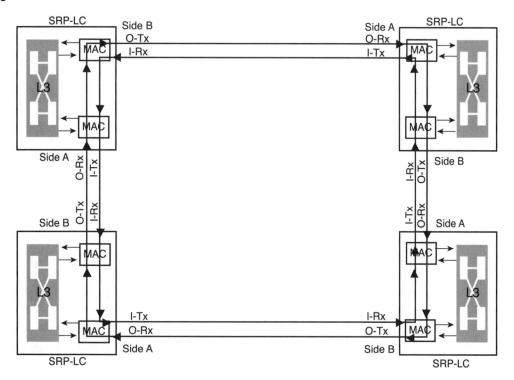

A CAM table is associated with each DPT node. Figure 11-13 displays the format of the DPT CAM table. The following information describes all the fields in the CAM table.

Figure 11-13 *CAM Table Layout*

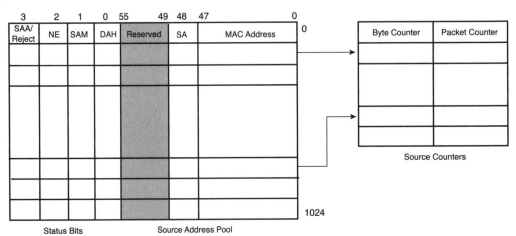

The CAM table contains an SA bit and the MAC address. The SA bit indicates whether the associated MAC address is a source or destination address. If the bit is set to 1, the associated MAC address is the source address. Status bits included in the CAM are used to make forwarding decisions. These bits include the following:

- **SAA/reject bit**—Used for source address accept/reject. If this bit is not set, an incoming packet with the associated MAC address is rejected. By default, no MAC addresses are rejected, but this might be configured through the IOS for security reasons.

- **Network element (NE)**—This bit is used to indicate that the MAC address entry is a DPT network element. A source or destination device's MAC address that's connected to the DPT ring through Ethernet would not have this bit set.

- **Source address monitor (SAM)**—This bit does not affect forwarding decisions. An associated set of byte and packet counters can be activated for specific source addresses. This bit is used to determine whether byte and packet counters should be activated for this address.

- **Destination address to host (DAH)**—This bit is used to indicate that the associated MAC address is a destination address destined for the host connected to the DPT ring. This packet needs to be removed from the ring and processed.

Packet Handling (Incoming)

Six things can happen to an incoming packet:

- **Stripped**—Packet is removed from the ring.

- **Received and stripped**—Packet is sent to host (Layer 3) and removed from the ring.

- **Received and forwarded**—Packet is a multicast packet. It is sent to the host (Layer 3) and transit buffer.

- **Forwarded**—Packet is sent to the transit buffer.

- **Wrapped**—Packet is sent back in the reverse direction.

- **Pass through**—All packets, including control packets, are sent to the transit buffer.

For all incoming packets, the SRP control information, including TTL, RI, and mode, is extracted from the incoming packet and the following process occurs:

1   The packet-processing software associated with the DPT interface checks the Mode field of the incoming packet header to determine whether it is a control or data packet. If it is a data packet, which is determined by a Mode field value of 0x111, processing starts at item two in this list. If it is a control packet, the type of control packet is determined by looking at both the Mode and Control Type fields.

2   The control packet is stripped and sent to the appropriate processing routine. The path taken differs depending on the setting of the Mode field, as follows:

   — Topology discovery packets have a control type of 0x01. If the Mode field is set to 0x100 (pass to host), the stripped packet is placed in the receive buffer (like data packets), and from there it is passed to the topology discovery processing routine.

   — IPS packets have a control type of 0x02. If the Mode field is set to 0x101 (locally buffered for host), the stripped packet is buffered by the SRP MAC, and an interrupt is sent to the processor to retrieve the packet and pass it on to the IPS processing routine.

   — Usage packets have a Mode field set to 0x110. The stripped packet is forwarded to the mate, which further passes it on to the SRP-fa routine for processing.

3   The RI field is checked to ensure that it matches the incoming ring ID. An RI of 0 should only be in packets received on the outer ring, and an RI of 1 should only be in packets received on the inner ring. The exception to this is if the node is wrapped. In this case, packets can be accepted regardless of the RI value as long as there is a DA match. If the RI does not match the incoming ring ID and the node is not wrapped, the packet is forwarded to the transit buffer.

4   Perform CAM lookup to see whether the packet is bound for the node.

   A   If the source MAC address is the same as the node's MAC address, strip the packet from the ring. This means that the packet has traversed the ring back to the source and needs to be removed.

   B   If the destination MAC address is equal to the node's MAC address or it has a matching entry in the CAM, accept the incoming packet and perform a CRC check. If the CRC check passes, forward the packet to the routing process for delivery to the host. If it does not pass the check, discard the packet and increment the CRC error counter. If the packet is a unicast packet, strip it from the ring.

5   The packet process software then checks the TTL field. Although the packet should have been removed by either destination or source stripping, there remain cases where both the source and destination nodes might be removed (either intentionally or otherwise) from the ring while a packet is still in circulation. This is handled by the TTL field in the MAC header. The packet processing software decrements the TTL field by 1 each time a node forwards the packet. When the TTL reaches 0, the packet is stripped from the ring.

6   If the packet passes all of these steps and is still not processed, it is forwarded to the transit buffer. Figure 11-14 displays a DPT network element's queues. In each DPT node, there is a set of high- (Hi) and low- (Lo) priority queues (buffers) for the following:

- Transmit traffic
- Receive traffic
- Transit upstream traffic
- Transit downstream traffic

Figure 11-14 *DPT Queues*

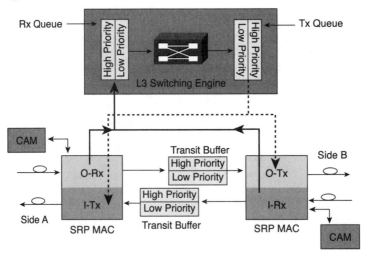

Packet Handling (Transmit)

When a packet is to be transmitted, it also needs to be processed. The transmit side is responsible for the following:

- Determining the priority of locally sourced packets
- Selecting the next packet to be sent on to the ring
- Managing the packet flow

Packets are classified as either transmit or transit traffic. Transmit packets are processed and then placed into either high- or low-priority queues based on the header Priority field. Transit traffic is processed and placed in high- or low-priority transit buffers (again defined by the Priority field in the packet's header). Traffic that is placed in a queue (buffer) is held for a configurable length of time. If traffic is backed up too long, fairness signaling or some other type of action results. Figure 11-15 displays the transmit processing and shows the two different queues for both transmit and transit traffic. The traffic meets at an X in the diagram where a scheduling function takes place. Schedulers are on a platform-by-platform basis. The Cisco 10720 router uses the latest in scheduling technology, the Versatile Traffic Management System (VTMS). Additional information on scheduling is available through Cisco Press QoS book offerings.

Figure 11-15 *Transmit Processing*

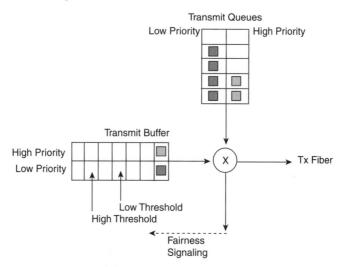

By separating the packets into high and low priority, SRP is able to provide packet prioritization and expedited priority handling for the transmit queues and transit buffers. This provides expedited handling for real-time and mission-critical applications that have strict delay bounds and jitter constraints. Control packets are put into a dedicated control queue for Layer 3 processing.

The node that is sourcing the packet onto the ring sets the Priority field in the SRP MAC header. The node automatically maps the value of the IP precedence bits in the ToS field of the IP header to the Priority field of the SRP MAC header.

Because there are only two priority queues (high and low) in SRP, a configurable priority threshold value is used to determine whether the packet should be placed in the high- or low-priority transmit or transit queues. It is recommended that the threshold point be set to limit contention for priority treatment. If too much traffic is considered high priority, the SRP fairness algorithms will not be able to manage traffic appropriately.

Transiting packets are checked for the value in the Priority field in the SRP MAC header and then compared to the configurable priority threshold to determine whether the packet is to be placed in the high- or low-priority transit buffer.

The next step is to determine which packet to transmit. This is accomplished with the help of a scheduler, which must choose between high- and low-priority transit packets and high- and low-priority transmit packets. The following rules are used in making the decision:

- Schedule high-priority packets before low-priority packets.

- Enforce ring packet conservation by avoiding discarding packets that are already in circulation around the ring. It is better to transfer traffic that has already made it onto the ring than to add traffic to the ring and drop the existing traffic. This could cause a lot of contention on the ring.

The following high-level, packet-handling hierarchy is used in making the scheduling decisions:

1 High-priority transit packets

2 High-priority transmit packets from host

3 Low-priority transmit packets from host

4 Low-priority transit packets

The system might modify this order if the low-priority transit buffer gets too full. This is done to ensure that the transit buffer does not overflow or have to wait extended periods of time if the node has an abundance of locally sourced traffic.

DPT Rings

DPT uses a ring architecture, which allows it to utilize existing cable infrastructures in MAN environments. The dual counter-rotating ring topology allows the concept of ring wrapping to provide protection by routing around failures. The inner and outer rings have symmetrical and equivalent roles and are composed of unidirectional P2P links. Traffic might be placed on either ring as determined by SRP.

SRP Ring Selection

SRP is responsible for discovering all nodes on the ring and building and maintaining a topology map of the network. SRP uses the topology map to determine the most optimal ring to place traffic on to get to the destination. Ring selection is provided by an ARP mechanism. The version 2 implementation of SRP just prefers the path with the least number of hops to get to the destination.

Rings transport both data and control packets. When data is placed on one ring, control packets are placed on the ring in the opposite direction. All the packet types have a common SRP header, thus allowing the MAC chips to handle all traffic in a similar way.

Some control packets are for general ring administration and are therefore sent in the appropriate direction for the specific function; for example, topology discovery control packets are always placed on the outer ring.

When data packets cause congestion, the usage information control packets, which are used to transmit available bandwidth, are sent on the ring in the opposite direction of the congestion to force the upstream node to throttle back.

Ethernet and the DPT Ring

There are some differences between passing traffic on an Ethernet segment and passing traffic around a DPT ring. Delivering IP traffic to a destination is simple on a standard Ethernet segment. Based on ARP, the destination MAC address is determined. An Ethernet frame is created with this MAC address as the destination, and the IP packet is encapsulated into the frame for transport onto the underlying physical media. In this example, there is only one media to place traffic onto, so the decision is simple.

In DPT, ARP has to be augmented to work on a ring. The goal is not to change the ARP protocol, but to provide the extra functionality needed to make ARP work in a ring environment. These procedures apply to unicast packets only because multicast packets are sent to all nodes on the ring and do not require ARP. The changes needed are as follows:

- When the source sends an ARP request packet, SRP uses a hashing algorithm on the destination address to determine which ring to use. The DA is hashed, and if the result is 0, the outer ring is selected; otherwise, the inner ring is selected.

- Upon receiving the ARP request, the responding node looks at the topology map (discussed in the section "Automatic Configuration of DPT Rings") and chooses, based on hop count, the best ring on which to send the ARP response. Figure 11-16 shows this process.

In Figure 11-16, node 1 needs to send traffic to node 3. Node 1 only knows the destination IP address of the packet. Node 1 hashes the broadcast destination MAC address. If the result is a 0, the traffic is put on the outer ring. If the result is a 1, the traffic is put on the inner ring. Each node around the ring sees the packet, but only node 3 responds because it is the destination of the unicast frame. Node 3 looks at its topology map and selects the ring with the shortest hop count to place the response packet.

Figure 11-16 *ARP on a DPT Ring*

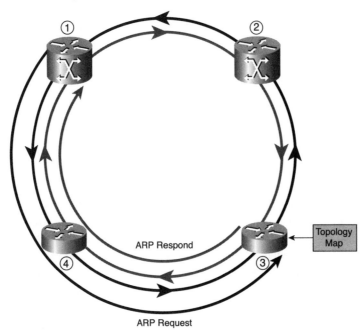

Note that in the version 2 SRP implementation, only the shortest hop count is used as the metric for ring selection. In the future, a more sophisticated approach could be used. Shortest hop-count metrics is a good selection criteria for DPT because this selection optimizes spatial reuse.

If a wrap is present in the path between the source and the responding node, the responding node chooses the opposite ring from which the ARP request came in, to send back the ARP response. This is because in this case, each node can receive ARP responses only from a single direction; one of the alternative directions is blocked.

A tie happens when the inner and outer rings are equal hops to the source, as in Figure 11-16. In this case, the hashing algorithm is used as a tiebreaker.

The source node receiving the ARP response checks the ring identifier where the ARP response is received. The opposite ring is used to send data packets.

When a node wraps or unwraps, the ARP cache of all nodes in the ring is flushed to trigger ring selection. This is done so that traffic traveling on the suboptimal path can be correctly forwarded on the most optimal path.

NOTE   This solves a long-time debate over wrap versus steering approaches to automatic failover. Wrap is faster but does not provide route optimization. The ARP flushing triggers a steering operation after the wrap. The DPT architecture uses the advantages of both solutions. Suboptimal routing happens but for a shorter time than the alternative. RPR uses the steering function.

For multicast packets, the destination address is hashed using the hashing algorithm. If the result is 0, the outer ring is selected; otherwise, the inner ring is selected.

As in any network design, there are times that you have a better idea of how you want certain traffic to flow through the ring than might be available to the dynamic procedures just discussed. Overriding dynamic decisions of the system always carries with it user responsibility to ensure traffic continues to flow smoothly and efficiently. In these situations, static ARP is used to override the dynamic ring-selection process. This is useful to SPs that need to manually engineer bandwidth utilization.

Knowledge of the ring topology can speed packet processing. Automatic ring topology discovery and configuration is accomplished through the use of the topology discovery packets. This process is described in greater detail in the next section.

Automatic Configuration of DPT Rings

Topology discovery is a key feature of DPT rings that provide automatic ring configuration. Topology discovery is performed as nodes are added and removed from the ring. The ring automatically reconfigures itself and updates each node through topology discovery packets. This is a sharp contrast to the provisioning and traffic-engineering needs of SONET ring networks. The key component of topology discovery is the topology discovery packet.

Each node on the ring sends out topology discovery packets on the outer ring. Each node that receives this packet appends its own MAC address and MAC type information (ring ID and wrap status) to the control packet. The topology discovery packet used for collecting node information has a variable size, and as the packet travels around the ring, its size increases. Figure 11-17 displays this operation.

Figure 11-17 *Topology Discovery Packet*

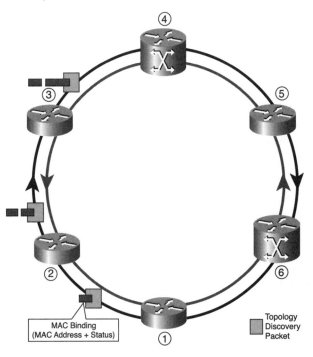

MAC Binding
(MAC Address + Status)

Topology
Discovery
Packet

NOTE   Control packets are always point to point, so the next node at Layer 2 always strips discovery packets. The packet needs to traverse the ring so that the discovery packets are not removed from the ring until reaching the source. The source populates this information into the topology table when it receives two identical topology discovery packets. The MAC chip does not do the stripping; rather, the software host does it. If the software host fails after it has placed packets onto the ring, the TTL field is used to determine the removal of those topology discovery packets.

In the current implementation (version 2), discovery packets are sent out on the outer ring, so they should also arrive back to the originating node on the outer ring.

If the node is wrapped because of a failure, the discovery packet just follows the wrap, which moves the packet from the outer ring to the inner ring. Topology discovery is based on building a sequence of node information to reflect node locations on the ring. Wrapping should logically provide a full ring topology. Because of this requirement, the topology discovery packet does not change when it enters into a wrap. The packet is again updated when it comes out from the wrap and returns to the outer ring. This results in a topology list that excludes only the nodes inside the failure domain.

If the failure that caused the wrapping changes quickly, the DPT ring topology becomes too unstable to use for route optimization. Changing failure conditions might also destroy control packets.

The solution to the instability problem is to compare two consecutive topology packets. If the packets are identical, the ring has reached a relatively stable status. It makes sense to re-optimize ring-selection decisions only when the topology is no longer changing.

When the node finally receives two consecutive identical topology packets, it begins to build a topology map from the information. Each node stores the learned topology map, which can also be read by the management console. The topology map includes the hop-count information, the MAC address, the wrap status, and the host name of each node in the same sequence that they appear connected on the ring. The topology map is the basis for answers to ARP requests. Management applications might also use the discovered topology to help fault management and troubleshooting.

Topology discovery is done periodically (every 5 seconds by default) but is also triggered by protection switching events. The periodic execution provides some independence from the IPS functions and also provides a basis for improving stability in a case of quick changes in failure status.

With the ring discovered and ring-selection mechanisms in place, you must then understand how traffic engineers and the SRP fairness algorithm can manage bandwidth. Traffic engineers have an important role because poor implementation can override even the best of automated systems, as you learn in the next section.

DPT Quality of Service

With the queuing capabilities supported by DPT, quality of service (QoS) issues can be addressed for traffic placed on the ring. While helping to manage traffic, it is important not to not overburden DPT with too much traffic management as opposed to traffic delivery. If you require DPT to perform excessive QoS duties, you can mitigate the burden by ensuring QoS processing is handled as much as possible at the IP layer.

DPT is designed to transport upper-layer services. DPT nodes typically do not originate traffic at Layer 2, except for control traffic. Control traffic is always considered by DPT to be high priority. The priority of data packets is determined at the upper layers and placed in the IP headers, so DPT does not have to make decisions on traffic priority but merely to pass traffic at the priority already specified.

Although SRP provides automatic mapping from the IP ToS field into the SRP priority field, you also can manually configure the mapping. This Priority field might then be used by the SRP-fa to provide different handling for different-priority traffic.

SRP Fairness Algorithm (SRP-fa)

A ring topology provides for shared bandwidth on a link with multiple source-destination traffic flows. It is possible to overload a link between two nodes. In such a case, congestion management is needed to ensure that higher-priority traffic gets priority over lower-priority traffic. The SRP-fa has the following high-level goals:

- **Global fairness**—Each node on the ring gets its fair share of the ring bandwidth by controlling the rates at which packets are forwarded onto the ring from upstream or downstream sources in relation to packets that are sourced by the node. The goal is to ensure that neighbors cannot act as bandwidth hogs and create either node starvation or excessive delay conditions.

- **Local optimization**—This ensures that nodes on the ring maximally leverage the spatial reuse properties of the ring so that they can use more than their fair share on local ring segments, so long as other nodes are not adversely impacted due to traffic locality.

- **Scalability**—It is expected that customers will generally be building large rings of up to 128 nodes over geographically distributed areas. The algorithms used for rings of this size must rapidly adapt to changing traffic conditions and accommodate propagation delays associated with interbuilding, metropolitan, and long-haul distances.

The fairness algorithm is designed so that it can adapt to changing traffic patterns quickly. The next section describes how SRP-fa handles packets.

Understanding SRP-fa Packet Handling

Figure 11-18 illustrates SRP-fa functionality. Node 1 begins transmitting packets to node 3 at full line rate on an OC-12 ring. Node 2 now has traffic to send on the ring but cannot because node 1 is using all the ring resources. Node 2 sends a fairness message to node 1. Node 1 throttles back its traffic rate to share the ring bandwidth with node 2. Node 3 now has traffic for node 1, while nodes 1 and 2 are consuming all the available bandwidth on the ring. Node 3 sends fairness messages to nodes 1 and 2. Nodes 1 and 2 throttle back their traffic rate. This allows the ring bandwidth to be shared among the three DPT network elements on the ring. This is a great simplification of the fairness algorithm, which crunches the usage parameters to allow the proper amount of bandwidth to each node.

Figure 11-18 *SRP-fa*

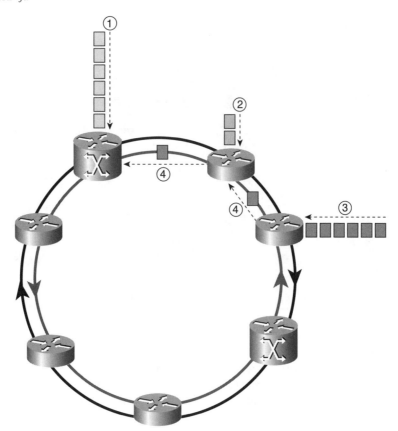

Figure 11-19 shows this operation from the vantage point of node 1 through a modeling program. Notice that the bandwidth on the vertical access changes for the different time points (1s, 2s, and 3s) on the horizontal access. The bandwidth changes are a reflection of the node throttling its bandwidth consumption so that node 2 and 3 get fair access to the ring.

To better understand the SRP-fa, it is beneficial to see how traffic is managed as it gets ready to be placed onto the DPT ring.

Figure 11-19 *SRP-fa Modeling*

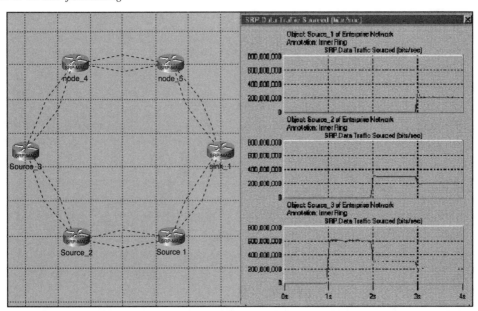

Low-priority packets are transmitted at any time as long as there is sufficient transit buffer space. If high-priority packets need to be controlled, the SRP-fa is not involved, and this needs to be handled with features such as committed access rate (CAR) at Layer 3. If a node becomes congested with high-priority traffic, the most SRP-fa can do is generate a fairness message to its upstream neighbors to throttle down the traffic rate.

If the node must choose whether to transmit high-priority packets sourced locally or transit packets moving through the node, the node forwards packets that are transiting the node.

Because ring integrity must be applied to all frames, the transit queue is given an increased priority prior to reaching a point where packets could overflow the buffer. At a certain threshold, the packet scheduling is changed to allow the transit buffer to empty out to an acceptable level. With a modification of the thresholds, it is possible to control the node delay and jitter on the ring.

Implementing SRP-fa Packet Handling

The packet-handling process is implemented by the scheduler functionality in the node. The scheduler follows the rules of the SRP-fa. This scheduler enforces priority levels, first giving higher priority to transit traffic. The following hierarchy shows the order of precedence for the priority levels that SRP-fa enforces under noncongested circumstances:

1 High-priority transit packets

2 High-priority transmit packets

3 Low-priority transmit packets

4 Low-priority transit packets

Fairness processing includes the generation and interpretation of congestion signaling messages and the updating of the SRP-fa traffic counters when a packet is transmitted onto the ring.

These traffic counters are maintained locally and are exchanged in usage control packets between nodes to manage bandwidth access and make forwarding decisions. The main counter values are as follows:

- My_usage tracks only low-priority packets that originate on the node and that are inserted onto the ring. My_usage does not measure high-priority transmit packets.

- Fwd_rate measures low-priority transit packets forwarded by the node. Excess transit traffic is not rate limited by the node. Instead, it generates a fairness message to the source of the congestion. Fwd_rate is adjusted periodically to allow packet-forwarding credits to accumulate over time and reflect the most recent usage conditions on the ring. Fwd_rate does not measure high-priority transit packets.

The main threshold values are as follows:

- Allow_usage is the maximum number of locally sourced packets that the node could transmit onto the ring as determined by the fairness algorithm. Nodes trying to source more than their current fair share of the traffic are rate limited by this parameter. The current value of allow_usage is set according to bandwidth data received from neighbors. This value can increase up to max_usage.

- Max_usage is a per-node parameter that places an upper limit on locally sourced low-priority packets, regardless of the availability of ring bandwidth as communicated by the SRP-fa. A node's my_usage is always less than or equal to the allowed_usage, which cannot exceed max_usage.

As usage packets are received at a node, the usage information is extracted and the fairness algorithm is used to calculate new values for the local counters and thresholds. The amount of change is factored by a time value called the decay interval, which helps to provide for fair access to the bandwidth. These values are then sent in usage packets to the upstream and downstream neighbors to be used in their calculations.

SRP-fa and Congestion

Several factors can congest the network. If congestion is due to too much high-priority traffic, an external mechanism such as CAR should be put into place. Proper network design requires that high-priority traffic inside the network be kept to a manageable level. Congestion levels affect what types of traffic are transmitted and transited. The resulting hierarchy is as follows:

1

    A　High-priority transit packets

    B　Low-priority transit packets (high congestion)

2

    A　High-priority transmit packets

    B　Low-priority transit packets (low congestion)

3

    A　Low-priority transit packets (if low-priority transmit packets are not allowed)

    B　Low-priority transmit packets

4　Low-priority transit packets

The scheduler is the part of SRP that controls access to the DPT ring. It works through a simple procedural checklist to determine which traffic should be placed onto the ring as follows:

1　The scheduler always checks the high-priority transit queue first. The scheduler moves on to serve other queues only if this queue is already empty.

2　If a high-congestion situation is declared, the low-priority transit queue is scheduled next. Then the scheduler returns to the top of the list.

3　If a high-congestion situation is not declared, the scheduler consumes packets from the high-priority transmit queue of the software host. The scheduler moves on to other queues only if this queue is already empty.

4　If the scheduler still does not have to source packets onto the ring, it checks whether a low-congestion situation is declared. If a low-congestion situation is declared, the scheduler forwards low-priority transit packets. After this operation, it goes back to the top of the list.

5　If the scheduler does not find anything else to transfer and a low-congestion situation is not declared, it checks, based on the allow_usage threshold, whether locally sourced packets should be inserted into the ring. If local sourcing is disabled because of allow_usage value, the scheduler starts forwarding packets from the low-priority transit queue.

6 If there is no congestion situation and allow_usage is not imposing limitations, the scheduler checks the low-priority transmit queue of the software hosts and puts the packets from this queue onto the ring.

7 Finally, the scheduler forwards low-priority transit packets.

At each checkpoint, only one packet is put onto the ring. After this packet is sent out, another scheduling cycle starts from the top of the list. The scheduler must go back to the top of the list after each packet is sent because of the strict-priority nature of the high-priority queue. This minimizes the blocking of higher-priority packets by low-priority traffic in the middle of a transfer. This process continues on in an infinite loop.

SRP-fa Monitoring

Monitoring the fairness operations details in a production environment might be done only by special remote monitoring (RMON)-style functions. Classic Simple Network Management Protocol (SNMP)-based Management Information Base (MIB) variable polling for a 100-ms process might provide limited useful information due to the speed of changes in DPT environments.

The stabilized traffic counter values after an SRP-fa settlement provide a good basis for spatial reuse analysis. A high fwd_rate value signals a traffic pattern that does not allow proper spatial reuse.

A remote network management station can run computations on this collected data to determine an optimum ordering of the nodes in the ring to maximize bandwidth utilization. However, this computed information might be meaningful only for stable traffic patterns.

Fairness operations can also be simulated with tools such as those provided by OPNET Technologies, Inc. OPNET specializes in management and modeling tools. You can find more information about this company at www.mil3.com.

With OPNET software, you can put collected traffic counter data into a model and use it to analyze your ring. A scenario analysis helps to determine the optimum reconfiguration of the ring. Because it is difficult to compute the exact behavior of the ring on paper, simulation is the only good way to validate new designs against such requirements as providing enough transit buffers for the time of SRP-fa convergence. Figure 11-19 shows OPNET output.

At this point, you should understand the capabilities of the DPT ring and how traffic can be moved and managed around the ring. The last step is to understand how the ring reconverges after a failure.

Maintaining Network Reliability

You want to minimize network congestion to keep traffic flowing. When making changes to the network, you must avoid as much as possible taking the network out of service to implement the changes, which makes the network unavailable to users. DPT rings have many features to help maintain the network's availability to users, as discussed in the following sections.

Automatic Reconfiguration and Rerouting

The DPT switching mechanism used for protection switching is called *intelligent protection switching* (IPS). IPS is similar to the protection switching capabilities of SONET. The DPT ring does not need to have the dedicated protection bandwidth used in SONET. In addition, although it does not rely solely on the overhead bytes of SONET, IPS can use the error information contained in the overhead bytes to determine loss of frame, loss of signal, or bit error rates over a prescribed threshold.

Protection against single points of failure is based on a hybrid concept of reconfiguration. The nodes that detect the failure wrap the ring access controller around the failure, and then a steering process optimizes the path selection.

Multiple link failures might segment the ring (similar to FDDI). The best way to avoid multiple concurrent failures is to ensure that multiple segments of the ring are not carried in the same conduit or as separate wavelengths on a single DWDM fiber. Ensuring that the segments take divergent paths is also helpful. An event hierarchy is used to make a distinction between the various levels of failures.

During a fiber failure, changes to the topology map need to occur. To verify that each node has the correct topology information, an Address Resolution Protocol (ARP) flush and a topology rediscovery are required to ensure accurate ring selection and optimized ring bandwidth use. Even though this might sound like a major operation that could take some time to complete, the operation is still transparent to Layer 3, which maintains its routing table information. When a new IP packet is ready to be placed on the ring, the ARP table is empty, so new ARP requests are sent out. This results in a newly optimized ring-selection map that passes the IP traffic around the ring. The ARP flushing process implements the steering requirement in the automatic failover concept.

IPS Failure and Recovery

IPS provides functionality that is analogous to automatic protection switching (APS) for SONET rings but with several important extensions. IPS does not require the allocation of protection bandwidth, so there is no unused slack. IPS is even better than extra traffic arrangements in SONET because no artificial distinction exists between protected traffic and preempt-able traffic. IPS automatically recovers from fiber facility failures by wrapping the transmit and receive functionality at the MAC layer.

IPS does not rely on configured knowledge of ring topology, such as node identities and node adjacencies, to execute restoration and recovery procedures. This capability also enables network operators to insert and remove nodes from the ring, and to merge independent rings with minimal provisioning and configuration. The SRP ring has extensive topology-discovery capabilities to facilitate this process.

IPS protection switches are rapidly executed to minimize packet loss after a fiber facility or node failure. *Wrapping* is the IPS method of protection switching because of its speed and the capability to deliver packets in transit to the original destination.

IPS is fully independent from SONET layer protection mechanisms. SONET overhead bytes can be monitored to speed up the detection of link problems because DPT is using the SONET framer. DPT does periodical link-integrity checking to detect problems and activate protection switching.

IPS Control Messages

Proactive fault and performance monitoring is based on the SONET overhead bytes. Event detection and reporting is done according to the same principles as in SONET. All nodes process these signals and propagate them appropriately. This communication scheme facilitates rapid recovery and restoration. IPS includes procedures minimizing IPS-related signaling traffic under normal conditions, while expediting IPS signaling during failure detection and recovery periods.

IPS supports a protection switching event hierarchy. This allows the handling of concurrent multiple events without partitioning the ring into separate subrings or islands.

IPS Event Hierarchy

IPS provides a protection switching event hierarchy (modeled after the SONET APS hierarchy) to handle multiple, concurrent events. The event priorities, listed from highest to lowest priority, are as follows:

- **Lockout of protection (LO)**—Disables protection switching functions.

- **Forced switch (FS)**—Allows the network operator to manually force a protection switch. This command is useful during procedures when adding or removing a new node on the ring.

- **Signal fail (SF)**—Automatically initiated by detection of the following: loss of signal (LOS), loss of frame (LOF), line bit error rate (BER) above a specific threshold, line alarm indication signal (AIS), or excessive cyclic redundancy check (CRC) error.

- **Signal degrade (SD)**—Automatically initiated by detecting line BER above a specific threshold or excessive CRC errors.

- **Manual switch (MS)**—Similar to the forced switch but of lower priority. MS is typically used for testing survivability because no switchover is performed before the successful dynamic testing of the ring.

- **Wait to restore (WTR)**—Prevents protection switching oscillations. The node does not unwrap immediately after a failure condition is solved. When the failure situation has been cleared, the node waits for a configured period of time before unwrapping in case the link is oscillating.

- **No request (IDLE)**—The lowest-priority event sent out during normal working conditions with no failure conditions.

If a ring is already in Wrap mode due to an SF condition between nodes A and B, a manual ring switch could not be performed at either node A or B because of its lower priority. An FS could be performed because an FS is of higher priority in the event hierarchy. Because an MS is an operator-initiated function, the FS needs to be cleared before automatic IPS recovery mechanisms can take over again.

IPS Signaling

IPS event requests are initiated either automatically (after the node detects a triggering event) or manually through one of various IOS commands entered by the network administrator.

IPS sends control packets describing the state of the ring. IPS trigger events can be categorized based on severity. The most serious problems are hard failures. A hard failure is a complete node or card failure, which would make it impossible to send anything through the failure domain. Signal failure is a type of hard failure. The SD failure is an example of a soft failure. There are also operator-initiated events. These events are not failures in the network, but events initiated by an administrator. These events are normally caused by maintenance procedures. The operator-initiated events include MS, FS, and LO. MS and FS provide the same functionality, but they have different priority levels in the event hierarchy. An MS also tests the rings for existing failures before performing a switch. If there are any failures in the ring, an MS will not result in a switch. An FS always forces a switch regardless of whether there is an existing wrap somewhere in the ring. If the administrator is not careful, an FS could result in a segmented ring.

Node States and Ring Operation

A type of failure can occur in which the Layer 3 software is no longer able to route traffic but the physical hardware is still up and operational at Layer 2. In this situation, the goal is to avoid ring wraps or ring partitioning due to node failure. When a DPT node is wired into an existing DPT ring with the interface shut down, the interface is in *Pass-through mode* to protect the ring from segmentation. The Layer 3 processing is disabled at this point. The node cannot transmit traffic on the ring, but all traffic transiting the node is forwarded. Figure 11-20 displays the operation of a node that is in Pass-through mode. Pass-through mode is used to protect against Layer 3 failures and to disable the ring sanity check when there is a wrap somewhere in the ring. Notice that the Layer 2 functionality of the MAC layer (Ring Access Controller) is functioning normally, whereas Layer 3 communication is not functioning.

Figure 11-20 *DPT Node in Pass-Through Mode*

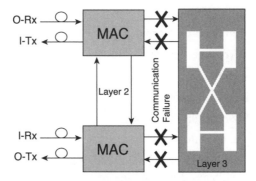

While in Pass-through mode, the following occurs:

- SRP-fa does not execute on this node.
- TTL is passed through and not changed.
- The fwd_rate counter is passed through and not changed.
- The node cannot source packets onto the ring.
- Control and data packets are passed to the transit buffer. No processing to determine packet type is done, and no CAM lookups are performed.
- The transit buffer acts as a simple buffer that is unable to discern between high- and low-priority queues.

Normal Working DPT Ring

IPS operations can be described by a four-tuple message format {R, S, W, P} that is sent out periodically by each node. The following is a description of this tuple:

- R is the IPS request type.
  - (LO)—Lockout of protection
  - (FS)—Forced switch
  - (SF)—Signal fail
  - (SD)—Signal degrade
  - (MS)—Manual switch
  - (WTR)—Wait to restore
  - (O)—IDLE No request
- S is the source address of the IPS message.
- W is the wrap status of the source.
  - (W)—Traffic is wrapped
  - (O)—Idle
- P is the Message path indication.
  - (S)—Short path
  - (L)—Long path

Initially, all nodes are in a Pass-through state. When the DPT interfaces on the node, all nodes transmit idle IPS messages to the upstream and downstream neighbors. In Figure 11-21, all nodes are in the Idle state and there are no issues in the ring. Node A is transmitting the idle message to its neighbors with the (0, A, 0, S) information. The information means the following to the routines programmed into the DPT routers:

- O is the no switch request.
- A is the source address of the transmission. In reality, this would be the MAC address of the node's DPT interface. A is used here for simplicity to denote node A.
- O is the no wrap.
- S is a short path message.

Short path messages are IPS control messages that are only transmitted to the next neighbor. Every node in the ring would be sending these idle messages. Figure 11-21 simplifies the operation by focusing on node A.

Figure 11-21 *DPT Normal Working Operation*

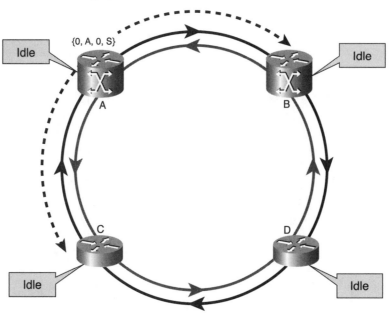

The failure and recovery situations in the following sections use this same normal ring to show the failure and recovery IPS event propagation throughout the ring.

One-Fiber Failure

Single-fiber failures create a situation that can be difficult to troubleshoot because only one node will have an indication of the failure. Figure 11-22 depicts an environment with a one-fiber failure. Node B detects the problem through an LOS on the receive interface. Immediately, node B generates two simultaneous IPS messages to both neighbors. Both IPS messages contain a request type SF, a source address of B, and a status of WRAP. What differs is whether a short or long message is sent. DPT nodes always transmit short messages in the direction of the failure and a long message in the direction away from the failure. The long message is propagated through each DPT network element so that the ring can converge around the failure. The short message is sent in the direction of the failure so that the other side of the link will be aware of the condition in the event of a one-fiber failure.

Figure 11-22 *Single-Fiber Failure*

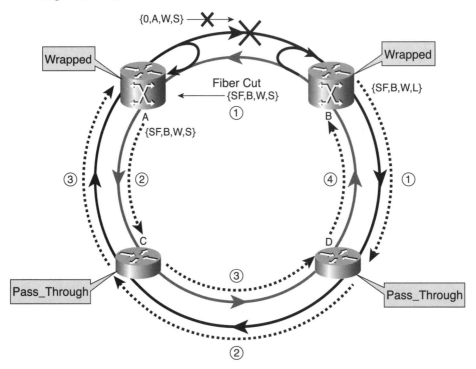

Figure 11-22 is numbered to display the order of the IPS messages, as follows:

1 Node B detects the failure and generates a short SF/WRAP message toward the failure
  and a long SF/WRAP message away from the failure. Node A receives the short
  message from node B. Node A identifies the message as a short IPS message with an
  SF/WRAP action. Node A is now aware that there is an issue in that direction and
  wraps the MAC interface on that side of the DPT interface. Both node A and node B
  flush their ARP cache.

2 Node C receives one IPS long message from each of its neighbors. Node C identifies
  the message as a long IPS message with an SF/WRAP condition. Node C enters the
  Pass-through state to turn off the ring identifier sanity check. The sanity check is
  turned off because packets can now potentially be sourced from the inner ring and be
  found on the outer ring due to the wrap. Nodes C and D forward the IPS long message.
  Both nodes flush their ARP cache.

3 Node A and node D receive an IPS long message from their respective neighbors
  with an SF/WRAP condition. Both of these nodes already know about the failure
  condition. No action results from the receipt of this frame. Node A removes the IPS

message from the ring because the outgoing interface to the next neighbor is wrapped. Node D forwards the message and makes no changes because the node is already in the Pass-through state.

4 Node B receives an IPS long message from node D with an SF/WRAP condition. Node A already knows about the failure and removes the IPS message from the ring.

When the issue between nodes A and B is fixed, neither node unwraps the ring. Instead, both nodes start a restoration timer known as the wait to restore (WTR) timer. The WTR timer is 60 seconds by default but might be configured between 10 and 600 seconds. It is advisable to set the WTR timer to its lowest value in lab environments so that testing can be performed.

When a node starts the WTR timer, short IPS messages indicating that the node is still wrapped and is in a WTR state are sent in the direction of the resolved failure. A long message to the same effect is sent in the opposite direction of the failure. This allows the ring to have knowledge of the process in place. Figure 11-23 shows this process.

Figure 11-23 *DPT Wait to Restore Timer*

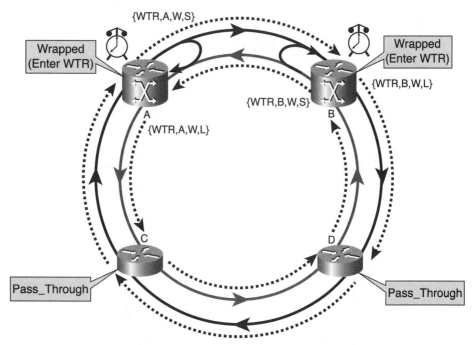

After the WTR expires, node A and B unwrap the ring and transition into the Idle state. Nodes C and D also transition into the Idle state. Each node sends out the short idle messages discussed earlier in this chapter. Figure 11-24 shows this process.

Figure 11-24 *DPT Wait to Restore Timer Expiration*

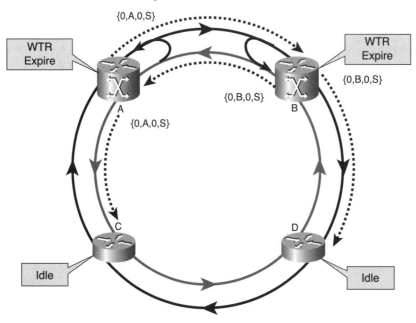

All nodes at this time change to the idle state as seen in Figure 11-25.

Figure 11-25 *DPT Idle State*

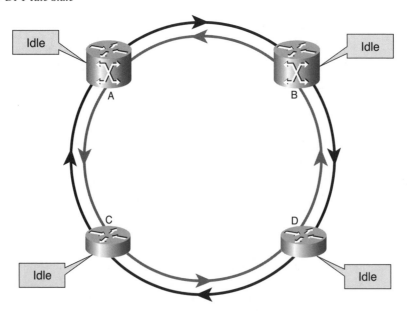

Two-Fiber Failure

Figure 11-26 displays a two-fiber failure between nodes A and B. The steps that were covered in the one-fiber failure are the same as those in a two-fiber failure. In a two-fiber failure, it is not important that each node send a short message in the direction of the failure, but they perform the operation anyway. This functionality is built in to the SRP protocol because each node has no way of knowing whether the failure is a one-fiber or two-fiber failure. Each node is only aware that an LOS occurred on its receive interface.

Figure 11-26 *Two-Fiber Failure Signaling*

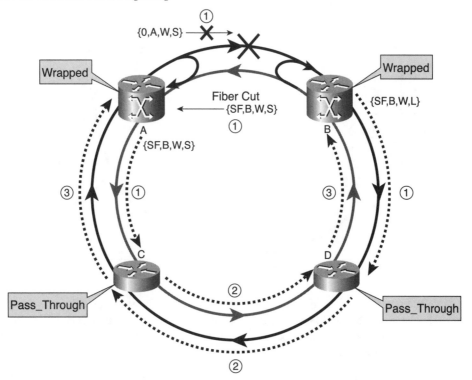

Signal Degrade and Signal Failure

Figure 11-27 displays the same DPT ring with an SD issue between nodes A and B. The steps in which the IPS signaling take place is the same as the one laid out in the "One-Fiber Failure" section. The only difference is that instead of carrying an SF message, the nodes send an SD message in their IPS signaling.

Figure 11-27 *DPT Ring with a Signal Degrade Condition*

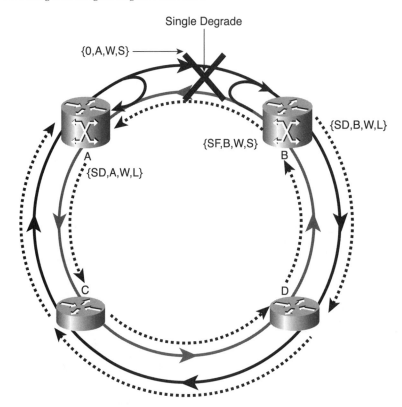

While the SD condition is still present between nodes A and B, an SF condition occurs between nodes B and D. Nodes B and D wrap their ring, and node B now becomes segmented from the rest of the ring due to the dual-failure nature. Figure 11-28 shows this network failure state.

IPS messages are sent around the ring regarding the SF condition. Node B is directly aware of the SF condition through an LOS on the receive interface. Node B determines that there is an SD and SF in the ring. An SD is lower in the IPS hierarchy than an SF, so node B unwraps the side of the ring directed toward node A. Figure 11-29 depicts this scenario.

Figure 11-28 *DPT Ring with a Signal Degrade and Signal Failure*

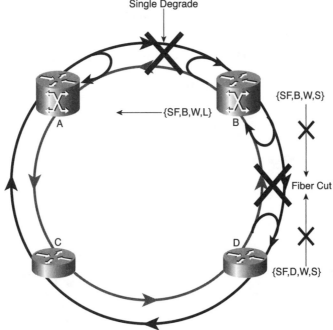

Figure 11-29 *DPT Ring/IPS Hierarchy Action*

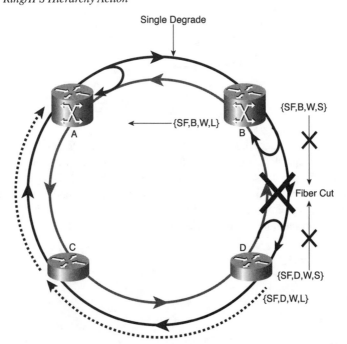

Node A receives the long message regarding the signal failure condition elsewhere in the ring, and like node B, node A unwraps the ring. It is better to receive traffic with a higher BER due to an SD than not to receive any traffic due to a segmented ring. Figure 11-30 depicts this scenario.

Figure 11-30 *IPS Hierarchy Result in a Ring with a Signal Degrade and Signal Failure*

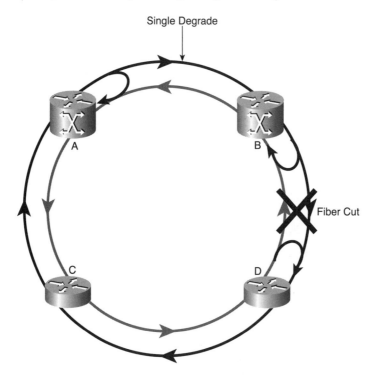

Summary

This chapter introduced DPT, SRP, IPS messaging, the IPS hierarchy, DPT failure, and failure recovery. DPT can co-exist with existing SONET infrastructures, dark fiber, and DWDM. DPT provides rings today with speeds up to OC-192 capable of supporting up to 128 network elements in a ring. SRP optimizes fairness around the ring. Destination-based packet stripping allows maximum efficiencies when compared to LAN-based ring architectures. DPT protection switching is as fast as that found in SONET networks.

Review Questions

1 Which three of the following statements are true?

    A DPT is a general solution based on the principles of multiaccess networks standardized by IEEE.

    B DPT is optimized for IP only and other Layer 3 technologies are not considered for support.

    C The DPT proposal borrows from both IEEE 802.3 and SONET.

    D DPT significantly increases configuration management requirements beyond that required for full-mesh topologies.

2 A being the highest and D being the lowest, letter the following switch states in the IPS hierarchy.

    A Signal failure _____

    B Signal degrade_____

    C Manual switch _____

    D Forced switch _____

3 Initial implementation of DPT is based on a _____ frame encapsulation for easy reuse of optical components.

    A APS

    B IPS

    C SONET

    D FDDI

4 Which two of the following are reasons why DPT is typically used in campus environments?

    A DPT uses a static point-to-point topology-based configuration.

    B Physical ring topology is typically already used, making the migration to DPT easy.

    C DPT rings can support multiple subnets, allowing the transport of many networks on a single optical backbone.

    D DPT provides an easy migration from LAN technologies because it provides a similar MAC layer interface as there was previously.

5 Which of the following is true?

   A It is not possible to overload a link between two nodes.

   B The SRP-fa controls only low-priority packets.

   C The rapid adaptation of the fairness algorithm causes instabilities.

   D Control of high-priority packets is left to Layer 4 or upper-layer functions.

6 Which two tasks are handled by SRP control packets?

   A Synchronization

   B Topology discovery

   C Protection switching

   D FDDI

7 Which one of the following statements is true?

   A The Layer 2 functions can see the DPT ring as a single subnet.

   B The DPT ring is a Layer 2 protocol.

   C DPT internals are mostly visible from the upper layers.

   D The Layer 3 functions see the DPT ring as multiple subnets.

8 What type of packet is indicated when setting the destination MAC address to 0?

   A Data

   B Control

   C Forwarding

   D Discovery

9 In a fully functional DPT ring, multicast packets are stripped by the _____.

   A Local host

   B Receiving node

   C Active monitor on the ring

   D Source node

10 What is the Mode field of data packets set to?

   A 0101

   B 010

   C 111

   D 1XX

11 Under what condition does a node accept a control packet but not forward it?

   A Control TTL > 1

   B Control TTL = 0

   C Control TTL = 1

   D Control TTL < 0

12 What is the goal of global fairness according to the SRP fairness algorithm?

   A The node with the highest-priority traffic gets the most bandwidth.

   B One node controls the bandwidth usage of its neighbors.

   C No node can excessively consume network bandwidth and create node starvation or excessive delay.

   D A high-priority node can retain as much bandwidth as needed, although this might create node starvation or excessive delay.

13 During a congestion situation, which type of traffic is more likely to cause harm to ring integrity?

   A High-priority

   B Control

   C Low-priority

   D Discovery

14 What is the proper transmit decision hierarchy during periods of no congestion?

   1 – High-priority transit packets

   2 – Low-priority transit packets

   3 – High-priority transmit packets

   4 – Low-priority transmit packets

   A 1, 2, 3, 4

   B 2, 1, 4, 3

   C 2, 3, 1, 4

   D 1, 3, 4, 2

15 Ring selection is controlled by what procedure?

   A AARP

   B MAC

   C ARP

   D AURP

16 In a DPT ring, the node has to decide whether to place traffic on the inner or outer ring for transport to the destination. What is ring selection based on?

   A Topology information

   B Ethernet frames

   C MAC address

   D Number of subnets

17 When is the ARP cache flushed?

   A When a new node is added

   B When a ring wrap or unwrap condition occurs

   C When the TTL expires

   D At regular intervals

18 What algorithm is used to determine which ring to send a packet on when the number of hops on the inner and outer rings is the same?

   A Fairness

   B Transmit

   C Hashing

   D Limiting

19 Which is not an IPS event priority?

   A FS

   B SM

   C SD

   D MS

20 Which of the following statements about WTR is true?

   A The request is generated when the wrap condition clears.

   B The request is generated when the wrap condition begins.

   C The timer can be configured in the 10- to 1200-second range.

   D The default value of the WTR timer is 100 seconds.

21 Which is the proper interpretation of the message {SF, B, W, L} when sent by node B to node D.

   A Request type = Signal failure, Source MAC = Node B, Node status = Wrapped, Path = Long, enter Pass-through Mode

   B Request type = Signal failure, Source MAC = Node B, Node status = Wrapped, Path = Long, enter Wrapped state

   C Request type = Signal failure, Source MAC = Node A, Node status = Wrapped, Path = Long, enter Pass-through Mode

   D Request type = Signal failure, Source MAC = Node B, Node status = Wrapped, Path = Short, enter Wrapped state

22 Intelligent protection switching (IPS) was designed for self-healing according to SONET terminology. IPS automatically recovers from fiber facility failures by _____.

   A Wrapping the ring

   B Reversing traffic flow

   C Stopping the flow of packets

   D Cutting off bandwidth from the damaged area

This chapter covers the following topics:

- Cisco Dynamic Packet Transport Platforms
- Metropolitan Access Rings with Dynamic Packet Transport
- Configuring Dynamic Packet Transport
- Verifying and Troubleshooting Dynamic Packet Transport
- Interconnecting Dynamic Packet Transport Through a SONET UPSR Ring

CHAPTER 12

Configuring Dynamic Packet Transport

This chapter provides the information necessary to configure a Dynamic Packet Transport (DPT) solution. By the end of this chapter, you should be able to perform the following tasks:

- Identify the hardware required to build a Cisco DPT network.
- Understand the basics of the Cisco 12xxx series routers and the cards that are needed to implement a DPT ring.
- Implement DPT over an existing SONET ring.
- Configure a circuit for use with SONET using the ONS 15454.
- Understand the commands necessary to configure DPT interfaces.

Reviewing the Cisco DPT Platforms

A variety of hardware platforms can support DPT within Cisco product lines. At the time of this writing, these platforms include the Cisco 12xxx, 7x00 10720, 1519x, 10000, 6500, and 7600 series routers and switches. Which product or set of products is correct for your setup depends on the network and equipment in place.

If a SONET network is already in place, it is possible to deploy DPT, leveraging the existing infrastructure. The DPT cards in the Cisco 12xxx series router can run over SONET, dense wavelength division multiplexing (DWDM), or dark fiber.

High-speed DPT core rings are normally found on Cisco 12xxx series routers because they support up to OC-192 DPT rings. The 12xxx series router has a number of high-speed interfaces through which it can bring diverse traffic onto the ring, including DS3/E3, ATM, Gigabit Ethernet, and Packet over SONET (PoS). Additional traffic types, such as cable IP, Frame Relay, and ATM, can be added onto the ring with lower-cost network edge devices such as the 7200 and 7500 routers. The 10720 router is a small form-factor DPT router optimized for metropolitan networks providing customers with Ethernet drops.

The ONS 15194 enables the creation of logical rings from physical star networks. The 15192 supports enhanced intelligent protection switching (E-IPS), which allows multiple ring breaks without ring segmentation. The 15194 reroutes traffic across the backplane to restitch the ring. The 15194 offers high port density (64 OC-48ports in a 9-RU chassis) and a broad range of management and monitoring functions. Each interface on the 15194 can be configured for either PoS or DPT operation.

If the distances between DPT nodes is too great, the ONS 15104 IP-manageable optical regenerator can be used to extend single-mode fiber transmission. The 15104 is an OC-48 optical regenerator device supporting 3R (re-amplification, reshaping, and retiming) functionality. The 15104 can extend any SONET framed network.

NOTE Although the Cisco product line contains many products that support DPT, this chapter focuses on the 12xxx series routers working with ONS products.

Metro IP Access Ring with DPT

A common implementation of DPT technology is in shared metropolitan or campus rings, as shown in Figure 12-1. These rings provide access to multiple-tenant units. A router in the building basement provides access for multiple building tenants to a range of robust, high-bandwidth IP services, including virtual private networks (VPNs) and Internet access as well as low-cost Voice and Video over IP services, which can be sold by IP service providers.

Figure 12-1 *Metro Ring with DPT-Supported Platforms*

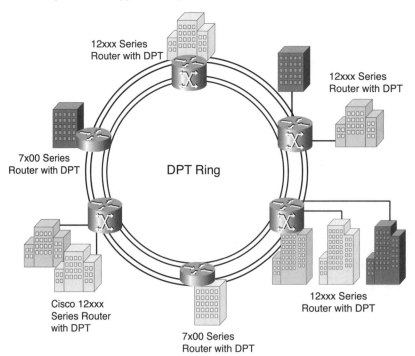

DPT rings run on a variety of transport technologies, including SONET, DWDM, and dark fiber. DPT provides carriers with the flexibility to operate packet rings over their existing fiber transport infrastructure as well as an evolution path to packet-optimized transport for high-bandwidth IP networks.

Cisco 12xxx Series DPT Line Cards

Cisco offers various cards to support DPT on the different 12xxx model routers. For instance, one Cisco DPT offerings is the two-port DPT line card for the 12xxx series router. Keep in mind that two ports are necessary to attach to a DPT ring due to the ring topology of DPT. You can order this card with a variety of optical connectors to support multimode or single-mode fiber with options for different optics. The card is available in intermediate, long, or extra long reach optics. Table 12-1 lists the card specifications.

Table 12-1   *OC12c DPT Line Card Specifications*

	Multimode	Single-Mode Intermediate Reach	Single-Mode Long Reach	Single-Mode Extra Long Reach
Connector type	SC duplex	SC duplex	SC duplex	SC duplex
Operating wavelength	1310 nm	1310 nm	1310 nm	1550 nm
Transmit power	−14dBm (max.)	−8 dBm (max.)	+2 dBm (max.)	+2 dBm (max.)
	−20 dBm (min.)	−15 dBm (min.)	−3 dBm (min.)	−3 dBm (min.)
Receive power	−14 dBm (max.)	−8 dBm (max.)	−8 dBm (max.)	−7 dBm (max.)
	−26 dBm (min.)	−28 dBM (min.)	−28 dBm (min.)	−28 dBm (min.)

Cisco also offers a single-port OC-48c DPT line card. An OC-48c DPT node must have two line cards connected via an external mate cable. The mate cable is necessary for the 12000 series routers because each DPT ring can run at a speed of OC-48 for an aggregate speed of 5 Gbps, whereas the line card's connectivity to the switch fabric matrix is only 2.5 Gbps. This limitation is no longer an issue with the 12400 series routers, which have line-card connectivity at up to 10 Gbps. The Cisco 12400 series router OC-192 card also has a mate interface because the aggregate port speed of two DPT interfaces is 20 Gbps. Figure 12-2 shows a mate cable. The mate cable is the cable that links the two cards.

Figure 12-2 *Two DPT Line Cards with a Mate Cable*

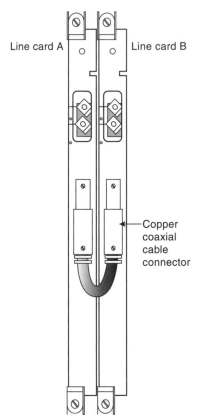

The one-port OC-48 DPT line card also provides the choice of single short- or long-reach optics to meet different network requirements. Table 12-2 shows the optical specifications of the OC-48c DPT line card.

Table 12-2 *OC-48c DPT Line-Card Specifications*

	Single-Mode Short Reach	**Single-Mode Long Reach**
Connector type	SC duplex	SC duplex
Operating wavelength	1310 n	1550 nm
Transmit power	–3 dBm (max.)	+3 dBm (max.)
	–10 dBm (min.)	–2 dBm (min.)
Receive power	0 dBm (max.)	–9 dBm (max.)
	–18 dBm (min.)	–28 dBm (min.)
Worst-case reach	2 km	80 km

All DPT line cards offer a number of features and benefits that are common to the cards. Table 12-3 describes these features and benefits.

Table 12-3   *Features and Benefits of the DPT Line Cards*

Feature	Benefit
Spatial Reuse Protocol (SRP) fairness and spatial reuse	Maximizes ring capacity, cost effectiveness, and service stability via spatial reuse, statistical multiplexing, and distributed, internodal fairness
Intelligent protection switching (IPS)	Maximizes ring robustness via self-healing around ring node or fiber failures and intelligent handling of multiple concurrent trouble events; provides fast IP service restoration without Layer 3 reconvergence to minimize impact on revenue-producing traffic
Multicast support	Provides efficient support for new revenue-producing multicasting applications in LAN, MAN, and WAN environments
Packet prioritization	Provides expedited handling of packets generated by mission-critical applications as well as delay-sensitive real-time applications such as Voice and Video over IP
Dual working fiber rings	Maximizes ring robustness and bandwidth capability using two interconnected line cards
Topology-discovery and routing procedures	Minimizes configuration requirements, optimizes routing decisions for ring bandwidth maximization, and aids in network monitoring and management
Network monitoring and management	Maximizes ring robustness and operational efficiency by providing SONET support, SRP MIB support, and MAC-layer counters for proactive monitoring and recovery and effective traffic-engineering capabilities
Pass-through mode support	Maximizes ring robustness and bandwidth availability by avoiding ring wraps caused by soft, recoverable failures in router hardware or software
Transport flexibility	Maximizes deployment flexibility by operating via dedicated fiber, WDM wavelength, or as SONET, thus matching both embedded and evolving infrastructures
Optics options	Maximizes application versatility and deployment flexibility by supporting single-mode short- and long-reach optics

DPT with SONET

DPT relies on SRP to transport traffic. SRP is a media-independent Layer 2 protocol. DPT is physical layer agnostic, which allows DPT to run over any Layer 1 protocol. In its current implementation, DPT uses SONET framing for transport. DPT can run across any SONET circuit that matches the speed of the DPT interface. The ONS 15454 and ONS 15327 can be used as transport in a DPT ring. The ONS 15454 and ONS 15327 currently do not have DPT interfaces and cannot participate in the DPT ring. They are used for transport only.

Building a DPT Ring

To build a DPT ring, you first must establish proper cabling. Figure 12-3 shows a ring with four DPT 12xxx series routers. Notice that the interfaces are identified as A and B. Similar to FDDI, A ports must be connected to B ports and vice versa. If the interfaces are incorrectly cabled, ring wraps will result. Depending on the platform, there might be a wrap LED to verify ring status. The Cisco IOS provides the most robust wrap identification by using the topology table. IOS commands are covered later in this chapter.

Figure 12-3 *DPT Ring Construction*

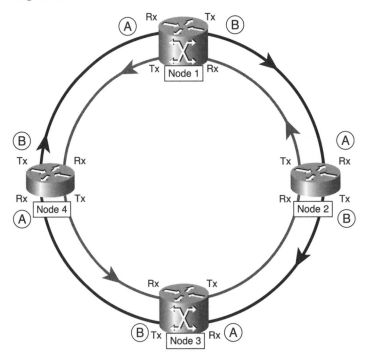

Figure 12-4 shows the fiber connectivity between the four nodes. Notice that the top ports on the two-port OC-12 DPT line card are connected to the bottom ports.

Figure 12-4   *Cabling the 2-port OC-12c DPT Cards*

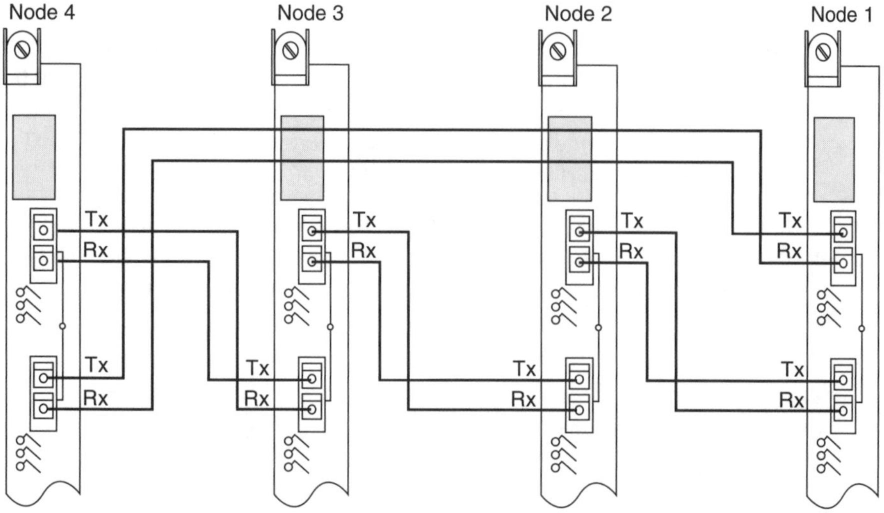

Table 12-4 describes the cable connections for a four-node ring that would be used in the DPT configuration.

Table 12-4   *Fiber Connectivity for a Four-Node Ring*

From Node / Connector	To Node / Connector
Node 1 / Tx side B	Node 2 / Rx side A
Node 1 / Rx side B	Node 2 / Tx side A
Node 2 / Tx side B	Node 3 / Rx side A
Node 2 / Rx side B	Node 3 / Tx side A
Node 3 / Tx side B	Node 4 / Rx side A
Node 3 / Rx side B	Node 4 / Tx side A
Node 4 / Tx side B	Node 1 / Rx side A
Node 4 / Rx side B	Node 1 / Tx side A

NOTE   Cisco recommends that you configure a node before the fibers are connected to it so that you avoid inserting an incorrectly configured node onto an SRP ring.

After the physical layer issues of cabling have been worked out, the DPT interfaces need to be configured. The next section covers configuration.

DPT Configuration

DPT and SRP have been covered as though they are separate entities thus far. DPT is the name of the technology, and SRP is the Layer 2 protocol that makes the technology possible. After you establish a Telnet or console connection with the device to be configured, you can use the **show ip interface brief** to obtain a listing of all the interfaces on the device. Although one might expect to see interfaces labeled DPT, all DPT interfaces are addressed as SRP interfaces in the Cisco IOS Software.

Example 12-1 outlines the initial configuration steps to assign an IP address to the DPT interface. Remember that all interfaces on a DPT ring must belong to the same IP network; and although there are two physical ports on the line card, there is only one addressable interface in the Cisco IOS Software.

Example 12-1 *IP Address Assignment*

```
Router>enable
Router#configure terminal
Router(config)#interface srp 1/0
Router(config)#ip address 10.1.1.1 255.255.255.0
```

You have several other interface configuration options that you can enter at this point, depending on how the DPT ring will be configured. The commands that include an "a" or "b" in brackets ([]) enable the administrator to choose which side of the ring the command applies to. If the [a] or [b] option is left out, the command applies to both sides. The side of the node that has the outer ring receive fiber is identified as side A. The side of the node that has the inner ring receive fiber is identified as side B, as shown previously in Figure 12-3. The requirements to use these commands are determined by your individual installation. Table 12-5 lists command options that enable you to configure a DPT ring.

Table 12-5 *DPT Configuration Commands*

Command	Purpose
srp clock-source [**line**|**internal**] [*a*	*b*]	Used to enter the **srp clock-source** as either from side A, B, or both.
	SRP clocking is similar to that of PoS. When connecting to existing SONET rings, the interface is normally line timed. When setting up a DPT ring over dark fiber, one side of the connection should be internally timed, whereas the other side of the connection is line timed.
	Each DPT interface has an internal Stratum 3 oscillator.
	A connection will work with both sides A and B internally timed, but BER will increase as the clocks drift over time.

Table 12-5   *DPT Configuration Commands (Continued)*

Command	Purpose
mac-address value	Used to manually configure the MAC address of an interface. The MAC address must be unique within the confines of the ring. Each interface has a burned-in MAC address that is normally used.
srp reject value	To force the SRP interface to reject packets sent to it by a specified source MAC address. This command uses the SAA/Reject field in the CAM table that was covered in the preceding chapter.
srp flag [*c2* | *j0*] value [*a* | *b*]	Used to specify SONET overhead values for the frame header. **c2**—Path signal label byte (default is 0x16) **j0**—Section trace byte (default is 0xCC)
srp framing [*sdh* | *sonet*]	Used to specify framing for the frame header and trailer to ensure synchronization and error control **sdh**—SDH framing s1s0 set to 2 **sonet**—SONET framing s1s0 set to 0 (default) Although SDH framing is a valid option, SDH is not covered by this book. The default framing is SONET.
srp ips timer [*1-60*] [*a* | *b*] value in seconds (default 1)	Used to control the frequency of the transmission of IPS requests on an SRP line card. SRP values should be the same for all nodes on the ring.
srp ips wtr-timer [*10-600*] value in seconds (default 60)	Used to set how long after a failure is resolved to unwrap the interface. The Wait To Restore (WTR) timer should be configured for the same value for all nodes on the DPT ring.
srp topology-timer [*1-600*] value in seconds. (default 10)	Used to set how frequently topology discovery messages are sent around the ring to identify the current nodes on the SRP ring. It is recommended that the **srp topology-timer** value be the same for all nodes on a ring.
srp ips request [*forced switch* | *manual switch*] [a | b]	Used to force a wrap on a ring. If a forced switch has been performed, a manual switch cannot be performed because it is lower on the IPS hierarchy. The IPS hierarchy was covered in the preceding chapter.
srp priority-map {**receive**} {*high* | *medium* | *low*} {**transmit**} {*high* | *medium*}	The SRP interface classifies traffic on the transmit side into high- and low-priority traffic. High-priority traffic is rate shaped and has higher priority than low-priority traffic. You have the option to configure high- or low-priority traffic and can rate limit the high-priority traffic.

Table 12-6 outlines some of the optional DPT configuration commands that you can use in a DPT configuration.

Table 12-6   *Configuration Commands*

Command	Purpose
Router1(config-if)# srp topology timer 20	Configures the topology timer frequency to 20 seconds.
Router1(config-if)# srp ips wtr-timer 10	Configures the WTR timer to 10 seconds, rather than the default 60-second timer.
Router1(config-if)# srp ips timer 30	Configures the frequency of the transmission of IPS requests on a DPT line card to 30 seconds.
Router1(config-if)# srp clock-source line a	Configures the DPT interface to take timing from the A side of the ring because the A side is connected to a service provider in which Stratum 1 timing is being provided.
Router1(config-if)#srp clock-source internal b	Configures the DPT interface to internally time on the B side because the B side is connected to another DPT node over dark fiber. The other DPT node will be line timing from this node.

In addition to the DPT-specific commands, you might need other interface configuration options to allow the interfaces to participate in routing or multicast transmissions. However, a discussion of these commands is beyond the scope of this book.

After the interface has been configured, you are ready to activate the interface by executing the **no shut** command from interface configuration mode. Make sure you have plugged in your cables before activating the interfaces.

After activating the DPT ring, you need to ensure that you have configured the ring correctly by using the appropriate **show** and **debug** commands. The next section discusses **show** and **debug** commands in detail.

Verifying and Troubleshooting DPT

To ensure that the DPT ring is configured correctly, use the appropriate **show** commands. You can use these commands in conjunction with **debug** commands to troubleshoot.

DPT show Commands

Many **show** commands enable you to display information about DPT. Table 12-7 lists some of these commands.

Table 12-7   *DPT show Commands*

Command	Description
show interfaces srp	Shows information about all SRP interfaces on the device.
show srp	Shows the current IPS source counter and topology status of SRP interfaces on the ring.
show version	Displays the IOS versioning running on the system.
show diag	Displays specific hardware information for all installed line card. Useful to see whether the IOS has the microcode necessary for the particular line card.
Show controllers srp	Displays the currently running SRP controller.
show srp topology	Command output shows the number of hops between nodes and identifies the status of the nodes. Very useful in troubleshooting A and B misconnections due to the wraps.

The following sections describe each command in more detail.

show interfaces srp Command

Purpose: Use the **show interfaces srp** command to show information about SRP interfaces. The syntax for the **show interfaces srp** command is as follows:

```
show interfaces srp slot/port
```

Syntax Description:

Parameter	Description
slot/port	Identifies the router slot and port number for the SRP line card

Example 12-2 shows a screen capture of an SRP interface. The srp2/0 interface is up and the line protocol is up. The IP address confirms the address assigned to the port and SRP discovery is reporting four nodes on the ring. Notice that SRP interfaces support loopbacks, but they are currently not set. The MAC address has been changed. This is apparent because the burned-in address (bia) is 0050.e28c.5440, whereas the current MAC address is 0012.3456.0001. Neither the A nor B side is wrapped. The IPS messages generated locally and received on both the A and B interfaces are all in the Idle state.

Example 12-2   **show interfaces srp** *Command*

```
Router#show interfaces srp 2/0
SRP2/0 is up, line protocol is up
  Hardware is SRP over SONET, address is 0012.3456.0001 (bia 0050.e28c.5440)
  Internet address is 10.1.2.1/24
  MTU 4470 bytes, BW 622000 Kbit, DLY 100 usec, rely 255/255, load 1/255
  Encapsulation SRP,       Side A loopback not set   Side B loopback not set
```

continues

Example 12-2 show interfaces srp *Command (Continued)*

```
      4 nodes on the ring   MAC passthrough not set
      Side A:not wrapped    IPS local:IDLE          IPS remote:IDLE
      Side B:not wrapped    IPS local:IDLE          IPS remote:IDLE
   Last input 00:00:00, output 00:00:00, output hang never
   Last clearing of "show interface" counters 18:26:08
   Queuing strategy:fifo
   Output queue 0/40, 0 drops; input queue 0/75, 0 drops
   5 minute input rate 20000 bits/sec, 1 packets/sec
   5 minute output rate 9000 bits/sec, 3 packets/sec
      111517 packets input, 184059367 bytes, 0 no buffer
      Received 0 broadcasts, 0 runts, 0 giants, 0 throttles
      0 input errors, 0 CRC, 0 frame, 0 overrun, 0 ignored, 0 abort
      203428 packets output, 78234051 bytes, 0 underruns
      0 output errors, 0 collisions, 0 interface resets
      0 output buffer failures, 0 output buffers swapped out
      Side A received errors:
      0 input errors, 0 CRC, 0 runts, 0 giants, 0 ignored, 0 abort
      Side B received errors:
      0 input errors, 0 CRC, 0 runts, 0 giants, 0 ignored, 0 abort
Router#
```

show srp Command

Purpose: Use the **show srp** command to show the current IPS source counter and topology status of SRP interfaces on the ring.

The syntax for the **show srp** command is as follows:

```
Router#show srp slot/port
```

Syntax Description:

Parameter	Description
slot/port	Identifies the router slot and port number for a specific SRP interface; otherwise, SRP information for all interfaces is shown.

Example 12-3 shows the results of the **show srp** command. This command enables you to see the results of the topology discovery process and configured values. Directly connected neighbors on the ring appear in the topology table next to their appropriate side. The topology table also includes the current IPS state and WTR period. The contents of the IPS messages display, indicating their Idle state. At the bottom of the output, the topology map displays all nodes that constitute the DPT ring. The details in this section include MAC address, IP address, wrap status, and the host name.

Example 12-3 *show arp Command*

```
Router# show srp
__IPS Information for Interface SRP2/0
  MAC Addresses
     Side A (Outer Ring RX) neighbor 0010.1b41.2dd0
     Side B (Inner Ring RX) neighbor 0010.f21e.d1c0
  ISP State
     Side A not wrapped
     Side B not wrapped
  IPS WTR period is 10 sec. (timer is inactive)
     Node ISP State IDLE
  IPS messages received
     Side A (Outer Ring RX) {0010.1b41.2dd0, IDLE,S}, TTL 128
     Side B (Inner Ring RX) {0010.f21e.d1c0, IDLE,S}, TTL 128
  Topology Map for Interface SRP2/0
     Topology pkt. Send every 5 sec. (next pkt. After 3 sec.)
     Last received topology pkt: 00:00:01
  Nodes on the ring:4
  Hops (outer ring)  MAC         IP Address    Wrapped Name
      0          0012.3456.0001  10.1.2.1      No      Router1
      1          0012.3456.0002  10.1.2.2      No      Router2
      2          0012.3456.0003  10.1.2.3      No      Router3
      3          0012.3456.0004  10.1.2.4      No      Router4
Router#
```

If the **show srp** command does not provide the expected output and you think your configuration is accurate, it is best to ensure that you are running the proper system hardware and
IOS software. Two main commands can help you determine whether the system recognizes
the hardware and is configured with the appropriate software: **show version** and **show diag**.

show version Command

Purpose: Use the **show version** command to display the configuration of the system
hardware, number of each line and card type installed, the Cisco IOS version, the names
and sources of configuration files, and boot images.

The syntax for the **show version** command is as follows:

```
Router#show version
```

Syntax description: This command has no parameters or arguments.

show diag Command

Purpose: Use the **show diag** command to view specific hardware information for any line card installed in the system.

The syntax for the **show diag** command is as follows:

```
Router#show diag slot
```

Syntax Description:

Parameter	Description
Slot	Identifies the slot number for which to display information

If the line card is not recognized, it will not appear in the **show diag** command output. Other commands can provide additional information about the operation and configuration of SRP. These commands include **show controllers srp** and **show srp topology**.

show controllers srp Command

Purpose: Use the **show controllers srp** command to display the SRP controller information.

The syntax for the **show controllers srp** command is as follows:

```
Router#show controllers srp [slot/port] [details]
```

Syntax Description:

Parameter	Description
slot/port	Identifies the router slot and port number for the SRP interface
Details	Provides additional information about the controller in the output

Example 12-4 shows output from this command. This command shows the status of the lines connecting this node to other nodes in the ring. The first part of the display refers to side A, which is transmitting on the inner ring and receiving on the outer ring. Errors are broken down into section, line, and path errors, with details of each type of possible transmission error displayed. Because the framing used is SONET, the contents of some of the internal frame bytes are shown. (The details of these bytes are discussed in Chapter 3, "SONET Overview.") In addition, information about configured clocking on the line, the remote host name, and IP address is available. This output gives you an overview of configuration or line errors that might be impacting ring operations.

Use the **show controllers srp** command to verify that no SONET errors exist on either side A or side B. If you discover errors, check the dBm levels of the DPT line card. If the dBm levels are higher than the published receive levels, too much power is being received. The interface needs to be attenuated until the levels are within the desired level. A low-level reading would need to be fixed as well but will involve troubleshooting where the additional loss is coming from.

Example 12-4 **show controllers srp** *Command*

```
Router1#show controllers srp 2/0

SRP2/0
SRP2/0 - Side A (Outer RX, Inner TX)
SECTION
 LOF = 0          LOS    = 0                      BIP(B1) = 0
LINE
  AIS = 0         RDI    = 0        FEBE = 0      BIP(B2) = 0
PATH
  AIS = 0         RDI    = 0        FEBE = 0      BIP(B3) = 0
  LOP = 0         NEWPTR = 0        PSE  = 0      NSE     = 0

Active Defects:None
Active Alarms: None
Alarm reporting enabled for:SLOS SLOF PLOP

Framing           :SONET
Rx SONET/SDH bytes:(K1/K2) = 0/0     S1S0 = 0  C2 = 0x16
Tx SONET/SDH bytes:(K1/K2) = 0/0     S1S0 = 0  C2 = 0x16  J0 = 0xCC
Clock source      :Internal
Framer loopback   :None
Path trace buffer :Stable
  Remote hostname :Router2
  Remote interface:SRP2/0
  Remote IP addr  :10.1.2.2
  Remote side id  :B

BER thresholds: SF = 10e-3  SD = 10e-6
TCA thresholds: B1 = 10e-6  B2 = 10e-6  B3 = 10e-6

SRP2/0 - Side B (Inner RX, Outer TX)
SECTION
  LOF = 0         LOS    = 0                      BIP(B1) = 0
LINE
  AIS = 0         RDI    = 0        FEBE = 0      BIP(B2) = 0
PATH
  AIS = 0         RDI    = 0        FEBE = 0      BIP(B3) = 0
  LOP = 0         NEWPTR = 0        PSE  = 0      NSE     = 0

Active Defects:None
Active Alarms: None
Alarm reporting enabled for:SLOS SLOF PLOP

Framing           :SONET
Rx SONET/SDH bytes:(K1/K2) = 0/0     S1S0 = 0  C2 = 0x16
Tx SONET/SDH bytes:(K1/K2) = 0/0     S1S0 = 0  C2 = 0x16  J0 = 0xCC
Clock source      :Internal
Framer loopback   :None
Path trace buffer :Stable
  Remote hostname :Router4
  Remote interface:SRP2/0
  Remote IP addr  :10.1.2.4
```

continues

Example 12-4   **show controllers srp** *Command (Continued)*

```
  Remote side id  :A

BER thresholds: SF = 10e-3  SD = 10e-6
TCA thresholds: B1 = 10e-6  B2 = 10e-6  B3 = 10e-6
Router2#
```

show srp topology Command

Purpose: Use the **show srp topology** command to see the number of hops between nodes and identify the nodes that are wrapped.

The syntax for the **show srp topology** command is as follows:

```
show srp topology
```

Syntax description: This is executed from Enable mode.

Example 12-5 shows the topology table, which also appeared at the bottom of the **show srp** command output. The **show srp topology** command is a subset of the **show srp** command. This command is normally used when the ring is first coming up to ensure all nodes have been discovered or to quickly determine whether any nodes are in a Wrap state.

Example 12-5   **show srp topology** *Command*

```
Router1#show srp topology

Topology Map for Interface SRP2/0
Topology pkt. sent every 60 sec.

(next pkt. after 2 sec.)

Last received topology pkt. 00:00:02
Nodes on the ring:4
Hops (outer ring)  MAC         IP Address Wrapped Name
0         0012.3456.0001   10.1.2.1    No     Router1
1         0012.3456.0002   10.1.2.2    No     Router2
2         0012.3456.0003   10.1.2.3    No     Router3
3         0012.3456.0004   10.1.2.4    No     Router4
```

DPT debug Commands

Table 12-8 lists several commands that you can use to debug configuration problems such as loss of packets, cyclic redundancy check (CRC) errors, card resets, alarms, and so on. Use **debug** commands with extreme care. Each character from a debug message generates an interrupt to the router and can bring down a router with ease (especially at multigigabit speeds). If your intention is to troubleshoot a busy production ring, typing in **debug srp packet** will probably aggravate the problem more. In many cases, the router would need to be rebooted. Table 12-8 shows a few of the **debug** commands that assist in troubleshooting a problematic DPT ring.

Table 12-8   *DPT* debug *Commands*

Command	Description
debug srp protocol error	Displays SRP interface protocol errors and error statistics
debug srp ips	Debugs an SRP interface on the ring
debug srp nodename	Displays node name packets by the source MAC address
debug srp packet	Displays information about how to debug a specific SRP packet
debug srp periodic activity	Debugs a specific periodic activity
debug srp topology	Examines ring topology information

debug srp protocol error Command

Purpose: Use the **debug srp protocol error** command to display SRP interface protocol errors and error statistics. When you want to see which types of errors are occurring within SRP, this command enables you to see the area in which errors are occurring.

The syntax for the **debug srp protocol error** command is as follows:

```
Router#debug srp protocol error
```

Syntax description: This command has no parameters or arguments.

The **debug srp protocol** error command generates the output shown in Example 12-6.

Example 12-6 *SRP Protocol Error Messages*

```
Lack of memory when attempting to originate packets
SRP version mismatches
Time To Live (TTL) problems. TTL problems should not affect the normal operation of
the ring.
Checksum failures
Incorrectly sized topology packets
Incorrect packet type
Internal software errors
```

The example shows that the ring is receiving Time To Live (TTL) mismatches. TTL mismatches can occur in a ring that includes older equipment with newer equipment. The older equipment may be using an older framer that does not support a TTL of 255. The single-port OC-48 line card for the Cisco 12000 series router comes in an A and B version. The A version has a framer with support of 64 nodes (TTL of 128), whereas the B version has a framer with support of 128 nodes (TTL of 255).

debug srp ips Command

Purpose: Use the **debug srp ips** command to debug an SRP interface on the ring that might not be correctly participating in the DPT ring. IPS is the intelligent protection switching mechanism. To see the IPS packets that control wrapping on the ring, use the **debug srp ips** command.

The syntax for the **debug srp ips** command is as follows:

```
Router#debug srp ips
```

Syntax description: This command has no parameters or arguments.

```
Date Stamp: srp_process_ips_packet: SRP1/0, checksum 64620, ttl 255, B
Date Stamp: srp_process_ips_packet: SRP1/0, checksum 14754, ttl 255, A

Date Stamp: Tx pkt node SRP1/0 side A {IDLE, SHORT}
Date Stamp: Tx pkt node SRP1/0 side B {IDLE, SHORT}
```

From the **debug srp ips** command output here, you can deduce that this node is sending and receiving IPS Idle messages on both the A and B side. Operation for this node looks good.

```
Date Stamp: Tx pkt node SRP12/0 side A {SF, LONG}
Date Stamp: Tx pkt node SRP12/0 side B {SF, SHORT}
```

From the **debug srp ips** command output here, you can deduce that this node has directly detected a failure on the B side of the ring. The node is sending out the appropriate messages: SHORT SF messages on the failed, wrapped side of the ring; and the SF, LONG messages on the operable side of the ring.

debug srp nodename Command

Purpose: Use the **debug srp nodename** command to display node name packets by source MAC address.

The syntax for the **debug srp nodename** command is as follows:

```
Router#debug srp nodename
```

Syntax description: This command has no parameters or arguments.

```
*May  9 08:28:39:srp_process_node_name_packet SRP4/0, len 27, 0048.1100.0002, M2305B

*May  9 08:28:39:srp_forward_node_name_packet:SRP4/0, len 27, 0048.1100.0002, M2305B

*May  9 08:28:39:srp_glean_node_name:SRP4/0, len 27 src 0048.1100.0002 data M2305B

*May  9 08:28:48:srp_process_node_name_packet SRP4/0, len 27, 0048.3300.0001, M2307A
```

debug srp packet Command

Purpose: Use the **debug srp packet** command and specify the MAC address of the SRP interface to display information contained within an SRP packet. SRP packets contain all the control information used by the DPT ring. Executing this **debug** command can result in a large amount of overhead on the router, which could affect performance. This command shows you all SRP activity and could be used to confirm any activity if SRP operation is in question.

The syntax for the **debug srp packet** command is as follows:

```
Router#debug srp packet
```

Syntax description: This command has no parameters or arguments.

debug srp periodic activity Command

Purpose: Use the **debug srp periodic activity** command to debug a specific periodic activity. This command checks the frequency of IPS requests and topology messages.

The syntax for the **debug srp periodic activity** command is as follows:

```
Router#debug srp periodic activity
```

Syntax description: This command has no parameters or arguments.

debug srp topology Command

Purpose: Use the **debug srp topology** command to examine ring topology information.

The syntax for the **debug srp topology** command is as follows:

```
debug srp topology
```

Syntax description: This command has no parameters or arguments.

Example 12-7 shows how to debug a specific periodic activity on an SRP interface.

Example 12-7 *debug srp topology Command*

```
*Jan 3 23:34:01.846: srp_input: pkt_hdr=0x0F002007, flags=0x00000002
*Jan 3 23:34:01.846: srp_forward_topology_map_packet: SRP12/0, len 20
*Jan 3 23:34:01.846: srp_input: pkt_hdr=0x0F002007, flags=0x00000003
*Jan 3 23:34:01.846: srp_forward_topology_map_packet: SRP12/0, len 20
*Jan 3 23:34:02.266: srp_send_topology_map_packet: SRP12/0 on side B - Not Wrapped
*Jan 3 23:34:02.266: srp_send_topology_map_packet: SRP12/0 on side A - Not Wrapped
*Jan 3 23:34:02.266: srp_input: pkt_hdr=0x0F002007, flags=0x00000002
*Jan 3 23:34:02.266: srp_consume_topology_map_packet: SRP12/0, len 34
*Jan 3 23:34:02.266: 0, src node_wrapped 0, src mac_addr 0001.c9ec.d300
!-- If the node is not wrapped, the node_wrapped bit should be zero (0)
*Jan 3 23:34:02.266: 1, src node_wrapped 0, src mac_addr 0000.5032.3037
*Jan 3 23:34:02.266: 2, src node_wrapped 0, src mac_addr 0006.d74a.f900
*Jan 3 23:34:02.266: 3, src node_wrapped 0, src mac_addr 0003.a09f.5700
topology changed = No
*Jan 3 23:34:02.266: 0, src node_wrapped 0, src mac_addr 0001.c9ec.d300
*Jan 3 23:34:02.266: 1, src node_wrapped 0, src mac_addr 0000.5032.3037
*Jan 3 23:34:02.266: 2, src node_wrapped 0, src mac_addr 0006.d74a.f900
*Jan 3 23:34:02.266: 3, src node_wrapped 0, src mac_addr 0003.a09f.5700
topology updated = No
*Jan 3 23:34:02.266: srp_input: pkt_hdr=0x0F002007, flags=0x00000003
```

continues

Example 12-7 *debug srp topology Command (Continued)*

```
*Jan 3 23:34:02.930: srp_input: pkt_hdr=0x0F002007, flags=0x00000002
*Jan 3 23:34:02.930: srp_forward_topology_map_packet: SRP12/0, len 13
*Jan 3 23:34:02.930: srp_input: pkt_hdr=0x0F002007, flags=0x00000003
*Jan 3 23:34:02.930: srp_forward_topology_map_packet: SRP12/0, len 27
*Jan 3 23:34:04.194: srp_input: pkt_hdr=0x0F002007, flags=0x00000003
*Jan 3 23:34:04.194: srp_forward_topology_map_packet: SRP12/0, len 13
*Jan 3 23:34:04.194: srp_input: pkt_hdr=0x0F002007, flags=0x00000002
*Jan 3 23:34:04.194: srp_forward_topology_map_packet: SRP12/0, len 27The ping
Command
```

With your DPT ring up and operational, you should test it to make sure it is carrying traffic properly. The most common test program is **ping**. To execute the **ping** command, follow these steps:

Step 1   Connect to the desired Cisco 12xxx series router.

Step 2   Type **ping** ip_address, where ip_address is the IP address of a different Cisco 12xxx series router that is connected to the DPT network. The example earlier created an SRP interface with an IP address of 10.1.2.1 and the examples show an SRP interface on router 2 with an IP address of 10.1.2.2. To verify connectivity to the next node, you should be able to ping its SRP port and see the following response:

```
Sending 5, 100-byte ICMP Echos to 10.1.2.2, timeout is 2 seconds:
!!!!!
Success rate is 100 percent (5/5), round-trip min/avg/max = 1/1/4 ms
```

Step 3   Repeat the **ping** command to verify connectivity to each Cisco 12xxx series router.

To show how DPT can leverage existing ONS 15454/15327 rings, this section shows how to configure a DPT ring over an existing SONET ring.

DPT over an ONS 15454/15327 SONET Ring

Figure 12-5 displays the physical connectivity of a DPT-based Cisco 12xxx router network intermixed with an ONS SONET ring. Figure 12-5 depicts the 12xxx series routers connecting to the optical cards in the ONS 15454 or ONS 15327s as appropriate. For the purpose of this book, the cards on both the 12xxx series router and the ONS nodes will be OC-12. Two fiber runs are needed for each connection between the ONS 15454 and 15327 combination because DPT interfaces have an A and B side for the ring creation. Each Cisco 12xxx uses the ports on the OC-12 DPT card to connect to two standard SONET OC-12 interfaces on the ONS 15454. An interconnecting set of circuits needs to be made from each ONS 15454 and 15327 OC-12 port, which must be routed across the SONET ring in the core.

Figure 12-5 *DPT Ring Running over SONET Core*

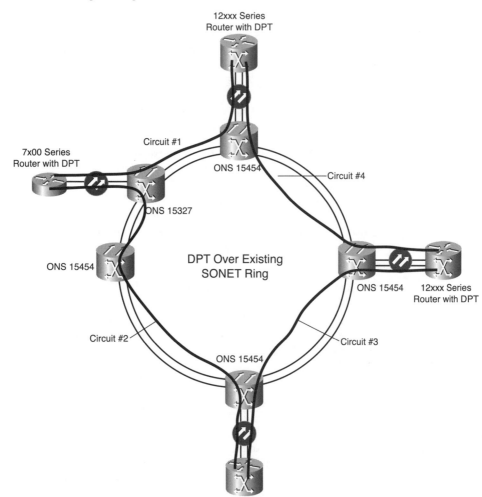

Figure 12-5 displays a DPT network over SONET. Notice that circuits are needed between all the DPT nodes in the ring.

NOTE   If the network will be running DPT over SONET, the SONET circuits should be provisioned and tested before the DPT nodes are connected to the network. This will assist with troubleshooting procedures.

ONS 15454 and 15327 DPT Circuit Provisioning

To provision the necessary circuit, first log in to the ONS 15454 or 15327 node using the Cisco Transport Controller (CTC). The OC-12 ports on the cards that you plan to connect the Cisco 12xxx series routers to must be put In Service and configured as necessary. This information was covered in Chapter 10, "Configuring Packet over SONET."

The circuit must be created next. To create a circuit, follow these steps:

Step 1   Choose the **Circuit** tab from the Node view in CTC and click the **Create** button. The Circuit Attributes dialog box displays.

Step 2   Give the circuit a descriptive name related to its purpose. The type will be STS. The size needs to be the DPT ring size. You will be using an OC-12 ring, so the STS size will be STS-12c. The bidirectional box should be selected because the circuit is bidirectional. (That is, it will transmit and receive traffic.) The number of circuits is 1. Do not select Revertive. Protection will be provided by DPT, so you will be provisioning an unprotected circuit. Click the **Next** button, and the Circuit Source dialog box displays.

Step 3   In the Circuit Source Node field, choose the first ONS 15454 node as the source. Choose the appropriate slot from the Slot drop-down menu. Choose the correct port from the Port drop-down menu. The STS number will be dynamically selected. Click the **Next** button to set the destination of the circuit. The Circuit Destination dialog box displays.

Step 4   Choose the ONS 15454 node at the other end of the circuit, and then choose the slot and port for the circuit to terminate on. Click **Next** to set the circuit routing preferences. The Circuit Routing Preferences dialog box displays.

Step 5   Make sure that the Route Automatically and the Fully Protected Path check boxes are not checked. Recall that this will be an unprotected circuit. Click **Next**. The Route Review and Edit window displays.

Step 6   Choose the span between the source ONS 15454 and the destination ONS 15454. When the link is selected, make sure the arrow is pointing away from your Cisco ONS 15454 and toward the other Cisco ONS 15454. (To change the direction of the arrow, click the span.) Proper configuration of these circuits is essential to proper operation of the DPT ring.

Step 7   When the circuit is properly configured, click **Add Span**. The arrow that indicates the span turns blue in color, and the span is added to the Included Spans field.

Step 8   Review the span you have created in the Included Spans list. You should see a contiguous span indicated by blue arrows pointing from the source node through the network to the destination node.

Step 9   Click **Finish** to complete the circuit. The Cisco Transport Controller window appears with the Circuits tab selected. You can verify the circuit by using the CTC circuit mapping features. Ensure that the Circuit State is Active.

You must repeat this process at each of the other nodes in the SONET ring.

Summary

After reading this chapter, you should have an idea of what it takes to configure a DPT ring using the Cisco 12xxx series router. You were exposed to the various **show** and **debug** commands that could help you to monitor and troubleshoot a DPT ring. Finally, you saw how DPT can run over an existing SONET ring and the provisioning process.

Review Questions

1 Which three speeds are supported by the Cisco 12xxx series DPT line cards?

    A   OC-3c/STM-1

    B   OC-12c/STM-4c

    C   OC-48c/STM-16c

    D   OC-192/STM-64c

2 What benefit does the Cisco 12xxx DPT line-card packet-prioritization feature provide?

    A   Maximizes application versatility and deployment flexibility

    B   Maximizes deployment flexibility by operating via dedicated fiber

    C   Maximizes ring robustness via self-healing around ring node or fiber failures

    D   Provides expedited handling of packets generated by mission-critical applications

3 Single-mode fiber, simplex or duplex, with SC connectors is used to connect Cisco 12xxx series routers to other routers for DPT connectivity. If the one-port OC-48 line card is used, how would the second line card be connected?

A Crossover cable

B Single-mode fiber

C Mate cable

D Straight-through cable

4 In the following table, indicate the order in which each Dynamic Packet Transport configuration and verification task is completed when configuring DPT over an external 15454 SONET network.

Enter the Step # Here	Task Description
	Using Cisco IOS Software, configure optional parameters on SRP interface.
	Use Cisco IOS **show** and **debug** commands to verify the configuration.
	Configure a circuit using the Cisco Transport Controller interface.
	Verify the DPT circuit using **ping**.
	Verify the circuit using the CTC circuit map display.
	Using Cisco IOS Software, assign an IP address to the SRP interface.

5 Which Cisco IOS Software command correctly configures the frequency at which topology messages are sent?

A srp ips timer 5 a

B srp ips time 5 b

C srp ips wtr-timer 10

D srp topology-timer 60

6 Which Cisco IOS Software command initiates the highest-priority ring wrap?

A srp ips wtr-timer value

B srp ips timer value [a | b]

C srp ips request forced-switch [a | b]

D srp ips request manual-switch [a | b]

7 Which Cisco IOS Software command correctly sets the frequency of the transmission of IPS requests on an OC-12c SRP line card to 30 seconds?

   A   **srp ips timer 3 a**

   B   **srp ips timer 30 a**

   C   **srp ips wtr-timer 30**

   D   **srp topology-timer 30**

This chapter covers the following topics:

- Steps for developing metro optical solutions
- Case study 1: SPE Enterprises
- Solution review: SPE Enterprises
- Case study 2: FC Enterprises
- Solution review: FC Enterprises

CHAPTER 13

Metro Optical Case Studies

In this book, you have reviewed the state of optical networks in the metro environment and the types of technologies that need to be supported. You were also introduced to some of the equipment and technology used in metro optical solutions. This equipment includes the following:

- The ONS 15454 and ONS 15327
- The Cisco 12000 series gigabit routers
- The 15216 optical amplifiers and EDFA DWDM equipment

The technologies supported by these pieces of equipment include the following:

- **SONET**—Synchronous Optical Network
- **SDH**—Synchronous Digital Hierarchy
- **PoS**—Packet over SONET
- **DPT**—Dynamic Packet Transport
- **DWDM**—Dense wave division multiplexing
- **Metro Ethernet/TLS**—Transparent LAN services
- **SAN**—Storage-area network
- **CWDM**—Coarse wave division multiplexing

This final chapter presents two case studies that represent some fictitious companies that want to establish metro services. Based on the equipment and technologies presented in the book, you are asked to design metro optical solutions around each business case.

After you have designed your solution, review the solution answers and see how close you came to the book solution. Your solution may differ, and that is all right as long as you think you can justify your answer.

Before introducing the case studies, this chapter reviews some of the material needed to understand design issues with attenuation, power-budget calculations, and optical signal-to-noise ratios.

Steps for Developing a Metro Optical Solution

When designing metro optical networks, you need to consider these factors:

- Customer requirements
- Component manufacturer requirements
- System manufacturer requirements
- Fiber manufacturer requirements
- Service provider requirements
- Physical technical specifications

Customer Requirements

Customer requirements are one of the key driving factors when building a metro optical solution because the customer services and needs have to be run over the designed solution. You need flexibility and depth to allow the addition of multiple solutions as customer or client needs arise.

Component Manufacturers

Component manufacturers realize that their products must integrate into the finished products. Therefore, component manufacturers create integrated solutions/products that bring together the capabilities of devices such as transmitters, modulators, receivers, multiplexers, and amplifiers in a single product. These players are companies such as Lucent, Nortel Networks, Corning, and others.

System Manufacturers

System manufacturers are engaged in the design of high-speed switches (wire speed) and routers (gigabit/terabit routers), including SONET/SDH/add/drop and terminal multiplexers, and DWDM (filters/switches). Key players in this area are Cisco, Nortel, Lucent, Ciena/ONI, Sycamore, Alcatel, and many others. All of these companies have positioned themselves to provide numerous products in different areas in the optical marketplace. To manufacture products in all different solutions, system manufacturers have had to acquire companies and integrate those products into an overall solution. For instance, Cisco has acquired companies such as PipeLinks, SkyStone Systems, Pirelli Optical Systems, Qeyton Systems, Cerent, Monterey Networks, and others. Currently, both Lucent and Nortel hold about 50 percent of the worldwide optical market, but Cisco and Alcatel are eroding into that percentage.

Cisco Systems, long considered the leader in the IP router/switching marketplace, has entered the optical marketplace through numerous acquisitions. Later in this chapter, you are introduced to two fictitious companies that want to establish metro optical services. You will build your optical solutions using the Cisco optical components discussed in this book (Cisco ONS 15454, Cisco ONS 15327, Cisco ONS 15216, Cisco 12*xxx* PoS/DPT/RPR interfaces, and others).

Fiber Manufacturers

Only a few companies manufacture fiber (for instance, Lucent and Corning). These companies are developing new fiber to overcome current limitations of the existing fiber.

Service Providers

Service providers can be classified into various categories, including the following:

- Incumbent local exchange carriers (ILECs / incumbent PTTs)
- Competitive local exchange carriers (CLECs)
- Interexchange carriers (IXCs)
- Cable carriers (MSOs)

Physical Technical Specifications

When dealing with the physical technical specifications, you must determine whether the equipment you will be using can send the signal the required physical distance. Because the data has to be gathered on the transmitters, the receivers, and the fiber in between the transmitter and receiver, you must also test how strong the transmitter is and how sensitive the receiver is to incoming photons. The most accurate way to test the power of the transmitting signal and the sensitivity of the receiving device is to use a power meter. Using a power meter may not always be possible. If it isn't possible, you must manually calculate the "span budget."

You determine the span budget by taking the power of the transmitters and subtracting power losses incurred because of the length of fiber, fiber type, number of splices, and the number and type of connectors. You compare that result to the sensitivity of the receiver. If the power is within the transmitter's specification, with some margin for the effects of time and additional splices, you can use the fiber. Otherwise, an optical amplifier or regenerator may be needed.

The case study designs focus on the customer requirements, system component requirements, and physical technical specifications.

Calculating Attenuation

Optical power budgets, or link-loss budgets, must be considered when you plan an optical network. Vendors must provide guidelines, or engineering rules, for their equipment. In general, signal loss occurs for many different reasons. The most obvious of these is the distance of the fiber itself; this tends to be the most important factor in long-haul transport. In metro-area networks (MANs), the number of access nodes, such as optical add/drop multiplexers (OADMs), is generally the most significant contributor to optical loss.

To determine minimum/maximum losses and maximum distances, you need to identify all the factors that contribute to fiber loss. Failure to identify even one of the factors can lead to potential problems. The ideal way to determine loss is to measure the actual loss after the fiber has been laid.

Calculating the signal strength on a cable is only half the job. To avoid overdriving a fiber receiver and to eliminate data-loss problems, it is equally important to calculate the maximum signal strength. Overdriving a receiver most commonly occurs when using single-mode products with low fiber attenuation. It is safe to assume average numbers for fiber loss, but you should measure the actual loss after the fiber has been laid, spliced, and terminated to verify your previous measurements and avoid performance problems.

To calculate fiber distance, you must consider not only loss variables described earlier, but also the launch-power and receive-sensitivity specifications on your fiber products. The equation can get a bit complicated because many vendors provide a launch-power range. Therefore, when calculating distance, you may want to use the lowest launch power to calculate the worst-case distance. Similarly, it is helpful to calculate the best-case distance by using the highest available launch power.

Fiber-Loss Variables

Fiber-loss variables may determine what type of equipment you need to design and build a metro network. Variables you need to consider include the following:

- **Age of the fiber**—Depending on the quality of the single-mode fiber deployed, the loss can vary. A typical value is .25 dB per km, although this can vary, especially on older fibers. If the fiber is old enough, there may be even larger issues (reflectance of old connector types, PMD, and so on) that make it unusable for DWDM or even 2.5-Gbps transmission.

- **Splices**—Although splices are small and often insignificant, there is no perfect loss-less splice. Many errors in loss calculations are made due to a failure to include splices.

- **Connectors**—Like splices, there is no perfect loss-less connector. It is important to note that even the highest-quality connectors can get dirty. Dirt and dust can completely obscure a fiber light wave and create huge losses. A .5- to 1-dB loss per connector is commonly the worst-case scenario, assuming a cleaned and polished connector is used. There is always a minimum of two connectors per fiber segment, so remember to multiply connector loss by two.

- **Safety buffers**—It is common to add a couple dB of loss as a buffer just to play it safe. Two dB of loss can take fiber aging, poor splices, and so on into account.

Fiber-Loss Information for ONS 15216 and 15454 Products

This section covers some of the technical specifications of the Cisco ONS 15216 and Cisco ONS 15454. You will need this information when developing and designing a metro network to find where the equipment can be placed to meet the proper specifications in the network.

ONS 15216 Product Specifications

Cisco Systems has numerous models that have the name ONS 15216, including the following:

- OADM
- Transparent filters
- Erbium-doped fiber amplifiers (EDFAs)

Table 13-1 shows the technical specifications of ONS 15216 OADMs. Pay close attention to the attenuation range and the one- or two-channel insertion loss. You will use these numbers when you design an optical network with this equipment.

Table 13-1   *15216 OADM Specifications*

ONS 15216 OADM	
Attenuation range 0 dB–35 dB	
One-channel OADM (end-to-end insertion maximums)	
Pass through	1.8 dB
Drop	2.1 dB
Add	3.2 dB
Two-channel OADM (end-to-end insertion maximums)	
Pass through	2.0 dB
Drop	2.6 dB
Add	4.0 dB

The following are the Cisco ONS 15216 DWDM filter specifications.

ONS 15216 DWDM Filters

DWDM loss ONS 15216 filter: Insertion Loss

9-wavelength base red filter:

- 4.5-dB multiplexing loss
- 4.5-dB demultiplexing loss

9-wavelength upgrade blue filter:

- 4.25-dB multiplexing loss
- 4.25-dB demultiplexing loss

18-wavelength red base filter + blue upgrade filters:

- 4.5-dB multiplexing loss
- 4.5-dB demultiplexing loss

Table 13-2 shows the Cisco ONS 15216 EDFA technical specifications.

Table 13-2   *15216 EDFA Technical Specifications*

Input Signal Wavelength	1530 nm to 1563 nm
Input power (total all channels)	–29 dBm to –6 dBm
Maximum input power	–6-dBm variable optical attenuation used to bring power < –6 dBm
Maximum output power	17 dBm

ONS 15454 Line Cards

Table 13-3 lists some of the ONS 15454 card specifications for fiber transmission and receipt.

Table 13-3   *ONS 15454 Line Card Specifications*

Optical Card	Rx	Tx
OC-3 IR 1310	–8 to –28 dBm	–8 to –15 dBm
OC-12 IR 1310	–8 to –28 dBm	–8 to –15 dBm
OC-12 LR 1310	–8 to –28 dBm	+2 to –3 dBm
OC-12 LR 1550	–8 to –28 dBm	+2 to –3 dBm
OC-48 IR 1310	0 to –18 dBm	0 to –5 dBm
OC-48 LR 1550	–8 to –28 dBm	+3 to –2 dBm
OC-48 ELR DWDM	–8 to –28 dBm	0 to –2 dBm

Table 13-4 lists some of the Ethernet line card specifications.

Table 13-4   *Ethernet 15454 Line Card Specifications*

Card	Distance
15454-GBIC-SX	1804 feet
15454-GBIC-LX	32,810 feet
Single-mode fiber	0.30, 0.40 dB/km

Table 13-5 shows other technical variables that you need to consider when designing a metro network.

Table 13-5   *Other Variables*

Fiber Connectors	0.75 dB per Pair
Fiber-splice loss	0.1 dB each
Mechanical splices	0.5 dB each
Fiber loss (attenuation)	Single-mode at 1310 nm: 0.40 dB/km Single-mode at 1550 nm: 0.30 dB/km
Fiber aging	3 dB over the life of the system; fiber life expectancy is 25 years

As you can see, each product and model usually has different variables that you must consider when designing a metro optical network. This section plugs these variables into different formulas to come up with the complete technical variables.

Calculation Formulas

The tables in the preceding section listed various technical specifications that are important when you are designing optical networks. Now it is time to consider these specifications in various scenarios.

Power Conversion (Aggregate to Per Channel)

One of the most fundamental concepts to understand when using DWDM is the relationship between the aggregate signal power and the per-channel signal power. You must be able to convert between the two to properly design and troubleshoot a DWDM network. The following equation shows you how to make the conversion:

Equation 1:
$$P_{total} \text{ (mW)} = P_{channel} \text{ (mW)} \times \text{Number of channels}$$

where the following is true:

- P_{total} is the total (aggregate) optical power in the fiber.
- $P_{channel}$ is the power per channel.
- Number of channels is the number of channels on the fiber.

This equation is intuitive. The confusion normally arises because power levels are usually expressed in terms of decibels (dB), and this simple multiplication does not apply when dealing with dBs.

To convert the preceding equation to dB, recall the equation used to convert power from milliwatts (mW) to decibels (dB):

Equation 2:
$$\text{Power (dB)} = 10 \times \log(\text{Power mW})$$

Applying that conversion to Equation 1 results in the following:

Equation 3:
$$10 \times \log(P_{total}) = 10 \times \log(P_{channel} \times \text{Number of channels})$$

Using log math, the preceding equation can be simplified as follows:

Equation 4:
$$10 \times \log(P_{total}) = 10 \times \log(P_{channel}) + 10 \times \log(\text{Number of channels})$$

Because $10 \times \log(\text{Power})$ is the same as the Power in dB (Equation 2), equation 4 can be simplified as follows:

Equation 5:
$$P_{total} \text{ (dB)} = P_{channel} \text{ (dB)} \times 10 \times \log(\text{Number of channels})$$

Using this equation, you can easily convert from total power to per-channel power, and vice versa. Out of all the equations in this discussion, Equation 5 is the only one you need to remember.

Power-Conversion Examples (Aggregate to Per Channel)

A simple rule of thumb to remember is that $10 \times \log(2)$ is about 3 dB (3.0103 dB). This means that every time you double the number of channels, the total power goes up by 3 dB. Therefore, if you have 1 channel at 0 dBm, the aggregate power is 0 dBm. If you add another channel, also at 0 dBm, the aggregate power goes up 3 dB to 3 dBm. Double the number of channels to 4, each at 0 dBm, and the aggregate power goes up 3 dB again, to 6 dBm. Table 13-6 uses Equation 5.

Table 13-6   *Demonstrating Equation 5*

Number of Channels	10 × log(Number of Channels)	Per-Channel Power (dBm)	Aggregate Power (dBm)
1	0.0	0	0.0
2	3.0	0	3.0
3	4.8	0	4.8
4	6.0	0	6.0
5	7.0	0	7.0
6	7.8	0	7.8
7	8.5	0	8.5
8	9.0	0	9.0
16	12.0	0	12.0
32	15.1	0	15.1

OADMs and Optical Power Levels

When calculating loss and gain across a DWDM, it is important to know when to use the aggregate and when to use the per-channel power formulas. Loss is always the same on both per-channel and aggregate power. If you have a fiber span with a loss of 10 dB, and 16 channels on the fiber, for example, you can apply the 10-dB loss the same to the aggregate or per-channel power.

To understand this, just calculate the result using both aggregate and per-channel power. Assume each channel has a power of 0 dBm entering the fiber.

Power into fiber: Per channel: 0 dBm
 Aggregate: 0 dBm + 10 t[s] log(16) = 12 dBm (see Table 13-6)
Power out of fiber: Per channel: 0 dBm – 10 dB = –10 dBm
 Aggregate: 12 dBm – 10 dB = **2 dBm**

If you compare the two calculation methods, you see that they gave the same results. Calculate the per-channel power back into aggregate power and see that it matches the aggregate power previously calculated:

Per channel out of fiber = –10 dBm
Aggregate out of fiber = –10 dBm + 10 × log(16) = –10 dBm + 12 dB = **2 dB**

So, when calculating loss, you can use either per-channel or aggregate and get the same result.

The only challenge is to know when a loss affects only a single channel, as in an OADM. The loss from dropping a channel in a two-channel OADM is 2.6 dB, from the specification at the beginning of this lab. This is the loss seen per channel (the one being dropped). So if the aggregate power into the OADM is –6 dBm, and there are 16 channels on the system, the power of the dropped channel is –20.6 dBm. This is calculated as follows:

Aggregate = Per channel + 10 × log(16)
Per channel = Aggregate – 10 × log(16) (simple algebra)
Per channel = –6 dBm – 12 dB (–6 dBm)
Per channel = –18 dB
Per channel out of OADM = –18 dBm – Drop loss = –18 dBm –2.6 dB
Per channel out of OADM = –20.6 dBm

NOTE   Always use the per-channel power when designing DWDM networks. You can see why it is important to get the proper technical specification numbers to apply them to the formulas and arrive at accurate numbers.

Optical Signal-to-Noise Ratio (OSNR)

Generally, the largest limiting factor when using EDFAs is not the loss, but the optical noise that accumulates. Accurately representing the characteristics of an EDFA is extremely complex and not practical (nor necessarily accurate) for every design; however, some general rules can aid in the design and estimation of amplifier placement. It is highly recommended that all designs stay within the guidelines set forth by the business unit supporting that amplifier, which are based on numeric and lab simulations. In addition, many other characteristics (such as gain tilt, saturation, noise variation with wavelength, and input power) are not easily estimated.

In general, a target OSNR of at least 21 dB (0.1 RBW) is required for OC-48/STM-16 signals, and this is the case with the ONS 15454. OSNR is measured in a certain resolution bandwidth (RBW), and the RBW used gives a different value. (0.5 RBW is another commonly used RBW, which provides a relatively lower OSNR reading; for example, the same signal measuring 14-dB OSNR at 0.5 RBW may be 21-dB OSNR at 0.1 RBW.)

An estimation of OSNR is not required in the lab exercise.

Exercises

It is time to apply this information and formulas to some real-world examples. Use the previously discussed formulas to complete these exercises. If you run into trouble, you can always check the answers at the end of this chapter.

Exercise Formulas

Use the following formulas for the exercises in this chapter:

> **Power budge calculation (dB)** = (Transmit power) – (Receive sensitivity)
> **Calculation of signal loss (dB)** = (Fiber attenuation * km) + (Splice attenuation * Number of splices) + (Connector attenuation * Number of connectors) + (Optical Filter / OADM) + (Buffer (fiber aging))
> **Distance calculation: (dB)** = (Fiber budget) – (Splice loss) – (Connector loss) – (Optical filter / OADM) – (Buffer (fiber aging)) / (Fiber attenuation)

Exercise 1

You are using two ONS 15454 OC-48 ELR cards through a DWDM network. One ONS 15216, a DWDM filter on each side of the connection, is used for the DWDM infrastructure. Refer to Figure 13-1 and review the technical information given to you to answer the following questions.

Figure 13-1   *Power-Budget Calculation*

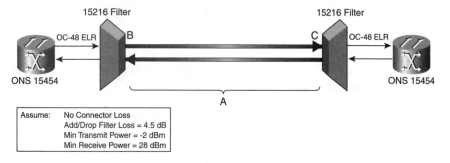

What is the maximum fiber-loss budget allowed using reference point A? _____

If you add seven more channels into the DWDM filter, what is the aggregate power out of the multiplexer using reference point B? _____

With eight total channels, what is the minimum aggregate power into the demultiplexer? Reference point C on Figure 13-1. _____

Figure 13-2 shows a power-budget calculation.

Figure 13-2 *Power-Budget Calculation*

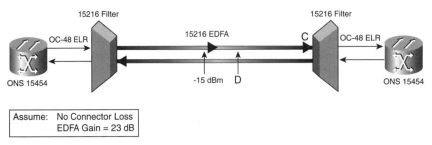

Suppose a single channel enters the 15216 EDFA at –15 dBm. When set to constant gain mode, what is the output power of the channel? (Refer to location D in Figure 13-2.) See Figure 13-3 and answer the following questions.

Figure 13-3 *Power-Budget Calculation*

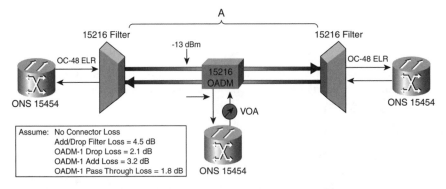

Perform the same calculations with the addition of a 15216 OADM to drop off a wavelength for a multiple-tenant unit at location A. _____

What is the maximum fiber-loss budget between end terminals?

_____

Assume a system with 8 channels enters the OADM1 with an aggregate power of –13 dBm. What is the power level out of the OADM of the dropped signal? _____

For the same scenario, what should the VOA be set to so that the added channel power equals the seven channels passing through at the output?_____

Exercise 2

You are connecting three ONS 15454 optical platforms to a DWDM infrastructure (Node A, Node B, and Node C). Each ONS 15454 is equipped with two OC-48 ELR 1533.47 wavelength line cards (East and West). Node A connects to Node B at a distance of 15 km. Clean fusion splices occur every 5 km in this span due to the size of the fiber spool that the provider could purchase. The span between Node B and Node C has the same distance and splice characteristics as the previous span. The span from Node C to Node A, however, is a 65-km span, and a splice is encountered every 5 km. Each node is connecting to the DWDM network using a one-channel ONS 15216 OADM. Fiber attenuation runs at 0.40 per km (see Figure 13-4).

Figure 13-4   *DWDM Power-Budget Calculation*

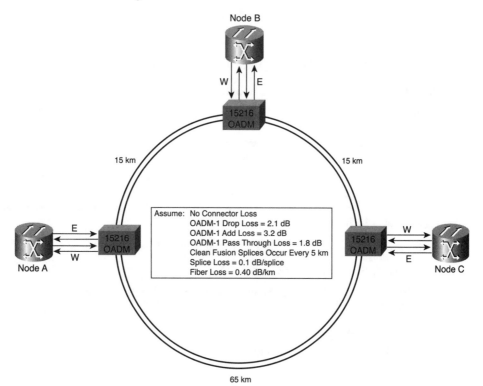

Calculate the fiber budgets for

Span 1:_____

Span 2:_____

Span 3:_____

Calculate the loss on

Span 1:_____

Span 2:_____

Span 3:_____

Calculate the number of ONS 15216 EDFAs that will be needed for

Span 1:_____

Span 2:_____

Span 3:_____

Exercise 3

Complete this exercise to practice what you learned regarding calculating power budgets. (See Figure 13-5.)

In this exercise, you explain the SONET architecture: Cisco ONS 15454:

- Identify fiber types: Cisco ONS 15454.
- Explain reasons for power budget loss and SONET-related calculations: Cisco ONS 15454.
- Perform specific SONET-related calculations.
- 15454 optical card specifications.

Figure 13-5 *ELR Power-Budget Calculation Network Diagram*

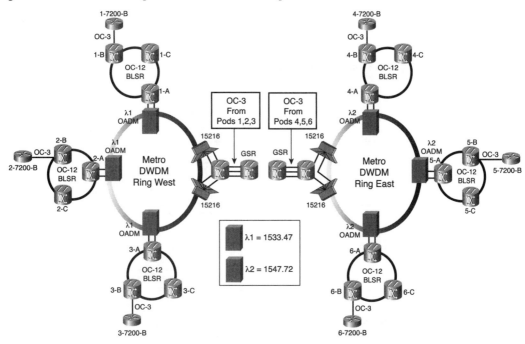

Table 13-7 provides the specifications for ONS 15454.

Table 13-7 *ONS 15454 Line Card Specifications*

Optical Card	Rx	Tx
OC-3 IR 1310	–8 to –28 dBm	–8 to –15 dBm
OC-12 IR 1310	–8 to –28 dBm	–8 to –15 dBm
OC-12 LR 1310	–8 to –28 dBm	+2 to –3 dBm
OC-12 LR 1550	–8 to –28 dBm	+2 to –3 dBm
OC-48 IR 1310	0 to –18 dBm	0 to –5 dBm
OC-48 LR 1550	–8 to –28 dBm	+3 to –2 dBm
OC-48 ELR DWDM	–8 to –28 dBm	0 to –2 dBm

Table 13-8 is a conversion table for single channel to aggregate power numbers.

Table 13-8   *Single Channel to Aggregate Power*

Number of Channels	10 × log (Number of channels)	Per-Channel Power (dBm)	Aggregate Power (dBm)
1	0.0	0	0.0
2	3.0	0	3.0
3	4.8	0	4.8
4	6.0	0	6.0
5	7.0	0	7.0
6	7.8	0	7.8
7	8.5	0	8.5
8	9.0	0	9.0
16	12.0	0	12.0
32	15.1	0	15.1

Table 13-9 shows passive optical component loss numbers.

Table 13-9   *Component Loss Table*

Device Name	Loss	Unit (per km)
SMF at 1550 nm	0.25	dB
SMF at 1310 nm	0.4	dB
DSF at 1550 nm	0.3	dB
Dispersion. Compensated Fiber	0.5	dB
Fusion splice	0.25	dB
SC-SC connection	1.0	dB
15216 200-GHz multiplexer	4.5	dB
15216 200-GHz demultiplexer	4.5	dB
Short fiber patch cable	0	dB

Use the information from Figure 13-6 to answer the following questions.

Figure 13-6   *ELR Power-Budget Calculation*

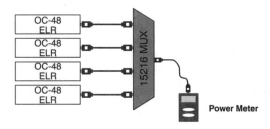

What would you expect the power meter to read if the OC-48 ELR cards are all outputting at their maximum power?

_____

_____

_____

Figure 13-7 shows the result if you now add the other side of the telecommunication link.

Figure 13-7   *ELR Power-Budget Calculation*

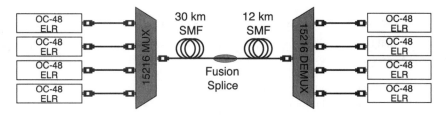

Each OC-48 ELR card's receiver will see what power (per-channel power) at the far right of this diagram?

_____

_____

_____

Figure 13-8 shows the network after you add another node in the middle. Drop the red lambda and add it again in the middle lambda. This means that going into the EFDA, the red lambda will have a much higher power than the other three.

Figure 13-8   *ELR Power-Budget Calculation*

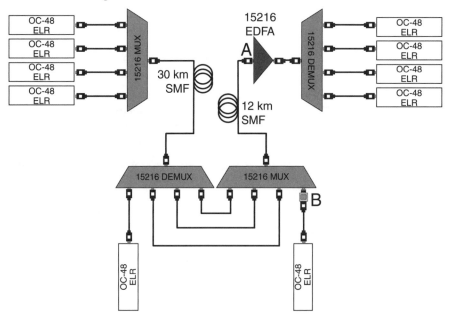

The 15216 EDFA has a maximum tolerance for differences in input power of ±1 dB (at point A).

To equalize the power going into the EFDA, the easiest thing to do is to add an attenuator to the red lambda (at point B).

What value of attenuator should you use?
Attenuator values do not include connector losses.

_____

_____

_____

Optical Amplification

If you are going to use DWDM, take into account additional concerns regarding fiber (such as optical regeneration and amplification). This subject is beyond the scope of this book. For the sake of this solution, the distances between nodes are close enough so that optical amplification or regeneration is not necessary.

Now you are ready to put the knowledge you learned from this book to work in the following two case studies.

Case Study 1: SPR Enterprises

Your customer, SPR Enterprises, is looking for an optical solution to facilitate its needs. SPR Enterprises has given you the following information.

Phase 1: Identify Requirements

Objective of Phase 1

Given a high-level customer problem statement, identify relevant questions that you must answer before you can identify proper solution options.

Challenge

Assume you have received the following e-mail:

> SPR Enterprises wants to link five sites via a high-speed optical fault-tolerant ring. One of the sites, called the admin site, has connections to the PSTN and the Internet. All the other sites (Sites 1 through 4) require a DS1 and a Gigabit Ethernet drop off. SPR's connection to the service provider is via an OC-3. This OC-3 has two DS3s for voice and one DS3 for Internet access. The requirement is to take these voice DS3s and break them down to DS1s that are then transported to the four remote sites. Between each site, SPR Enterprises can lease only a single OC-12 SONET circuit. (Figure 13-9 summarizes these requirements.)

Figure 13-9 *SPR Enterprises*

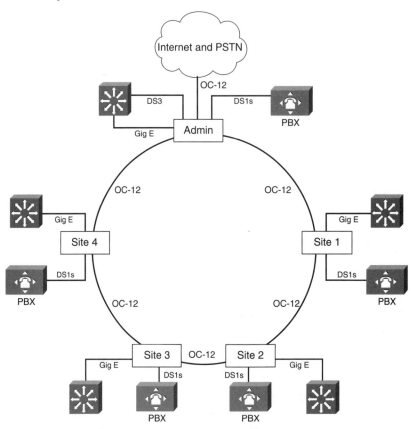

You will be meeting with SPR representatives next week. Prepare a list of follow-up questions you should ask at this meeting.

Your Task in This Phase

Formulate the customer's network challenges and decide what requirements must first be identified before you can effectively identify design options.

Write questions to uncover those requirements in a Word document.

Phase 2: Identify Options

In this phase, you identify design options and create a high-level topology diagram and rough cost estimate.

Objective of Phase 2

Given a clear understanding of customer requirements, design at least two suitable solution options for the customer to consider.

Your Task in This Phase

Following are some detailed requirements for this case study. Assume that you have received these in response to your follow-up questions.

Your task now is to analyze the detailed requirements and to identify 2–4 potential solutions to the customer's network challenge.

Detailed Customer Requirements

The customer has provided the following answers to your questions. Use this information to help you design your solution.

Customer Answers to Follow-Up Questions

1 Is the support for data multicast and broadcast required?

Answer: No need to support multicast today.

2 What is the expected data flow? Hub and spoke between admin and all other sites? Or distributed?

Answer: Hub and spoke between admin site and all other sites.

3 Is –48VDC power available at each site?

Answer: Yes.

4 How many DS1s are required per site?

Answer: A mix. But each site requires fewer than 14 DS1s.

5 What is the expected framing of the voice DS3s?

Answer: M23.

6 What is the required wavelength of the Gigabit Ethernet handoff? SX, LX?

Answer: SX.

7   Is fiber protection required for all circuits or only a subset?

Answer: All circuits must be fiber protected.

8   Is client-side protection required for Gigabit Ethernet? DS1s? DS3?

Answer: Client-side protection is not required.

9   What type of DS1 connector is required at each site? AMP champ or SMB?

Answer: AMP champ.

10   Can voice be transported over IP?

Answer: No.

11   Are 10/100 ports acceptable at the remote sites, instead of the Gigabit Ethernet ports?

Answer: No. Gigabit Ethernet handoff is a requirement at all the sites.

Phase 3: Design Solution

In this phase, you receive new information from the customer. Create an optimal design based on the following additional requirements.

Objective of Phase 3

Given detailed customer requirements, design the optimal optical network solution.

Additional Customer Service Requirements

A customer came back and said that his PBXs cannot provide timing. The current PBX can only receive timing.

Phase 4: Create a Solution Summary

After you have designed what you consider to be the optimal solution, you must build a presentation, called a *solution summary*, to deliver to the customer.

Objective of Phase 4

Present an optimal design solution to an informed audience.

Your Tasks in This Phase

You will create a 45-minute presentation to deliver to the customer on your solution.

Your aim in the presentation is to guide the IT staff through your design and to ensure they understand your reasoning and your process.

The presentation will be followed by a question-and-answer session, in which the customer will challenge your design solution and assumptions.

Be prepared to patiently explain each decision as if your audience were a tough customer team requiring a lot of convincing. The presentation should include the participant deliverables listed here.

Components of the solution summary need to include the following:

- Step-by-step review of the design process (Show your work!)
  - Customer requirements
  - Design options
  - Solution and rationale
- Detailed topology (final)
- Equipment list with detailed costs
- Ordering issues
- Required applications

Case Study 1 Solution: SPR Enterprises

This solution could be used for SPR Enterprises. Your answers might differ; however, if you can defend your reasoning, it is acceptable.

Option 1

The following solution for SPR Enterprises uses a SONET solution with ONS 15454s:

- A BLSR ring would provide a more efficient use of bandwidth. BLSR would allow an STS-3 circuit between each node for shared packet rings. If a UPSR ring is selected, the maximum bandwidth that can be used between the nodes for shared packet ring is an STS-1.
- The 12-port DS3 module terminates all DS3 connections at the admin site. These are the DS3s that are carried inside the OC-12 connection.
- The transmux card is used to break out the voice DS3s to DS1s. This is accomplished by the use of two external connections between the 12-port DS3 card and the transmux card.

- Build a shared packet ring between all the nodes. Another option is to create point-to-point circuits between all the remote nodes and the admin node. With a shared packet ring, the circuit between each node can be an STS-3; if point-to-point circuits are used, however, the maximum circuit size is an STS-1.

Figure 13-10 shows a conceptual diagram of the network.

Figure 13-10 *Solution for SPR Enterprises: Option 1*

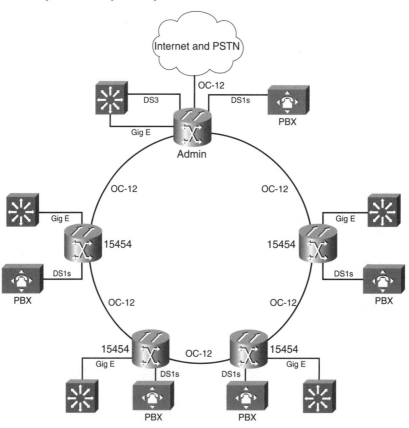

Figure 13-11 shows more details for the admin site.

Figure 13-11 *Admin Solution: SPR Enterprises*

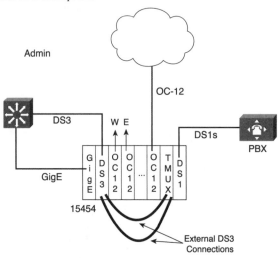

Table 13-10 shows a partial list for the admin site.

Table 13-10 *Partial Price List for the Admin Site*

Product	Description	Qty	Price
CISCO 15454	15454 ATO (assemble to order)	1	0
15454-EIA-BNC-A24	Elect I/F, 24 BNC, A side	1	1440.00
15454-EIA-AMP-B84	Elect I/F, 84 AMP, B side	1	1440.00
15454-TCC+	Timing Communications Control Plus	2	9000.00
SF15454-R3.3.0	Rel. 3.3.0 SW, Preloaded on TCC	2	0
15454-XC-VT	Xconn, 576 STS, 672 VT	2	9360.00
15454-FTA3-T	Shelf fan tray assembly, ANSI, 15454, HPCFM, ITEMP	1	720.00
15454-SA-ANSI	15454 SA NEBS3 ANSI w/RCA and ship kit	1	2700.00
15454-E1000-2-G	Gigabit Ethernet, 2 Ckt., GBIC - G	1	8910.00
15454-DS1-14	DS1, DSX, 14 CKT	1	3600.00
15454-DS3-12E	DS3, DSX, Enhanced PM, 12 Ckt.	1	8995.00
15454-DS3XM-6	DS3, DSX Transmux, 6 CKT	1	17,910.00
15454-OC121IR1310	OC-12, IR, 1310, 1 CKT, SC.	3	10,530.00
15454-GBIC-SX	1000Base-SX, MM, standardized for 15454/327	1	500.00
15454-BLANK	Empty slot filler panel	6	1350.00
Total price:			USD 76,455.00

Table 13-11 lists products and prices for the non-admin sites.

Table 13-11 *Pricing for Non-Admin Sites*

Product	Description	Qty	Price
CISCO 15454	15454 ATO (assemble to order)	1	0
15454-EIA-AMP-B84	Elect I/F, 84 AMP, B side	1	1440.00
15454-TCC+	Timing Communications Control Plus	2	9000.00
SF15454-R3.3.0	Rel. 3.3.0 SW, Preloaded on TCC	2	0
15454-XC-VT	Xconn, 576 STS, 672 VT	2	9360.00
15454-FTA3-T	Shelf fan tray assembly, ANSI, 15454, HPCFM, ITEMP	1	720.00
15454-SA-ANSI	15454 SA NEBS3 ANSI w/RCA and Ship Kit	1	2700.00
15454-E1000-2-G	Gigabit Ethernet, 2 Ckt., GBIC - G	1	8910.00
15454-DS1-14	DS1, DSX, 14 CKT	1	3600.00
15454-OC12IR1310	OC-12, IR, 1310, 1 CKT, SC.	2	7020.00
15454-GBIC-SX	1000Base-SX, MM, standardized for 15454/327	1	500.00
15454-BLANK	Empty slot filler panel	8	1800.00
Total price: USD 45,050.00			

Option 2

A SONET solution using the ONS 15454s in combination with a channel bank is as follows:

- A BLSR ring would provide a more efficient use of bandwidth.

- A channel bank would be used to break out the voice DS3s into DS1s.

- A shared packet ring would be used to transport data.

- ONS 15327 cannot be used because the Gigabit Ethernet module is not available.

This solution is similar to Option 1 except this solution uses an external channel bank rather than the transmux card (see Figure 13-12).

Figure 13-12 *Solution for SPR Enterprises: Option 2*

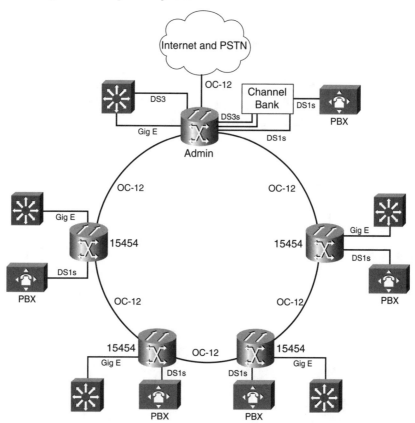

Because the PBXs cannot provide timing, and the ONS 15454s cannot provide timing on the DS1s, a CSU in front of the main PBX can provide this functionality. The ONS 15454s provide pass-through timing for electrical ports such as DS1s (see Figure 13-13).

Figure 13-13 *SPR Enterprises Option 1 Timing Solution*

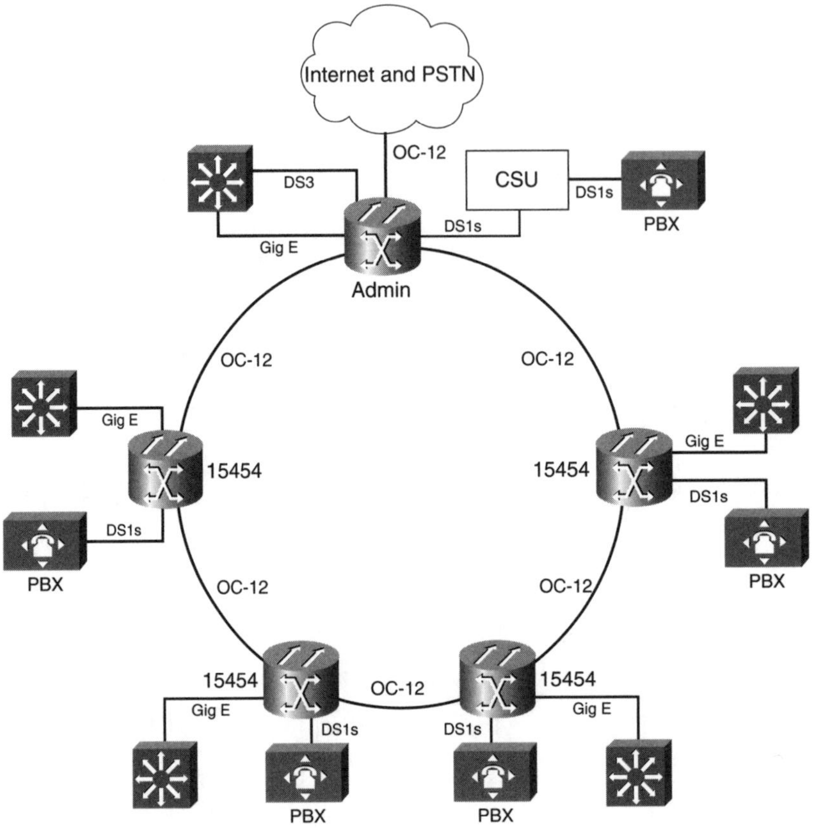

Table 13-12 shows a partial price list for admin site equipment.

Table 13-12 *Admin Site Pricing*

Product	Description	Qty	Price
CISCO 15454	15454 ATO (assemble to order)	1	0
15454-EIA-AMP-B84	Elect I/F, 84 AMP, B side	1	1440.00
15454-TCC+	Timing Communications Control Plus	2	9000.00
SF15454-R3.3.0	Rel. 3.3.0 SW, Preloaded on TCC	2	0
15454-XC-VT	Xconn, 576 STS, 672 VT	2	9360.00
15454-FTA3-T	Shelf fan tray assembly, ANSI, 15454, HPCFM, ITEMP	1	720.00
15454-SA-ANSI	15454 SA NEBS3 ANSI w/RCA and Ship Kit	1	2700.00

Table 13-12 *Admin Site Pricing (Continued)*

Product	Description	Qty	Price
15454-E1000-2-G	Gigabit Ethernet, 2 Ckt., GBIC - G	1	8910.00
15454-DS1-14	DS1, DSX, 14 CKT	5	18,000.00
15454-OC121IR1310	OC-12, IR, 1310, 1 CKT, SC.	2	7020.00
15454-GBIC-SX	1000Base-SX, MM, standardized for 15454/327	1	500.00
15454-BLANK	Empty slot filler panel	5	1125.00
Total price: USD 58,775.00			

Case Study 2: FC Enterprises

FC Enterprises has contracted you to build a solution to help with its current needs. What the company expects you to do is outlined here. Read the following information a couple of times to help prepare you for your design work.

Phase 1: Identify Requirements

Objective

Given a high-level problem statement, identify customer requirements through analysis and careful questioning.

Challenge

Assume that you have received the following e-mail:

FC Enterprises is building a secondary data center for disaster recovery. The company wants to link its servers at the main data center to the secondary data center via Fibre Channel. FC Enterprises is going to need four Fibre Channel connections on day one with the potential to go to six in the future. In addition to the Fibre Channel connections, FC Enterprises needs to support two Gigabit Ethernet connections today, growing to four in the future. Along with the two data centers, FC Enterprises wants to have three other sites connected on the same ring. The main data center would have a single wire-rate Gigabit Ethernet to each of these three sites, and four DS1s per site for PBX connections. Only two strands of fiber are available between each site.

You are meeting with customer representatives next week. Prepare a list of follow-up questions you should ask at this meeting.

Your Task in This Phase

Discuss and review the customer's network challenges and decide what requirements must first be identified before you can effectively identify design options.

Write questions to uncover those requirements in a Word document and compare them to the printed responses.

Phase 2: Identify Options

In this phase, you identify design options and create a high-level topology diagram and rough cost estimate.

Objective

Given a clear understanding of customer requirements, design at least two suitable solution options for the customer to consider.

Your Task in This Phase

Identify 2–4 potential solutions to the customer's network challenge.

Detailed Customer Requirements

The customer has provided the following answers to your questions. Use this information to help you design your solution.

Customer Answers to Follow-Up Questions

1 Is fiber protection required? How about client protection?

Answer: Only fiber protection is required for all channels.

2 What type of fiber is being used?

Answer: SMF-28.

3 What are the optical power losses between the sites? If not known, what are the distances between these sites?

Answer: Power losses are not known; however, the fiber lengths between the nodes are the following:

- 20 km between main data center and backup data center
- 25 km between backup data center and Site 1
- 20 km between Site 1 and Site 2
- 2 km between Site 2 and Site 3
- 30 km between Site 3 and main data center

4 Is –48VDC power available at each site?

Answer: Yes.

5 Would interconnecting remote storage via iSCSI be acceptable?

Answer: No.

Phase 3: Design Solution

In this phase, you receive new information from the customer. Create an optimal design based on the following additional requirements.

Objective

Given detailed customer requirements, design the optimal optical network solution.

Additional Customer Service Requirements

A customer came back and said that the only fiber that he can get is DSF (dispersion-shifted fiber). The customer also changed his mind about the Fibre Channel over IP requirement. Now he is saying that Fibre Channel over IP or iSCSI is acceptable.

Phase 4: Create a Solution Summary

After you design what you consider to be the optimal solution, you must build a presentation, called a *solution summary*, to deliver to the customer. Create a solution that you can present to the customer.

Objective

Present an optimal design solution to an informed audience.

Your Tasks in This Phase

Your aim in the presentation is to guide the class through your design and to ensure that the customer understands your reasoning and your process.

The presentation is followed by a question-and-answer session.

Be prepared to patiently explain each decision as if your audience were a mix of management and engineers. The presentation should include the participant deliverables listed here.

Components of the solution summary should include the following:

- Step-by-step review of the design process (Show your work!)
    — Customer requirements
    — Design options
    — Solution and rationale
    — Detailed Topology (final)
    — Equipment list with detailed costs
    — Ordering issues
    — Required applications

Case Study 2 Solution: FC Enterprises

This section present a solution that works for FC Enterprises. Your solution might differ, but that is acceptable if you meet the customer's needs and you can defend your solution.

Objective

Design an optical network solution to customer specifications using Cisco products and a 15454 solution.

Solution

The following solutions uses a combination of SONET and DWDM:

- SONET would transport DS1s.
- DWDM for Fibre Channel and gigabit connectivity.
- Because Fibre Channel is required, either the ONS 15540 or the ONS 15252 can be used.

- Because the total channel count is fewer than 16, the ONS 15252 fits the solution well.

- The ONS 15216 OADMs and terminal filters can be used in this solution. The nice thing about using these products is that they integrate well with the CLIPs of the ONS 15252 and the ELR modules of the ONS 15454.

- When designing this network, use the future channel-count requirement. This would mean that the composite power increases, and it affects the amount of attenuation needed to enter the EDFAs. Therefore, potentially more EDFAs will be needed (see Figure 13-14).

Figure 13-14 *FC Enterprises Solution*

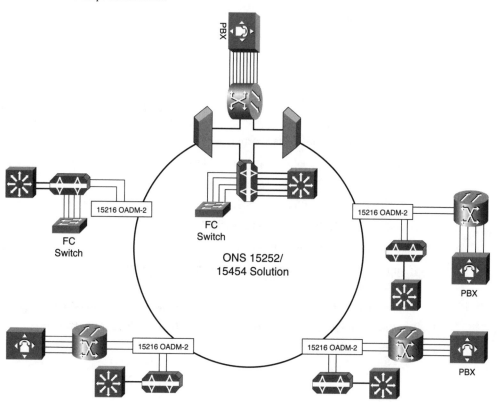

Table 13-13 shows the pricing that matches Figure 13-14.

Table 13-13 *Price List for FC Enterprises Solution*

		Line Card Req.	Line Card Req.	SONET Req.
454 Shelf	**Network**	**DS-1**	**GE**	**OC-192**
1	Master Data Ctr	12	9	2
2	Backup Data Ctr	0	6	2
3	Site 1	4	1	2
4	Site 2	4	1	2
5	Site 3	4	1	2
			Total Cost	1,128,285

Solution

FC Enterprises came back and said that the only fiber that it can get is DSF. FC Enterprises also changed its mind about the Fibre Channel over IP requirement. Now it is saying that Fibre Channel over IP or iSCSI is acceptable.

Because the new requirement is for DSF fiber, DWDM is not possible. So a SONET is the only solution. Because iSCSI is acceptable, the SN542*x* storage routers can be used.

- A SONET solution with the SN542*x* storage routers.
- An OC-192 SONET ring using the ONS 15454.
- Each node would provide DS1s.
- If you use the four-port G-series cards in each node, gigabit connectivity can be provided.
- The Cisco SN542*x* can be used to provide iSCSI connectivity to remote servers.
- Because wire-rate Gigabit Ethernet connectivity is required, only two ports in the four-port G-series cards can be used (see Figure 13-15).

Figure 13-15 *FC Enterprises Solution*

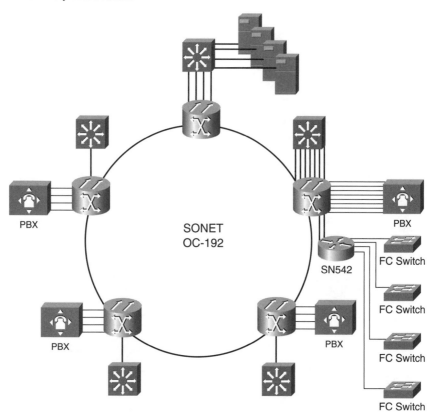

Table 13-14 shows the new pricing corresponding to Figure 13-15.

Table 13-14 *Prices for FC Enterprises*

		Line Card Req.	Line Card Req.	SONET Req.
454 Shelf	**Network**	**DS-1**	**GE**	**OC192**
1	Master Data Ctr	12	13	2
2	Backup Data Ctr	0	10	2
3	Site 1	4	1	2
4	Site 2	4	1	2
5	Site 3	4	1	2
			Total Cost	1,232,265

Answers to Review Questions

The answers to the chapter questions appear in boldface.

Chapter 1

1 Which of the following is not an optical metro region service provider service?

   a. Private-line aggregation

   b. Voice line switching

   c. Transparent lambda

   d. Data networking

2 Which portion of the network focuses on routing and caching?

   a. Metro network

   b. Service POP

   c. Core network

   d. None of the above

3 Which portion of the network is focused on connecting to the customer?

   a. Metro network

   b. Service POP

   c. Core network

   d. None of the above

4 What are the four main elements of a metropolitan environment?

   a. Metro network, core packet switching, web content, core network

   b. Access network, core packet switching, edge packet switching, service POP

   c. Access network, metro network, service POP, core network

   d. Metro network, DNS servers, VPN services, core network

5 Which four physical topologies are used in a metro network?

    a. Ring, mesh, fault, point-to-point

    b. Point-to-multipoint, bus, mesh, ring

    c. Point-to-multipoint, point-to-point, mesh, bus

    d. Bus, mesh, point-to-point, ring

6 Which one of the following functions is generally not associated with a service POP?

    a. Grooming of traffic

    b. Edge packet switching

    c. Core packet switching

    d. Accessing technologies

7 Which type of physical topology is predominantly used in metro networks?

    a. Point-to-point

    b. Mesh

    c. Bus

    d. Rings

8 How is time-division multiplex service typically sold?

    a. Demand basis

    b. Full-time basis

    c. Usage-based basis

    d. None of the above

9 Which of the following is *not* a service traffic pattern?

    a. Point-to-point, high-service density

    b. Point-to-point, low- to medium-service density

    c. Multipoint-to-point

    d. Multipoint-to-multipoint

    e. Point-to-multipoint

10 Wavelength service offering is equivalent to which of the following?

    a. Wireless

    b. Dark fiber

    c. Multiple fibers

    d. Gigabit Ethernet

11 SAN is defined as which of the following?

    a. Fiber Channel fabric connecting servers and servers

    b. Fibre Channel fabric connecting servers and storage

    c. Super high-speed access network

    d. None of the above

12 Metro and remote mirroring can be offered to fulfill _____ needs. Choose all that apply.

    a. Disaster-recovery

    b. Peak-loading

    c. Video-on-demand

    d. Bandwidth-protection

13 Transparent lambda wavelengths will be enabled by which of the following?

    a. SONET

    b. DWDM

    c. DPT

    d. PoS

14 What is the predominant broadband technology deployed by service providers in the local metro market?

    a. Wireless

    b. DSL

    c. SONET/SDH

    d. ATM

15 DPT is a _____-based technology.

a. Point-to-point

b. Hub-and-spoke

c. Ring

d. Point-to-multipoint

16 What are the capabilities of next-generation SONET multiservice?

a. It reduces the overall backbone cost.

b. It reduces the requirement for equipment.

c. It simplifies the provisioning process for backbone services.

d. All of the above.

Chapter 2

1 Engineers measure attenuation in an optical network with which unit of measure?

a. dB

b. dBμ

c. SNR

d. Reflection

2 Who invented the voice-sampling method of digital voice?

a. Sir Isaac Newton

b. Bill Gates

c. Einstein

d. Nyquist

3 Digital voice transmission compared to analog transmission uses a sampling method of what?

a. 8000 samples per second $\times$ 64 bits per sample = 256 bits per second

b. 8000 samples per second $\times$ 8 bits per sample = 64,000 bits per second

c. 6400 samples per second $\times$ 8 bits per sample = 256 bits per second

d. 64 samples per second $\times$ 8000 bits per sample = 64 bits per second

4 What invention took the voice signal and created a pattern of pulses, which could be construed as 1s and 0s?

a. FDM

b. PCM

c. QCM

d. CPM

e. TDM

5 What invention allowed up to 12 channels to be transmitted simultaneously across a single line?

a. FDM

b. PCM

c. QCM

d. CPM

e. TDM

6 What device was created to take the digital signals of the computer and put them into analog signals to be sent to the central office phone system to be converted back to digital signals?

a. Phone

b. Router

c. Switch

d. Modem

e. Multiplexer

7 How fast is a DS0 transmission rate?

a. 64 bps

b. 256 bps

c. 64 kbps

d. 256 kbps

e. 1.544 kbps

8 A DS1 has a transmission rate of what speed?

   a. 1.544 Mbps

   b. 43.008 Mbps

   c. 43.232 Mbps

   d. 44.736 Mbps

9 A DS3 with no overhead and no bit stuffing has a transmission of what speed?

   a. 1.544 Mbps

   b. 43.008 Mbps

   c. 43.232 Mbps

   d. 44.736 Mbps

10 A channelized DS3 with overhead and bit stuffing has a transmission rate of what speed?

   a. 1.544 Mbps

   b. 43.008 Mbps

   c. 43.232 Mbps

   d. 44.736 Mbps

11 An unchannelized DS3 with overhead and bit stuffing has a transmission rate of what speed?

   a. 1.544 Mbps

   b. 43.008 Mbps

   c. 43.232 Mbps

   d. 44.736 Mbps

Chapter 3

1 SONET electrical signals are represented by

   a. OC-N

   b. STM-N

   c. STS-N

   d. SDH-N

2 What is the line rate for an OC-12?

   a. 51.84 Mbps

   b. 155.52 Mbps

   c. 622.08 Mbps

   d. 2.488.32 Gbps

3 An OC-48 SONET frame is repeated how many times per second?

   a. 4000 times per second

   b. 8000 times per second

   c. 16,000 times per second

   d. 48,000 times per second

4 What do SONET TOH bytes carry? (Choose two.)

   a. Payload data

   b. Path overhead

   c. Line overhead

   d. Section overhead

   e. SPE data

5 Which of the following layers is responsible for framing and scrambling?

   a. Path layer

   b. Section layer

   c. Line layer

   d. Photonic layer

6 How many rows are in a SONET frame?

   a. 1

   b. 9

   c. 90

   d. 270

7 How many VTGs are in an STS-1 SPE?

a. 1

b. 2

c. 3

d. 4

e. 7

8 A VT1.5 encapsulates what type of traffic?

a. DS0

b. DS1

c. E1

d. DS3

9 Match these SONET layers to their network elements

a. Section   __**c** End to end

b. Line   __**a** Regenerator to regenerator

c. Path   __**b** Multiplexer to multiplexer

10 What is a SONET regenerator's primary role?

a. 1r

b. 2rs

c. 3rs

d. 4rs

11 Which of the following is true of a terminal multiplexer?

a. Exists at the edge of a SONET network

b. Exists in the core of a SONET network

c. Exists on the fiber span of a SONET link

12 A SONET add/drop multiplexer has at least how many ring and add/drop ports?

a. One ring port and two add/drop ports

b. Three ring ports and two add/drop ports

c. Two ring ports and one add/drop port

13 Which digital cross-connect system is capable of switching at the DS1 level between SONET OC-*n*s?

   a. Wideband DCS

   b. Broadband DCS

   c. Narrowband DCS

   d. Low-band DCS

14 A SONET broadband DCS is capable of switching which of the following?

   a. Minimum of DS1 levels and up

   b. Minimum level of DS3 levels and up

   c. Minimum level of STS-3 levels and up

   d. Minimum level of STS-1 levels and up

15 Which of the following devices is used to multiplex low-speed (56-kbps, voice) services onto a high-speed SONET network?

   a. Terminal multiplexer

   b. Add/drop multiplexer

   c. Digital loop carrier

   d. Digital cross-connect

16 Bidirectional automatic protection switching uses the ___ byte(s) in the line overhead.

   a. H1 and H2

   b. D1 and D2

   c. K1 and K2

   d. H3

17 Which of the following SONET ring topologies duplicates data on both the working and protect rings all the time?

   a. UPSR

   b. BLSR

   c. ULSR

   d. BPSR

18 Extra traffic is possible in a _____ ring.

  a. BLSR

  b. UPSR

  c. ULSR

  d. BPSR

19 What is the maximum traffic-carrying capacity of an OC-12 UPSR ring with four nodes?

  a. OC-3

  b. OC-6

  c. OC-12

  d. OC-24

  e. OC-48

20 What is the maximum traffic carrying capacity of an OC-12 BLSR ring with four nodes?

  a. OC-3

  b. OC-6

  c. OC-12

  d. OC-24

  e. OC-48

21 Which OC levels can support a BLSR ring?

  a. OC-1

  b. OC-3

  c. OC-12

  d. OC-48

  e. OC-192

22 Which type of ring can perform either a span switch or a ring switch?

  a. 2-fiber UPSR

  b. 4-fiber UPSR

  c. 2-fiber BLSR

  d. 4-fiber BLSR

23 An STS-12 frame is how many bytes?

a. 9700

b. 9720

c. 9740

d. 9780

24 In which section(s) can you find the data communication channel(s)?

a. Photonic

b. Section

c. Line

d. Path

e. Virtual tributary

25 How many DS1s are in a DS3?

a. 21

b. 24

c. 28

d. 32

Chapter 4

1 The largest ring size that the ONS 15327 can be a part of is

a. OC-3

b. OC-12

c. OC-48

d. OC-192

2 For 1:n protection, which slots on the ONS 15454 must be used for the protection cards?

a. 1,17

b. 2,16

c. 3,15

d. 6,10

3 How many DS3s are supported when the ONS 15327 is configured with the XTC-14 and MIC-28-3?

   a. 0

   b. 1

   c. 2

   d. 3

4 How many DS1s are supported when the ONS 15327 is configured with the XTC-14 and MIC-A/MIC-B cards?

   a. 3

   b. 14

   c. 28

   d. 672

5 How many STS cross-connect ports does the XC-10G support?

   a. 144

   b. 288

   c. 336

   d. 572

   e. 1152

6 The ONS 15327 supports: (Select all the correct answers.)

   a. OC-3

   b. Gigabit Ethernet

   c. OC-192

   d. OC-12

   e. OC-48

   f. 10/100 Ethernet

7 Which of the following fulfill the requirements of running the Cisco Transport Controller application?

    a. CTC does not require any special installation. Point any web browser to the IP address of the ONS device.

    b. Requires that it runs on a Sun workstation.

    c. Can run from a JRE-compliant web browser but requires installation of the JRE and policy file.

    d. Does not require a browser. CTC is a standalone application.

8 The three view levels of the CTC are

    a. Network view

    b. Node view

    c. Provisioning view

    d. Card view

Chapter 5

1 In a four-fiber BLSR network, _____ fibers are dedicated to working traffic and _____ fibers are used for protection bandwidth.

    a. two, two

    b. four, zero

    c. zero, four

    d. one, three

2 A four-fiber BLSR ring is best for what type of traffic patterns?

    a. Meshed

    b. Metro area

    c. Point-to-point

    d. Add/drop linear

    e. Centralized

3 The internal clock on the ONS devices is a

a. Stratum 1

b. Stratum 2

c. Stratum 3

d. Stratum 4

4 To receive timing from the central office via the access line, you should use

a. Internal clock

b. External clock

c. Line clock

d. Access clock

5 In a point-to-point network with no service provider, what clock type is preferred?

a. Internal clock

b. External clock

c. Line clock

d. System clock

6 In SONET, a typical network node located away from the central office is synchronized using _____.

a. Line timing

b. Internal timing

c. GPS

d. None of the above

7 Which subtabs enable you to set the node name?

a. Network

b. Ring

c. General

d. Node

8 Which subtabs enable you to create a UPSR ring?

a. General

b. Ring

c. SONET DCC

d. Timing

9 Which is the recommended way to interconnect fibers between nodes in a ring?

a. East to East

b. West to West

c. East to West

10 What needs to be configured the same for a BLSR ring to be created?

a. Ring ID

b. Node ID

c. West ports

d. East ports

11 What needs to be unique for a BLSR ring to be created?

a. Ring ID

b. Node ID

c. West ports

d. East ports

Chapter 6

1 ONS equipment can easily support metro Ethernet to the customer site through the addition of which of the following?

   a. SONET technology

   b. Transfigured LAN services

   c. Ethernet service cards

   d. DWDM technology

2 Customers have the advantage of advanced traffic shaping and flexibility in logical network design because of the native _____ and _____ support in Ethernet.

   a. QoS, VLAN

   b. VLAN, MPLS VPN

   c. Point-to-point, Gigabit Ethernet

   d. QoS, Gigabit Ethernet

3 Metro Ethernet allows the service provider to offer high-speed _____ connectivity in a native _____ technology that the customer understands and already supports.

   a. MAN/WAN, LAN

   b. LAN, MAN/WAN

   c. SONET, LAN

   d. SONET, MAN/WAN

4 CPE equipment would most likely be found where?

   a. Central office

   b. Customer's site

   c. Point of presence

   d. User's desktop

5 Traffic carried between POPs will most likely run at which speeds? (Choose all that apply.)

   a. DS1

   b. DS3

   c. 10 Mbps

   d. OC-3

   e. OC-12

   f. OC-48

   g. OC-192

6 Which three IEEE standards do the E-series ONS Ethernet cards support?

   a. 802.1h

   b. 802.1c

   c. 802.1d

   d. 802.1p

   e. 802.1r

   f. 802.1Q

7 Which two IEEE standards do the G-series metro Ethernet cards support?

   a. 802.3x

   b. 802.1d

   c. 802.1r

   d. 802.1p

   e. 802.1q

8 What is the amount of bandwidth that can be provisioned if the E-series Ethernet card is configured as a Single-card EtherSwitch?

   a. STS-3c

   b. STS-6c

   c. STS-12c

   d. STS-48c

9 What amount of bandwidth can be provisioned if the E-series Ethernet card is configured as a Multicard EtherSwitch group?

a. STS-3c

b. STS-6c

c. STS-12c

d. STS-48c

10 How many VLANs can be configured on the E-series cards?

a. 500

b. 590

c. 509

d. 8192

11 What does 802.1d specify?

a. VLANs

b. Flow control

c. Priority queuing

d. Spanning tree

12 The maximum bandwidth available on a G-series metro Ethernet card is which of the following:

a. STS-6c

b. STS-12c

c. STS-24c

d. STS-48c

13 What is the maximum bandwidth available on an ONS 15454 on a G-series card configured as a Multicard EtherSwitch Group?

a. OC-48c

b. OC-24c

c. OC-12c

d. OC-3c

e. You cannot configure the card with this configuration.

14 The bandwidth provisioning capabilities of the G1000-2 metro Ethernet card is which of the following?

   a. STS-12c

   b. STS-24c

   c. STS-36c

   d. STS-48c

15 Which terminology is associated with Multicard EtherSwitch group operation? (Choose all that apply.)

   a. Flow control

   b. Shared packet ring

   c. Point to point

   d. Shared bandwidth

Chapter 7

1 Which two tools enable you to create a metro Ethernet circuit between Cisco ONS 15454/15327 E-series cards?

   a. Cisco Transport Manager (CTM)

   b. TL-1

   c. Cisco IOS

   d. Cisco Transport Controller

2 Which three tools enable you to create a metro Ethernet circuit between Cisco ONS 15454/15327 G-series cards?

   a. Cisco Transport Manager (CTM)

   b. TL-1

   c. Cisco IOS

   d. Console Cable

   e. Cisco Transport Controller

3 In which three places must E-series card VLANs be configured?

a. Circuit

b. Fiber

c. Customer-side port

d. Backplane port

e. Customer premises equipment (CPE)

4 What operation does the native/management 802.1Q default to?

a. --

b. Untagged

c. Tagged

d. Pop Tag

5 Traffic entering a VLAN-configured metro Ethernet port from a customer would be categorized as which of the following?

a. Untagged traffic

b. Tagged traffic

c. 802.1Q

d. 802.1p

6 What tab must you choose to assign Single-card EtherSwitch?

a. Port

b. VLAN

c. Card

d. Circuits

7 If a customer does not have 802.1Q-compatible equipment, how should you configure your circuits' VLANs?

a. All ports should be tagged as VLAN 1.

b. All ports should be tagged as VLAN 2.

c. No ports should be tagged except for the port the customer is coming in on.

    d. The port that connects to the customer should be left as the default Untag selection.

    e. The port that connects to the customer should have a VLAN configured, and that VLAN should be set to Untag.

8 True or False: The E1000-2 card manages congestion by initiating pause messages.

    a. True

    b. False

9 True or False: Ethergroup is used to create a Single-card EtherSwitch circuit.

    a. True

    b. False

10 10.802.1Q-trunked VLANs should be set to which VLAN type?

    a. - -

    b. Untag

    c. Tagged

    d. d.Pop Tag

Chapter 8

1 The two largest problems facing service providers today in the metro environment are _____ and _____.

    a. Bandwidth scalability, fiber exhaustion

    b. Bandwidth scalability, wavelength multiplexing

    c. Fiber exhaustion, virtual dark fiber support

    d. Fiber exhaustion, wavelength multiplexing

2 Which of the following allows for a composite optical signal on a fiber to be strengthened simultaneously by one element?

    a. Fiber

    b. Microfilter

    c. Splitter

    d. Optical amplifier

3 Generally for a DWDM system, an optical signal can be sent how far before an OA is needed?

a. 1–10 km

b. 20–30 km

c. 40–60 km

d. 100–500 km

4 The wavelength separation between two different channels is known as

_____.

a. Dispersion

b. Attenuation

c. Channel spacing

d. Overlap

5 What is the function of the optical amplifier in DWDM?

a. Demultiplex wavelengths

b. Individually regenerate a specific wavelength

c. Amplify all wavelengths simultaneously

d. Multiplex wavelengths

6 Which specification defined the standard channel-spacing wavelengths for a DWDM system?

a. ITU-T G.253 wavelength grid

b. ITU-T G.962 wavelength grid

c. ITU-T G.692 wavelength grid

d. ITU-T G.893 wavelength grid

7 _____ are used to equalize the output signal when it leaves the demultiplexer stage.

a. Post-amplifiers

b. Mid-span amplifiers

c. Pre-amplifiers

d. Wavelength amplifiers

8 Which two components comprise the Cisco DWDM solution?

    a. Cisco ONS 15216 optical filter

    b. Cisco ONS 15327 with OC-12 ELR ITU optics

    c. Cisco ONS 15454 with OC-48 ELR ITU optics

    d. Cisco 12000 GSR router with OC-48 ELR ITU optics

9 What is the maximum network bandwidth over a single fiber provided by the Cisco ONS 15454 OC-48 ELR ITU card with 200-GHz spacing?

    a. 1.544 Mbps

    b. 24 Gbps

    c. 45 Gbps

    d. 80 Gbps

10 Into what slots on the Cisco ONS 15454 can you install the OC-48 ELR ITU optics cards?

    a. Into slots 1 through 6

    b. Into any high-speed card slot

    c. Into any multispeed card slot

    d. Into any high-speed or multispeed card slot

11 How many wavelengths in the red and blue band does the ONS 15216 with 100-GHz spacing support?

    a. 9 in blue optical band, 9 in red optical band

    b. 16 in blue optical band, 16 in red optical band

    c. 9 in blue optical band, 18 in red optical band

    d. 18 in blue optical band, 18 in red optical band

12 When connecting the Cisco ONS 15216 red base unit to the blue upgrade unit, what port on the base unit connects to the Common In port on the upgrade unit?

    a. To Upgrade

    b. Monitor Out

    c. Common In

    d. Common Out

Chapter 9

1   True or False: Packet over SONET was developed because there was no other way to transport data over a SONET network.

    a. True

    b. False

2   PoS can be directly encapsulated onto the network media. Which of the following is *not* a method for connecting PoS to network media?

    a. Connectivity to SONET/SDH ADMs

    b. Connectivity to transponders in a DWDM system

    c. Dark-fiber connectivity

    d. ATM connectivity

3   Which of the following are the three requirements for data to be successfully transported over SONET/SDH?

    a. The use of high-order containment is required.

    b. The PoS frames must be placed inside of the SONET containers aligned on frame boundaries.

    c. The $x \wedge 43 + 1$ scrambler must be used in addition to SONET native scrambling.

    d. The PoS frames must be placed inside of the SONET containers aligned on the octet boundaries.

4   What is the hex value of an HDLC delimiter flag byte?

    a. 0x7D

    b. 0xE7

    c. 0x7E

    d. 0xFF

    e. 0x03

5   What does the protocol field inside of the PPP frame indicate?

    a. The protocol that is carrying the PPP frame

    b. The protocol used to decode the FCS field

c. The protocol used to detect the number of padding bytes found in this frame

d. The protocol used to format the data in the Information field

e. None of the above

6 The C2 byte value of a PoS interface that is using the payload-scrambling function is set to which two of the following values?

a. 0xCF%

b. 0x16

c. 22

d. 207

e. 0xFF

7 It is recommended that the Layer 3 protocol and the SONET protocol configurations should _____.

a. Have both Layer 3 and SONET in a bidirectional ring configuration

b. Have Layer 3 in a point-to-point configuration and SONET in a bidirectional ring configuration

c. Both be in a point-to-multipoint configuration

d. Both be in a point-to-point configuration

8 HDLC frames in a PoS environment contain which four fields?

a. Address, Data, Destination, and Frame Check Sequence

b. Location, Data, Destination, and Protocol

c. Location, Control, Information, and Protocol

d. Address, Control, Information, and Frame Check Sequence

9 What is the purpose of PGP?

a. Transport data packets across SONET/SDH links

b. Overcome routing problems between Layer 3 and the SONET network layer

c. Reliable end-to-end communication and error-recovery procedures

d. Achieve adequate transparency, protection against malicious attacks, and enough zero-to-one transitions to maintain synchronization between adjacent SONET/SDH devices

10 What is the ideal configuration for APS 1+1 to reduce the need for routing updates due to a failure?

a. One SONET line between the private network router and the service provider ADM.

b. Two SONET lines between one private network router and the service provider ADM

c. Two routers with one line each to the service provider ADM

d. PoS Reflector mode

11 PoS Reflector mode is used for what purpose?

a. By the working router to keep the distant router up to date

b. By the protect router to notify the distant router when it takes over for the working router

c. By the Layer 3 protocol to send routing updates

d. To send AIS downstream

12 In which three ways can you interconnect PoS interfaces?

a. SONET

b. Dark fiber

c. Gigabit Ethernet

d. DWDM

e. Bidirectional path switched rings (BPSRs)

13 One of the advantages to PoS is that when there is a network failure, _____ can restore the network connection before the Layer 3 routing protocol even realizes that there is a problem.

a. ATM

b. SDH

c. APS

d. IPS

Chapter 10

1 Which three features do the Cisco 12000 series PoS interface cards support?

   a. 128-Mb burst buffers

   b. **ASIC-based queuing**

   c. **Multiple virtual output queues**

   d. **Quality of service (QoS) support**

   e. Cisco Express Forwarding (CEF) table that accommodates up to 14 million forwarding entries

2 The Cisco OC-3c PoS line cards are available in three different versions, each offering a different number of ports. How many ports are supported by the OC-3c/STM-1 PoS line cards?

   a. 1 port

   b. 2 ports

   c. **4 ports**

   d. **8 ports**

   e. **16 ports**

3 What two IETF RFCs specify PPP in HDLC-like framing?

   a. RFC 1615

   b. **RFC 1619**

   c. RFC 791

   d. RFC 1918

   e. RFC 2615

4 What command is used to configure the APS protect interface?

   a. **aps protect**

   b. **aps protect "IP address of protect interface"**

   c. **aps protect "IP address of working router"**

   d. **aps protect "IP address of working router"**

   answer: d

5 Which Cisco IOS command sets framing to SONET STS-3c?

   a. **pos framing 3**

   b. **pos framing SDH**

   c. **pos framing SONET****

   d. **pos scramble-atm**

answer: c

6 Which Cisco IOS command displays status information for the active network protocols?

   a. **debug interfaces**

   b. **show protocols pos**

   c. **show interfaces pos**

   d. **show controllers pos**

answer: b

7 Which Cisco IOS command displays information about SONET alarm and error rates divided into Section, Line, and Path sections?

   a. **debug interfaces**

   b. **show protocols pos**

   c. **show interfaces pos**

   d. **show controllers pos**

answer: d

Chapter 11

1 Which three of the following statements are true?

   a. **DPT is a general solution based on the principles of multiaccess networks standardized by IEEE.**

   b. **DPT is optimized for IP only and other Layer 3 technologies are not considered for support.**

   c. **The DPT proposal borrows from both IEEE 802.3 and SONET.**

   d. DPT significantly increases configuration management requirements beyond that required for full-mesh topologies.

2 A being the highest and D being the lowest, letter the following switch states in the IPS hierarchy.

   a. Signal failure_____B_____

   b. Signal degrade_____C_____

   c. Manual switch_____D_____

   d. Forced switch_____A_____

3 Initial implementation of DPT is based on _____ frame encapsulation for easy reuse of optical components.

   a. APS

   b. IPS

   c. SONET

   d. FDDI

4 Which two of the following are reasons why DPT is typically used in campus environments?

   a. DPT uses a static point-to-point topology-based configuration.

   b. Physical ring topology is typically already used, making the migration to DPT easy.

   c. DPT rings can support multiple subnets, allowing the transport of many networks on a single optical backbone.

   d. DPT provides an easy migration from LAN technologies because it provides a similar MAC layer interface as there was previously.

5 Which of the following is true?

   a. It is not possible to overload a link between two nodes.

   b. The SRP-fa controls only low-priority packets.

   c. The rapid adaptation of the fairness algorithm causes instabilities.

   d. Control of high-priority packets is left to Layer 4 or upper-layer functions.

6 Which two tasks are handled by SRP control packets?

   a. Synchronization

   b. Topology discovery

   c. Protection switching

   d. FDDI

7 Which one of the following statements is true?

   a. The Layer 2 functions can see the DPT ring as a single subnet.

   b. The DPT ring is a Layer 2 protocol.

   c. DPT internals are mostly visible from the upper layers.

   d. The Layer 3 functions see the DPT ring as multiple subnets.

8 What type of packet is indicated when setting the destination MAC address to 0?

   a. Data

   b. Control

   c. Forwarding

   d. Discovery

9 In a fully functional DPT ring, multicast packets are stripped by the
   _____.

   a. Local host

   b. Receiving node

   c. Active monitor on the ring

   d. Source node

10 What is the Mode field of data packets set to?

   a. 0101

   b. 010

   c. 111

   d. 1XX

11   Under what condition does a node accept a control packet but not forward it?

a. Control TTL > 1

b. Control TTL = 0

c. Control TTL = 1

d. Control TTL < 0

12   What is the goal of global fairness according to the SRP fairness algorithm?

a. The node with the highest-priority traffic gets the most bandwidth.

b. One node controls the bandwidth usage of its neighbors.

c. No node can excessively consume network bandwidth and create node starvation or excessive delay.

d. A high-priority node can retain as much bandwidth as needed, although this may create node starvation or excessive delay.

13   During a congestion situation, which type of traffic is more likely to cause harm to ring integrity?

a. High-priority

b. Control

c. Low-priority

d. Discovery

14   What is the proper transmit decision hierarchy during periods of no congestion?

1 – High-priority transit packets
2 – Low-priority transit packets
3 – High-priority transmit packets
4 – Low-priority transmit packets

a. 1, 2, 3, 4

b. 2, 1, 4, 3

c. 2, 3, 1, 4

d. 1, 3, 4, 2

15 Ring selection is controlled by what procedure?

   a. AARP

   b. MAC

   c. ARP

   d. AURP

16 In a DPT ring, the node has to decide whether to place traffic on the inner or outer ring for transport to the destination. What is ring selection based on?

   a. Topology information

   b. Ethernet frames

   c. MAC address

   d. Number of subnets

17 When is the ARP cache flushed?

   a. When a new node is added

   b. When a ring wrap or unwrap condition occurs

   c. When the TTL expires

   d. At regular intervals

18 What algorithm is used to determine which ring to send a packet on when the number of hops on the inner and outer rings is the same?

   a. Fairness

   b. Transmit

   c. Hashing

   d. Limiting

19 Which is *not* an IPS event priority?

   a. FS

   b. SM

   c. SD

   d. MS

20 Which of the following statements about WTR is true?

    a. The request is generated when the wrap condition clears.

    b. The request is generated when the wrap condition begins.

    c. The timer can be configured in the 10- to 1200-second range.

    d. The default value of the WTR timer is 100 seconds.

21 Which is the proper interpretation of the message {SF, B, W, L} when sent by node B to node D.

    a. Request type = Signal failure, Source MAC = Node B, Node status = Wrapped, Path = Long, enter Pass-through Mode

    b. Request type = Signal failure, Source MAC = Node B, Node status = Wrapped, Path = Long, enter Wrapped state

    c. Request type = Signal failure, Source MAC = Node A, Node status = Wrapped, Path = Long, enter Pass-through Mode

    d. Request type = Signal failure, Source MAC = Node B, Node status = Wrapped, Path = Short, enter Wrapped state

22 Intelligent protection switching (IPS) was designed for self-healing according to SONET terminology. IPS automatically recovers from fiber facility failures by _____.

    a. Wrapping the ring

    b. Reversing traffic flow

    c. Stopping the flow of packets

    d. Cutting off bandwidth from the damaged area

Chapter 12

1 Which three speeds are supported by the Cisco 12*xxx* series DPT line cards?

    a. OC-3c/STM-1

    b. OC-12c/STM-4c

    c. OC-48c/STM-16c

    d. OC-192/STM-64c

2 What benefit does the Cisco 12*xxx* DPT line-card packet-prioritization feature provide?

a. Maximizes application versatility and deployment flexibility

b. Maximizes deployment flexibility by operating via dedicated fiber

c. Maximizes ring robustness via self-healing around ring node or fiber failures

d. Provides expedited handling of packets generated by mission-critical applications

3 Single-mode fiber, simplex or duplex, with SC connectors is used to connect Cisco 12*xxx* series routers to other routers for DPT connectivity. If the one-port OC-48 line card is used, how would the second line card be connected?

a. Crossover cable

b. Single-mode fiber

c. Mate cable

d. Straight-through cable

4 In the following table, indicate the order in which each Dynamic Packet Transport configuration and verification task is completed when configuring DPT over an external 15454 SONET network.

Enter the Step # Here	Task Description
4	Using Cisco IOS, configure optional parameters on SRP interface.
5	Use Cisco IOS **show** and **debug** commands to verify the configuration.
1	Configure a circuit using the Cisco Transport Controller interface.
6	Verify the DPT circuit using **ping**.
2	Verify the circuit using the CTC circuit map display.
3	Using Cisco IOS, assign an IP address to the SRP interface.

5 Which Cisco IOS Software command correctly configures the frequency at which topology messages are sent?

a. **srp ips timer 5 a**

b. **srp ips time 5 b**

c. **srp ips wtr-timer 10**

d. **srp topology-timer 60**

answer: d

6 Which Cisco IOS Software command initiates the highest-priority ring wrap?

    a. srp ips wtr-timer *value*

    b. srp ips timer *value* [*a*|*b*]

    c. srp ips request forced-switch [*a*|*b*]

    d. srp ips request manual-switch *[a|b]*

7 Which Cisco IOS Software command correctly sets the frequency of the transmission of IPS requests on an OC-12c SRP line card to 30 seconds?

    a. srp ips timer 3 a

    b. srp ips timer 30 a

    c. srp ips wtr-timer 30

    d. srp topology-timer 30

Chapter 13

Answers to the case study questions.

Step 1

    a. Xmt = –2 dBm

       Rcv = –28 dBm

$$\text{Xmt} - \text{MuxLoss} - \text{FiberLoss} - \text{DmxLoss} = \text{Rcv}$$
$$\text{FiberLoss} = \text{Xmt} - \text{MuxLoss} - \text{DmxLoss} - \text{Rcv}$$
$$= -2 \text{ dBm} - 4.5 \text{ dB} - 4.5 \text{ dB} + 28 \text{ dBm}$$
$$= \underline{\mathbf{17\ dB}}$$

    b. PerChOutOfMux = Xmt – MuxLoss

$$= -2 \text{ dBm} - 4.5 \text{ dB}$$
$$= \underline{-6.5 \text{ dBm}}$$

$$\text{AggOutOfMux} = \text{PerChOutOfMux} + 10 \times \log(\text{QtyChs})$$
$$= -6.5 \text{ dBm} + 9 \text{ dB}$$
$$= \mathbf{2.5\ dBm}$$

c. PerChIntoMux – DmxLoss = Rcv

PerChIntoMux= Rcv + DmxLoss

$$= -28 \text{ dBm} + 4.5 \text{ dB}$$

$$= -23.5 \text{ dBm}$$

AggIntoMux= -23.5 dBm + $10 \times \log(8)$

$$= -23.5 \text{ dBm} + 9 \text{ dB}$$

$$= \underline{\textbf{–14.5 dBm}}$$

d. Constant Gain mode = 23 dB gain

PwrOut   = PwrIn + Gain

$$= -15 \text{ dBm} + 23 \text{ dB}$$

$$= \underline{\textbf{8 dBm}}$$

Step 2

a. Xmt – MuxLoss – FiberLoss – OADMThroughLoss – DmxLoss = Rcv

FiberLoss = Xmt – MuxLoss – OADMThroughLoss – DmxLoss – Rcv
$$= -2 – 4.5 – 1.8 – 4.5 +28$$
$$= \underline{\textbf{15.2 dB}}$$

b. PerChIntoOADM $= -13 – 10 \times \log(8)$

$$= -22 \text{ dBm}$$

PerChDrop $= -22$ – DropLoss
$$= -22 – 2.1$$
$$= \underline{\textbf{–24.1 dBm}}$$

c. AggOutOfOADM $= -13$ – OADMThroughLoss

$$= -13 – 1.8$$
$$= -14.8 \text{ dBm}$$

PerChOutofOADM $= -14.8 – 9 = -23.8$ dB

Set VOA so PerChOutofOADM = AddChOutofOADM
AddChOutofOADM = PerChOutofOADM= Xmt – AddLoss – VOA
$-23.8= -2 – 3.2$ – VOA
VOA = $\underline{\textbf{18.6 dB}}$

References

To review the latest advances in metro optical technologies and equipment, visit Cisco.com. Other books are also available to expand on those areas not covered in this book or only touched upon lightly. These books include the following:

- *SONET*, Second Edition, by Walter Goraliski. Osborne McGraw-Hill, May 2000. ISBN 0-07-212570-5.

- *Multiwavelength Optical Networks: A Layered Approach*, by Thomas E. Stern and Krishna Bala. Prentice Hall PTR, May 1999. ISBN 0-201-30967-x.

- *Optical Networks: A Practical Perspectives*, Second Edition, by Rajiv Ramaswami and Kumar N. Sivarajan. Morgan Kaufman Publishers, October 2001. ISBN 1-55860-655-6.

- *Next Generation Optical Networks: The Convergence of IP Intelligence and Optical Technologies,* by Peter Tomsu and Christian Schmutzer. Prentice Hall PTR, August 2001ISBN 0-13-028226-x.

- *Understand Optical Communication,* by Harry J. R. Dutton. Prentice Hall PTR, January 1999. ISBN 0-13-020141-3

- *Understanding SONET/SDH and ATM,* by Stamatios V. Kartalopoulos. Wiley IEEE Press, May 1999. ISBN 0-7803-4745-5.

- *Introduction to DWDM Technology,* by Stamatios V. Kartalopoulos. Wiley IEEE Press, November 1999. ISBN 0-7803-5399-4.

- *Fiber Optic Reference Guide*, Second Edition, by David R. Goffs. Focal Press, March 1999. ISBN 0-2408-0360-4.

INDEX

Numerics

A

B

D

E

I

J-L

M

N

O

Q-R

S

V

W-Z

Cisco Press

Learning is serious business.

Invest wisely.

Voice Solutions

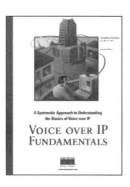

Voice over IP Fundamentals
Jonathon Davidson, James Peters, Brian Gracely
1-57870-168-6 • **Available Now**

Voice over IP Fundamentals explains the basic concepts of VoIP technology including how the modern telephone system works. Comparisons between today's PSTN and tomorrow's integrated network are included. Case studies show real-world examples of the technology in use as well as next-generation applications.

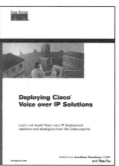

Deploying Cisco Voice over IP Solutions
Jonathon Davidson
1-58705-030-7 • **Available Now**

Deploying Cisco Voice over IP Solutions provides networking professionals the knowledge, advice, and insight necessary to design and deploy Voice over IP (VoIP) networks that meet customers' needs for scalability, services, and security. Beginning with an introduction to the important, preliminary design elements that need to be considered before implementing VoIP, *Deploying Cisco Voice over IP Solutions* also demonstrates the basic tasks involved in designing an effective service provider-based VoIP network. This book concludes with design and implementation guidelines for some of the more popular and widely requested VoIP services, such as prepaid services, fax services, and Virtual Private Networks (VPNs).

Integrating Voice and Data Networks
Scott Keagy
1-57870-196-1 • **Available Now**

Integrating Voice and Data Networks is both a conceptual reference and a practical how-to book that bridges the gap between existing telephony networks and the new world of packetized voice over data networks. Underlying technologies are explained in a nonproduct-specific manner that gives a holistic understanding of voice/data integration.

CCIE®

Cisco LAN Switching
Kennedy Clark, Kevin Hamilton
1-57870-094-9 • **Available Now**

This book is essential for preparation for the CCIE Routing and Switching exam track. As well as CCIE preparation, this comprehensive volume provides readers with an in-depth analysis of Cisco LAN Switching technologies, architectures, and deployments. Product operational details, hardware options, configuration fundamentals, spanning tree, source-route bridging, multilayer switching, and other technology areas related to the Catalyst series switches are discussed.

CCIE Routing and Switching Exam Certificatin Guide
Anthony Bruno
1-58720-053-8 • **Available Now**

CCIE Routing and Switching Exam Certification Guide is a comprehensive study and assessment tool. Written and reviewed by CCIEs, this book helps you understand and master the material you need to know to pass the test. The companion CD-ROM includes over 200 practice exam questions in a simulated testing environment, customizable so you can focus on the areas in which you need the most review.

CCIE Practical Studies, Volume I
Karl Solie, CCIE
1-58720-002-3 • **Available Now**

CCIE Practical Studies, Volume I, provides you with the knowledge to assemble and configure all the necessary hardware and software components required to model complex, Cisco-driven internetworks based on the OSI reference model-from Layer 1 on up.

Cisco CCNP BSCI Certification

CCNP® BSCI Exam Certification Guide (CCNP Self-Study), Second Edition

Clare Gough

1-58720-078-3 • **Available Now**

CCNP BSCI Exam Certification Guide, Second Edition, is a comprehensive exam self-study tool for the CCNP/CCDP/CCIP BSCI exam, which evaluates a networkers ability to build scalable, routed Cisco internetworks. This book, updated with more than 100 pages of IS-IS protocol coverage, addresses all the major topics on the most recent BSCI #640-901 exam. This guide enables readers to master the concepts and technologies upon which they will be tested, including extending IP addresses, routing principles, scalable routing protocols, managing traffic and access, and optimizing scalable internetworks. CCNP candidates will seek out *CCNP BSCI Exam Certification Guide* as timely and expert late-stage exam preparation tool and useful post-exam reference.

CCNP Self-Study: Building Scalable Cisco Internetworks (BSCI)

Catherine Paquet, Diane Teare

1-58705-084-6 • **Available Now**

CCNP Self-Study: Building Scalable Cisco Internetworks (BSCI) is a Cisco authorized, self-paced learning tool for CCNP, CCDP, and CCIP preparation. The book teaches readers how to design, configure, maintain, and scale routed networks that are growing in size and complexity. The book focuses on using Cisco routers connected in LANs and WANs typically found at medium-to-large network sites. Upon completing this book, readers will be able to select and implement the appropriate Cisco IOS® Software services required to build a scalable, routed network.

ciscopress.com

Cisco CCNP Certification

CCNP Preparation Library (CCNP Self-Study), Third Edition
Various Authors
1-58705-131-1 • **Available Now**

CCNP Preparation Library, Third Edition, is a Cisco authorized library of self-paced learning tools for the four component exams of the CCNP certification. These books teach readers the skills in professional level routing, switching, remote access and support as recommended for their respective exams, including for the new Building Scalable Cisco Internetworks (BSCI) exam.

Based on the four component exams of the CCNP certification, this four-book library contains *CCNP Self-Study: Building Scalable Cisco Internetworks* (BSCI), *Building Cisco Multilayer Switched Networks*, *Building Cisco Remote Access Networks*, and *Cisco Internetwork Troubleshooting*. These books serve as valuable study guides and supplements to the instuctor-led courses for certification candidates. They are also valuable to any intermediate level networker who wants to master the implementation of Cisco networking devices in medium to large networks.

Cisco CCNP Certification Library (CCNP Self-Study), Second Edition
Various Authors
1-58720-080-5 • **Available Now**

Cisco Certified Network Professional (CCNP) is the intermediate-level Cisco certification for network support. This is the next step for networking professionals who wish to validate their skills beyond the Cisco Certified Network Associate (CCNA®) level or who want to have a path to the expert level certification of CCIE. CCNP tests a candidates skill in installing, configuring, operating, and troubleshooting complex routed LANs, routed WANs, switched LANs, and dial access services. Where CCNA requires candidates to pass a single exam, CCNP requires candidates to pass four written exams, including 640-901 BSCI, 640-604 Switching, 640-605 Remote Access, and 640-606 Support.

The official exam self-study guides for each of these exams are now available in this value priced bundle. These books, *CCNP BSCI Exam Certification Guide*, *CCNP Switching Exam Certification Guide*, *CCNP Remote Access Exam Certification Guide*, and *CCNP Support Exam Certification Guide*, present the certification candidate with comprehensive review and practice of all the key topics that appear on each of the CCNP exams.

Learning is serious buisiness. **Invest wisely.**

Cisco CCNP Certification

Internetworking Technologies Handbook, Third Edition

Cisco Systems, Inc.

1-58705-001-3 • **Available Now**

Internetworking Technologies Handbook, Third Edition, is an essential reference for every network professional. *Internetworking Technologies Handbook* has been one of Cisco Press' best-selling and most popular books since the first edition was published in 1997. Network engineers, administrators, technicians, and support personnel use this book to understand and implement many different internetworking and Cisco technologies. Beyond the on-the-job use, *Internetworking Technologies Handbook* is also a core training component for CCNA and CCDA® certifications. It is a comprehensive reference that enables networking professionals to understand and implement contemporary internetworking technologies. You will master terms, concepts, technologies, and devices used in today's networking industry, and will learn how to incorporate internetworking technologies into a LAN/WAN environment.

This Third Edition features new chapters on cable technologies, wireless technologies, and voice/data integration. After reading this book, networking professionals will possess a greater understanding of local and wide-area networking and the hardware, protocols, and services involved. *Internetworking Technologies Handbook* offers system optimization techniques that will strengthen results, increase productivity, and improve efficiency--helping you make more intelligent, cost-effective decisions for your network environment.